CW01669696

Tinsley

and the Modernisation of

Sheffield's Railways

by Chris Booth and Alex Fisher

PLATFORM 5

Tinsley and the Modernisation of Sheffield's Railways

Published by Platform 5 Publishing Ltd, 52 Broadfield Road, Sheffield, S8 0XJ. England.

Printed in England by The Lavenham Press, Lavenham, Suffolk..
ISBN: 978 1 909431 24 9

Front cover: The start of the sorting process undertaken by thousands of wagons at Tinsley. The steep gradient of the hump can clearly be seen in this view taken along the hump line. The main yard control tower is in the centre of this 1966 image, taken not long after the yard opened. *John Pavitt collection, colourised by Stephen Arrandale*

Back cover: By June 2015, much of the area once occupied by Tinsley Yard has been taken over by the huge bulk of the Sheffield International Rail Freight Terminal. Just a handful of sidings remain. *John Webster*

Contents

Below: 46016 departs from Tinsley Yard with a mixed bag of wagons in 1981. *Steve Ryszka*

Preface

Tinsley was not only one of Britain's best-known yards, it was also one of Europe's most technologically advanced yards. October 2015 was the 50th anniversary of its opening, and March 2018 the 20th Anniversary of the closure of the Diesel Depot. Although the general story of Tinsley is essentially well known, the purpose of this book is to create the definitive story by bringing together as much information, from as many different sources as can reasonably be researched. The National Archives at Kew in London, Sheffield Archives, Rotherham Archives, the National Railway Museum and various other archives and sources have been visited; an extensive review of contemporary literature has been undertaken and reminiscences of people directly involved have been documented.

We hope that, as authors who are both professional railwaymen, we have been able to combine our working knowledge of the railway with our mutual desire to uncover more about the history and past times of Tinsley. We both felt that we could tackle Tinsley and hopefully produce a book befitting such an important piece of Britain's railway heritage.

Introduction

In writing this book we have tried to cover not only the history of Tinsley, but also the context of Tinsley's development, use and near ultimate demise. Thus the start of the book analyses the decade or so before Tinsley's opening and the creation of a unified Sheffield Division. The dieselisation era and the particular importance of Darnall is followed by two particular components - the Sheffield District Railway and Aldwarke Junction - without either of which Tinsley may well never have been built where it was; thus, their influence on how Tinsley came to be in that location is described.

Tinsley Yard's first plans, its construction and opening are examined closely, together with the type and construction of the yard buildings. Tinsley Traction Maintenance Depot is described in its own chapter, although mention of diesels cannot fail to include the famed "Siamese Twins", the Class 13 shunters which were synonymous with Tinsley Yard. Equally as synonymous was the Dowty Retarder System and this is fully described, as is the "Addo" Data Transmission System and the communication within the yard. In mentioning communication, the various types of signalling used, from traditional mechanical to then "state of the art" power signalling panels is detailed. Tinsley Yard was also integrated into the electrified Woodhead route and part of the yard was wired until the cessation of electric haulage.

With the yard and depot open, an account of operations starts with the 1960s and 1970s, with trip notices and actual railway material used to aid context. Of particular note is an Appendix with a full description of all the job roles involved in maintaining the smooth operation of a busy yard like Tinsley, along with a full Tinsley Yard simplifier which gives details of every booked train departure from Tinsley during late 1972. After entering the TOPS era, Tinsley's most turbulent decade – the 1980s – is followed by its downfall during the 1990s and how and why this came about. Once into the new millennium, the continued demise was met with a mini-renaissance thanks to a new stone terminal and the Sheffield International Railfreight Terminal warehouse building.

Detailed separately are Tinsley's much appreciated open days. A flavour of what could be found on Tinsley on a typical day in each year from the depot's opening to closure can be found in "On Shed", enhanced by a collection of personal recollections about Tinsley which together form an interesting record from those that were there, told in their own words. The book rounds off by looking at Tinsley 50 years on, and what remains today.

We hope you enjoy reading this book as much as we have enjoyed putting it together.

Chris Booth & Alex Fisher, 2019

Acknowledgements

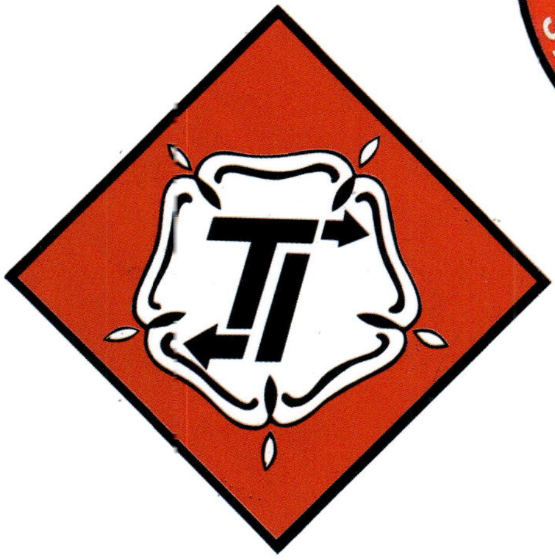

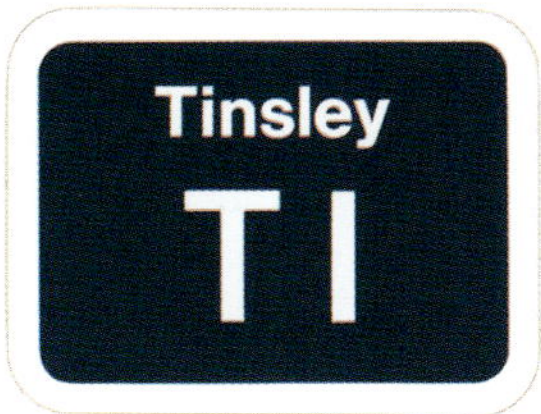

In collating this book we have been in contact with many people and as such the Authors would like to thank those for their help with this project. So in no particular order our thanks go to the following contributors.

David Higton and John Pavitt have helped immensely with their inside knowledge of the development of the yard and its buildings; there can be few others who have been involved with the yard from its first inception through to its opening. Thanks go to Owen Owens, who was the Wimpey site engineer for the construction of the diesel depot. Now in his 90s, his son John Owens forwarded the images for use. They were important contributors to the whole Tinsley project, but without the input of former staff members – drivers, signalmen and others interested in Tinsley – this book would have been all the poorer.

David Withers and Les Woad of the Breakdown Crane Association are thanked for information about the breakdown cranes, as are Alan Waterhouse, Alex Scott, Clive Greedus, Fred Kerr, Leonard Ball, Vince Minto, Andrew Booth, Brian Hancock, Keith Long, the Pennine Railway Society, along with the Shed Surprise and WNXX websites, and all others who permitted use of their spotting notes to give a good overview of locos to be seen on depot.

We are also grateful to the following who allowed use of their personal memories: James Greaves, John Miller, Walter Ford, Brian Hanley, Roger Wainwright and Alan McKenna. These memories would eventually have been lost had they not put them down for the readers of this book.

Special thanks go to the Network Rail Corporate Archive for allowing use of diagrams, to the National Railway Museum for access to images and to Sophia Brothers at the Science & Society Picture Library for expediting copies.

Dave Ball of Sheffield Hallam University is thanked for an image along with the following organisations: Kidderminster Railway Museum, Rail-Online, The Transport Interchange, English Heritage Aerofilms and GEC-Alstom.

Of course, without photographs the book would be very bland. Retired former Tinsley driver Roger Wainwright set the whole ball rolling with this project by allowing us privileged access to his extensive collection of images; it was through looking at his slides that the idea for the book was born.

The following photographers, listed in no particular order, have also allowed use of their images: Roger Wainwright, Phil Hodgkiss, John Miller, Alan Smalley, Andrew Overton, John Law, Adrian Wynne, John Gabbitas, Les Nixon, Paul Braybrook, Dave Stones, Alan McKenna, Ted Evans, Phillip Chrome, Andy Wills, John Byers, Allan Trotter, Dave Wragg, Michael Rhodes, Pete Churchman, Ivan Stewart, Mick Nicholson, Bob Wallace, Colin J. Marsden, Brian Daniels, Gordon Edgar, Alan Hazell, Paul Lindley, Barry Johnson, Arnie Furniss, David Ford, John Miller, Andy Hoare, Stewart Donohoe, Steven Gay, Andrew Johnson, David J. Smith, Paul Cooke, Peter White, Andrew Walker, Stuart Reeves, John Webster, Kim Collinson, David Ford, Nial Jordan, Keith Long, Mick Charlesworth, Brian Hancock, Ross Byers, Simon Judd, Steve Ryszka.

Chris must thank his wife Linda Booth for putting up with him disappearing out to London, York and Sheffield researching the book, and for the hours shut away in his little cubby hole working on the computer, compiling the text and photographs. Thanks go to Greg Sherlock for his abilities for proof reading and finding those hidden gaffes, and to Stephen Arrandale for his excellent colourisation of the front cover photograph. Finally, apologies go to anyone else who has helped with this project and is not named; your help was much appreciated.

Chapter 1

The Origins and Control of Sheffield's Railways

Like most of Britain's railways, the history of Sheffield's railways comprises a complicated web of lines from competing companies, who in many cases duplicated routes to serve the same destinations. The eventual amalgamation of companies lead to a rationalisation of resources and lines, and brought the railway we see today.

The Sheffield and Rotherham Railway (S&RR) opened in 1838 as the first railway to serve Sheffield, this running from Wicker station in Sheffield to Westgate station in Rotherham. Wicker station was comprised of a large shed supported on cast iron pillars which, in contemporary reports, was described as "commodious, handsome and spacious" when it opened. The S&RR was absorbed into the Midland Railway (MR) on 21st July 1845 and by 1870 Wicker station was being described in further contemporary reports as "wretched accommodation" and "miserable and dilapidated, being made to do duty as a station".

A new line skirting around the east of Sheffield, was originally surveyed and constructed by George Stephenson as part of his North Midland Railway (NMR) line from Derby to Leeds via Chesterfield and Rotherham. The section from Derby to Rotherham (Masborough) opened for traffic on 11th May 1840, with the section through to Leeds opening by the end of that June. The NMR was also and later absorbed by the MR and eventually become part of the MR main line from St Pancras to the north.

Following the opening of the NMR, a new through station at Masborough was opened and a spur at Holmes Junction was laid to connect the NMR to the S&RR, giving the NMR access to the Wicker terminal of the S&RR in Sheffield. Although the through station at Masborough was now open, Westgate station would remain in use as it was much more central.

From Manchester to Lincolnshire and Beyond

The route between Manchester and Sheffield (Bridgehouses) via Woodhead was opened in stages between 1841 and 1845 by the Sheffield, Ashton-under-Lyne & Manchester Railway. In conjunction with the Sheffield and Lincolnshire Junction Railway (S&LJR) and Great Grimsby and Sheffield Junction Railway, the company was renamed the Manchester, Sheffield and Lincolnshire Railway (MS&LR) in 1847, the line extending towards Woodhouse and onwards in 1849. A new station at Sheffield Victoria opened in September 1851, replacing

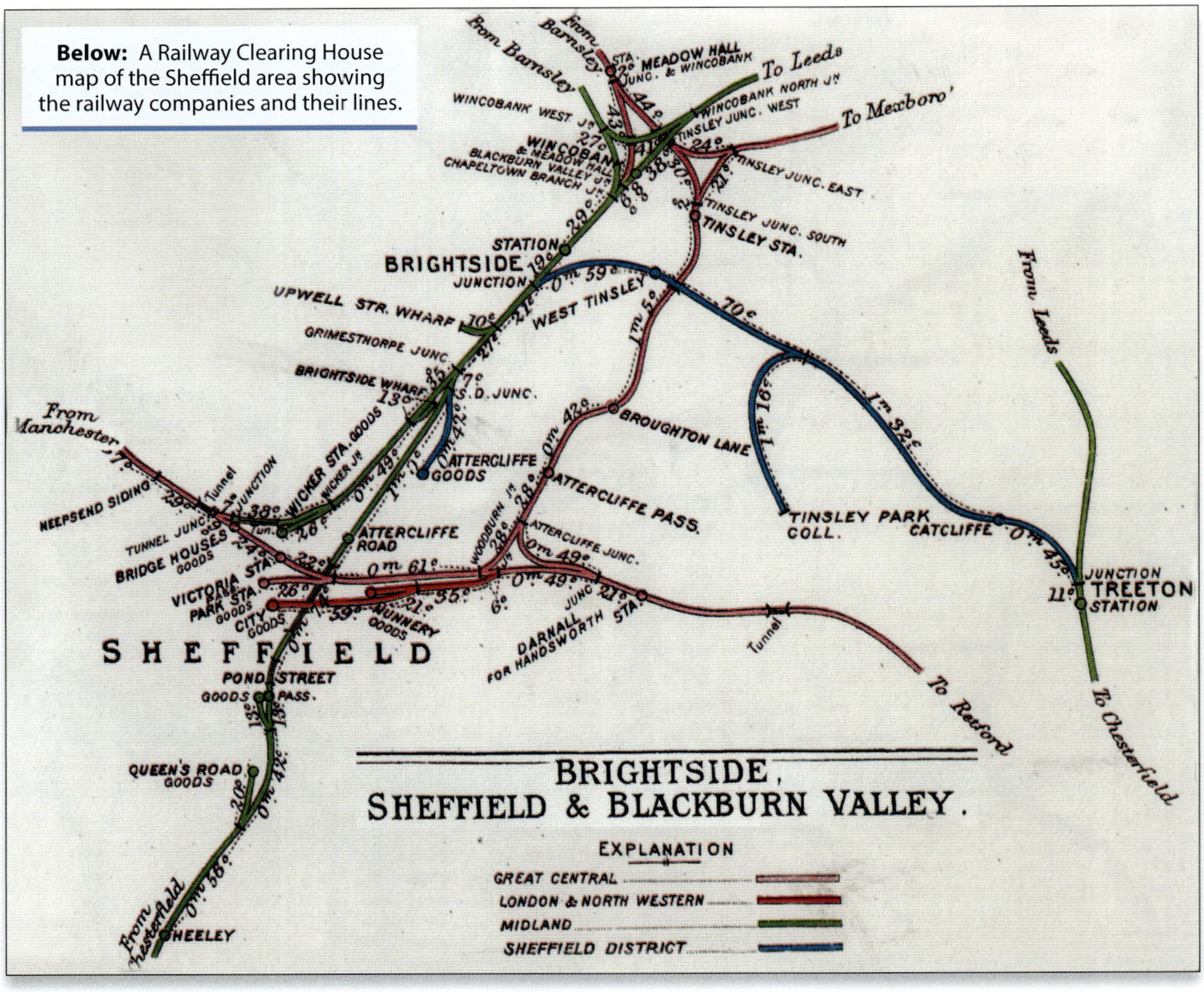

Below: A Railway Clearing House map of the Sheffield area showing the railway companies and their lines.

Above: Sheffield Wicker Goods Depot as seen from the Wicker Arches on 10th April 1962, 122 years on from when it opened as a passenger station. The road to the right is Saville Street, with a Sheffield Corporation No. 75 bus descending from Spital Hill on the left. The west gate is in the foreground, with the fire brigade, ambulance room and calling office block just inside the wall on the right. To the left of that, with a lorry being weighed, is the weighbridge and the large building just beyond is the three story grain warehouse. The main goods shed and warehouse building is to the left and rear of the grain warehouse.
National Railway Museum/Science & Society Picture Library DF396

Bridgehouses which became the company's terminal for goods and cattle traffic. The S&LJR line from Woodhouse to Beighton connected with the MR line initially, allowing passengers a route to Chesterfield via the "Old Road", before the MS&LR Derbyshire Lines to Chesterfield Central opened. The route onwards towards Annesley opened in 1892, eventually reaching London Marylebone in 1899 (On 1st August 1897, the name of the railway was changed to Great Central Railway (GCR)).

Parallel Lines

The GCR line from Woodburn Junction to Rotherham and Mexborough, including the north to east curve from Attercliffe Junction to Darnall, had opened throughout by 1864, paralleling the MR line from Sheffield to Rotherham and Mexborough. The two companies also had paralleling routes to Barnsley which served the same areas. The MR line running from Wincobank to Chapeltown opened in 1894, later extending to Barnsley, whilst the South Yorkshire Railway (later absorbed into the GCR) Blackburn Valley Line had opened in September 1854, originally connecting with the MR at Blackburn Junction and thus providing a link between Barnsley and Sheffield. This connection closed in August 1864 when the line from Woodburn Junction to Rotherham opened and a triangular junction was installed at Tinsley.

Sheffield finally became accessible from the south in 1870 when a line was opened from Chesterfield (Tapton Junction) through Dronfield, Bradway Tunnel and Dore & Totley, to a new station at Sheffield Pond Street (later Midland), joining the S&RR at Grimesthorpe Junction. The direct line to Sheffield then became the main line, being known as the "New Road" and most main line passenger traffic was henceforth routed that way; the line via Barrow Hill and Beighton was commonly known as the "Old Road". With the opening of Pond Street station in February 1870, Wicker then became a goods station. In the same year a new eastbound spur was built from Nunnery Main Line Junction, just north of Pond Street, to connect with the MS&LR line east of Sheffield Victoria. As it involved a reversal to access Victoria, it was rarely utilised.

After 1870 Rotherham Westgate station was still used for several Sheffield–Nottingham, Sheffield–Derby and Sheffield–Manchester local services which commenced their journey at Westgate and ran through to Sheffield Pond Street, calling at all stations. The station would remain in use through to nationalisation, but due to the parlous condition of the original 1830s vintage wooden bridge over the river, BR closed the station on 4th October 1952 with demolition coming in 1970. A short section of this line from Holmes Junction survives today, giving access to C.F. Booth's scrap yard. 1894 saw the opening of the Hope Valley

Line from Dore & Totley to Chinley and in 1900 a junction was constructed at Treeton with the new Sheffield District Railway (see later), which ran through to Brightside on the former S&RR.

Growth, Grouping and Nationalisation

As the lines grew amid fierce competition between the companies, so did the number of goods yards and private sidings to service customers. The duplication of routes meant that most of the larger firms and collieries had duplicated connections from competing railways, the GCR and the MR. Although the companies' lines were often situated adjacent to each other, the only physical exchange points between the former GCR and MR lines were at the two main junctions at Beighton and Swinton.

The advent of the grouping in 1923 saw the formation of the London Midland & Scottish Railway (LMS) and London & North Eastern Railway (LNER). It was from this point that a start was made on the rationalisation of the respective companies' services and facilities. The duplication of lines remained, but at least an agreement for the pooling of traffic at a number of locations had begun.

Prior to the creation of the newly-nationalised British Railways, the British Transport Commission (BTC) had deliberated whether the former LNER should remain as one area or whether it should be broken into two. The former LMS survived similar proposals of being split in two as it was thought it would split too many main lines, but with some diffidence the BTC accepted that, broadly speaking, the Southern and North Eastern areas of the LNER should be constituted into the Eastern and North Eastern areas of British Railways from 1st January 1948.

From 1949 the BTC were pressing for the rationalisation of the former GCR system but both the Eastern and London Midland Regions (LMR) successfully objected. The Railway Executive settled the argument however by segregating the commercial and operating management of the disputed line and other lines which penetrated the new regions. From 2nd April 1950 all ex-LMS routes in and around Sheffield became part of the Eastern Region along with the former GCR lines. However as the operating and motive power departments were left with previous owners, the former GCR lines now had to contend with two chief regional officers, two commercial managers, two civil and signal engineers and so on.

Although some small effort had been made in regard to pooling of wagons, by the end of WW2 and the nationalisation of the railways little else had been done to rationalise the situation despite there being a common central management. In 1950 the British Railways regional boundary changes saw the whole of the Sheffield area come under the control of the Eastern Region but, as if to emphasise the separation between the former company's lines, there were still two control offices – one at Rotherham for the Midland line and one at Sheffield Victoria for the GC lines.

Problems

The problem that this obviously caused was that it didn't give each area true control over all of the functions within its geographic area. In Sheffield especially this manifested itself by the continuation of duplicate facilities and a lack of connections between the two systems. What was needed was just one traffic manager per area and in the next re-organisation of the Eastern Region three territories were created, each with their own line traffic manager (LTM). In mid to late 1957 J.W. Dedman became the LTM for London, Tilbury and Southend (LT&S); from 18th November 1957 G.F. Fiennes took the post of LTM Great Northern (GN) and finally W.G. Thorpe became LTM for the Great Eastern

Below: Sheffield Midland Station on 29th May 1968. Sheaf House, the Sheffield Division headquarters, prominently overlooks the station on the site of Pond Street Goods Yard. The cars lined up in the foreground are parked on what is now the route of the South Yorkshire Supertram.
Chris Booth collection

(GE) from 30th December 1957. The posts and departments of commercial manager, motive power superintendent and operating superintendent were abolished.

Whilst J.W. Dedman had just the single area of the LT&S under his control, the GN LTM had four sub-districts under his control, each with their own traffic manager (TM), their districts being Kings Cross, Doncaster, Lincoln and Sheffield. This left the GE LTM to have control of the Liverpool Street, Cambridge and Norwich Districts. The motive power depots which came under each traffic area were then entirely under the control of the new LTMs.

Further Reorganisation

In 1958 further reorganisation saw the former LMR depots at Canklow, Grimesthorpe, Millhouses and Staveley (Barrow Hill) all transferred to the Eastern Region. The new combined Sheffield District under its own TM then covered from just north of Stairfoot Junction on the line to Cudworth, just north of Wath North signal box on the former Midland line and one chain north of Dearne Junction signal box on the Moorthorpe line. The eastern boundaries were just east of Denaby Crossing, (including the Silverwood branch) and Tuxford (on the Shirebrook to Lincoln line). To the south the district encompassed the Midland Main Line to Horns Bridge (Chesterfield) and the former Great Central Main Line to Heath, whilst the western boundaries were just west of Totley Tunnel East signal box, just west of Dunford Bridge and from Huddersfield to just west of Penistone station. Also included was all of the line between Dodworth and Barnsley Station Junction. As new offices for the Sheffield District were not ready at this time, the former Rotherham Control Office continued to function in the same way; however it was now answerable to Great Northern House and G.F. Fiennes for working to other districts within his area, and workings to either of the other two LTM's areas.

Under the unified control of the Sheffield Area TM, a start was made on rationalisation and modernisation. Duplicate facilities at Barnsley and Chesterfield were some of the first to be eliminated and a large proportion of the collieries lost one of their connections; some small stations and goods depots were also closed.

In 1962 the GE and LT&S LTMs were combined and from 6th July 1964 the Eastern Region abolished its two LTMs with their seven sub-areas. In their place came a new two-tier organisation based on a regional headquarters and divisional levels of management. The seven former areas were reduced to five divisions answerable directly to general management at regional headquarters. The Sheffield Division was managed by S.C. Webb who had previously been the Sheffield Area's Traffic Manager since January 1963. The other divisional managers were located at Liverpool Street, Norwich, Kings Cross and Doncaster.

What a Waste

The greatest waste of time and money in the Sheffield Area arose from the large number of small yards, nearly all of which were badly laid out. There was a great deal of duplication that persisted in train and trip working to different yards, and the handling and cartage of merchandise traffic was a logistical nightmare. The multitude of yards ensured that movement of trains from one part of the Sheffield Area to the other was a complex affair which often

Above: In a scene played out regularly in the chaotic Sheffield Area, Brush Type 2 D5844 waits at a signal at Grimesthorpe on 21st August 1962, waiting to proceed towards Mill Race Junction with a typical mixed goods train of the time. Grimesthorpe Junction No. 2 signal box is in the distance. The loco was allocated new to Darnall on 7th June 1962, moving to Tinsley on 2nd May 1964. Renumbered to 31310, and later to 31422 upon fitment of electric train supply, it would meet its end in C.F. Booth's scrapyard on 7th October 2014.
Alan Smalley

resulted in many reversals of direction and a lot of unnecessary mileage. It was not unusual for one wagon to be shunted several times before it was able to leave the district for its ultimate destination. A short BR film made in 1961 showed this to perfection, when a film unit spent a day with a wagon of 22 tons of steel reinforcing bars heading for the Manchester area from one of the steel works.

The wagon began its journey at the former GCR Attercliffe Goods Yard, where it was coupled to a train, complete with brake van at either end and with a trip number of 85. The train pulled out of Attercliffe Yard onto the Midland Line, then immediately set back into the former MR Engine Shed Sidings to pick up traffic for the former GCR Broughton Lane. Continuing out of the yard back on to the MR line and past the Wicker branch and Brightside Station Junction, it then turned off at Wincobank Station Junction along the former MR double-track connection (which by now was only used as a single-line connection) and joined the former GCR Blackburn Valley Line at Meadow Hall. There the engine ran round its train, hence the requirement for two brake vans. It then set off back beneath the MR line via Tinsley West and South Junctions to Broughton Lane, where it set back into the sidings there. The wagon then became part of trip No. 210 heading for the former GCR sidings at Bernard Road, hauled by a different locomotive and crew. Heading along the line from Broughton Lane to Woodburn Junction, the train joined the former GCR Worksop to Sheffield Victoria line and then eventually set back into the "up" sidings at Bernard Road, where it was now on the right line for forward dispatch to Manchester. This was the fourth yard that the wagon had been in and it was still only half a mile from where it had started!

Conclusion

Due to the amount of both passenger and the remaining freight traffic, the sight of one or more freight trains waiting at signals to gain access to yards was a common one. The system was described at the time as being "chaotic" with "signalmen being fully occupied trying to keep trains moving in their own little patch". Although steam had been replaced by diesel traction, operating methods had not changed at all and something drastic had to be done to join up Sheffield's "incoherent" railway system. So the seeds of Tinsley were sown.

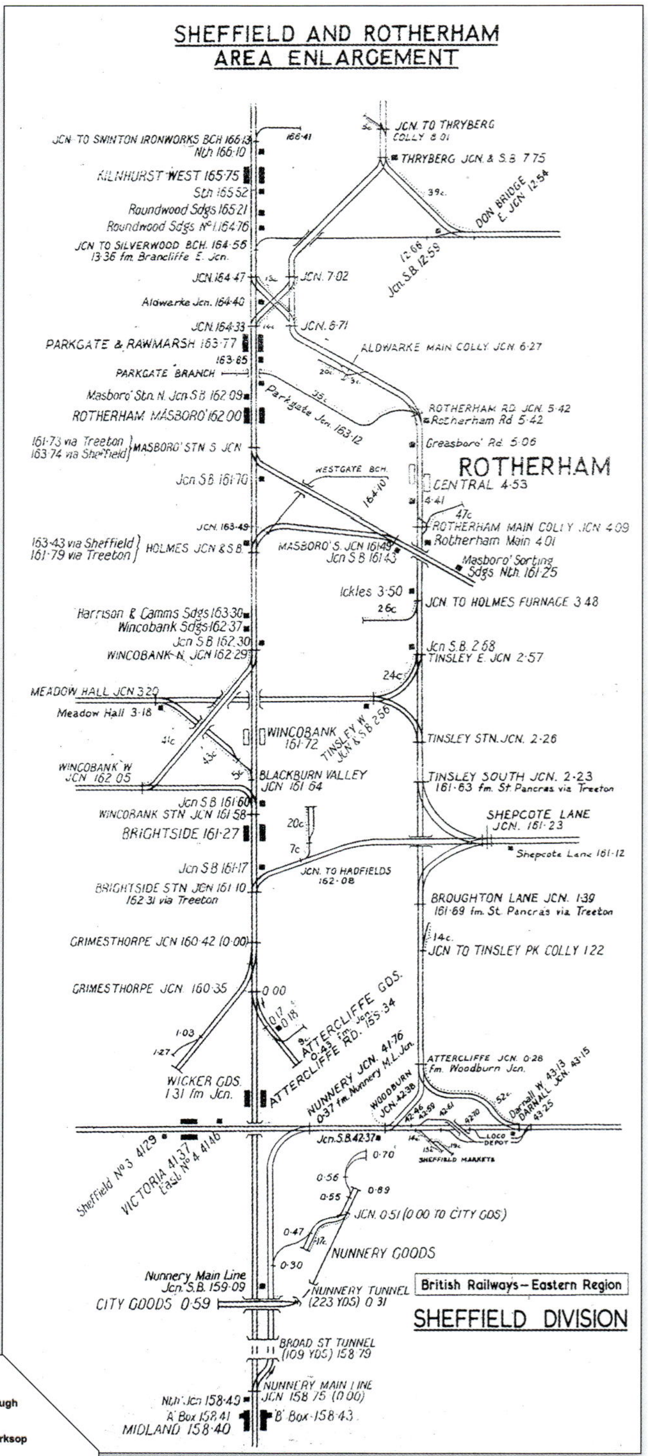

Above: Sheffield Division diagrammatic map enlargement after the construction of Aldwarke Junctions and Tinsley Yard.
Courtesy: Network Rail Corporate Archive

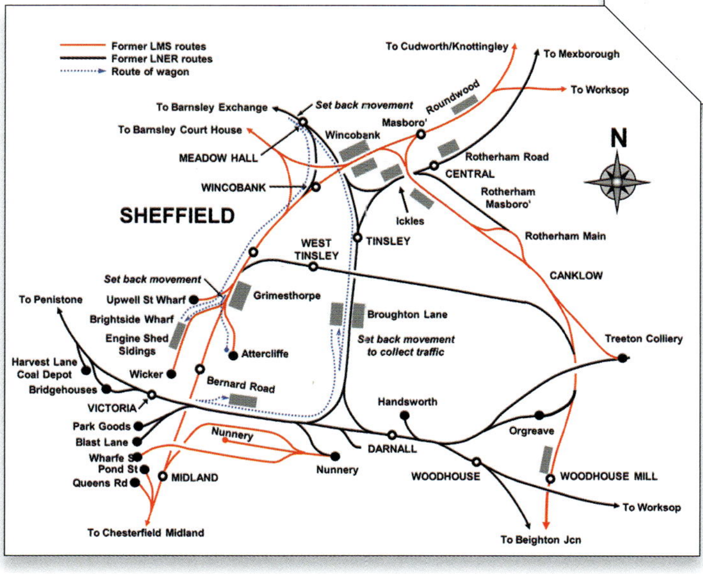

Left: Route taken by wagon of steel reinforcing rods as described in text.
Diagram by Alex Fisher

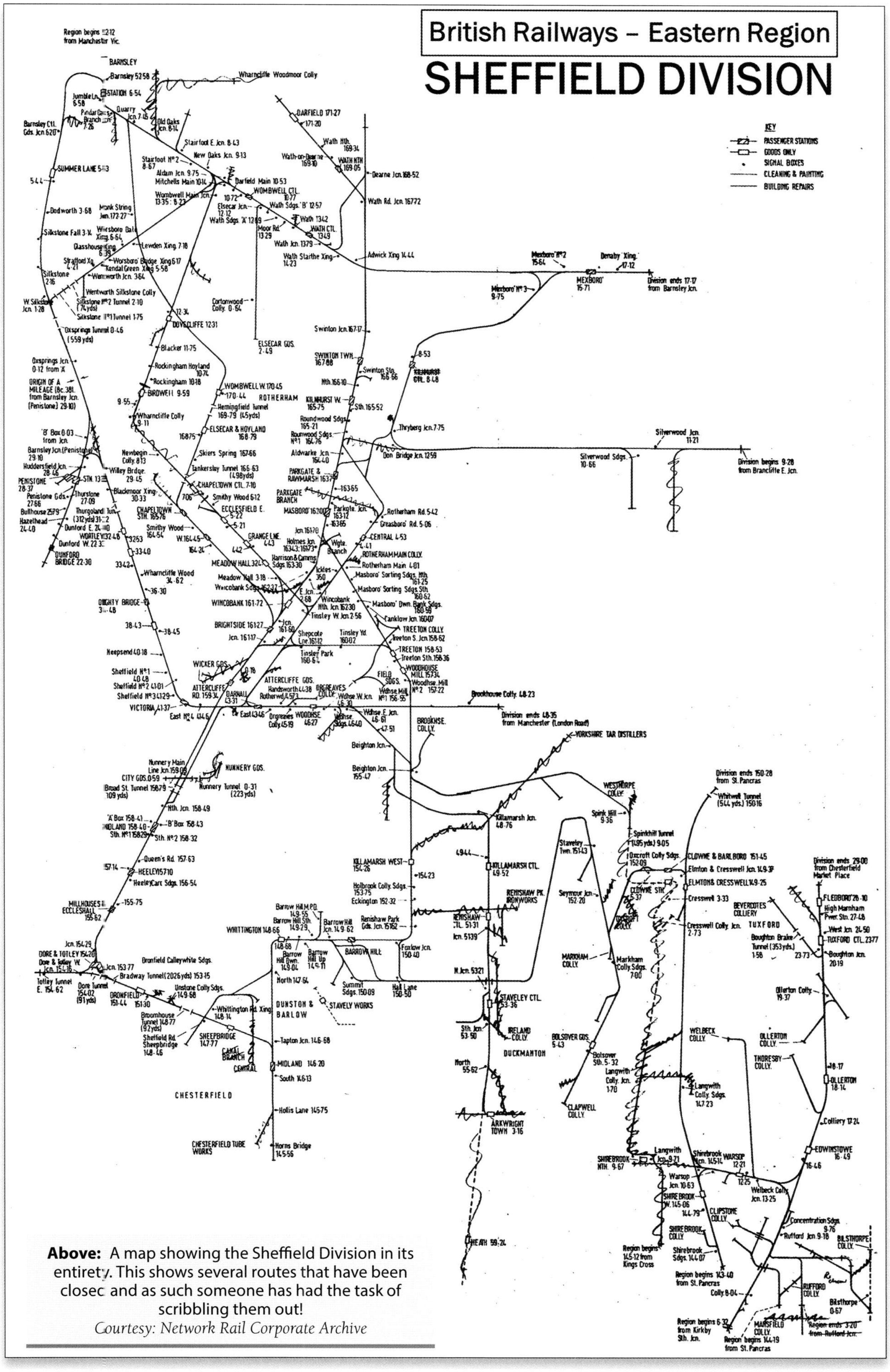

Above: A map showing the Sheffield Division in its entirety. This shows several routes that have been closed and as such someone has had the task of scribbling them out!
Courtesy: Network Rail Corporate Archive

Chapter 2

The Plan for the Modernisation of Freight Facilities in the Sheffield and Rotherham Area

By the early part of the 20th century, Sheffield had goods stations all around the city and surrounding areas which, even in the early years of BR, still saw considerable amounts of traffic. Table 1 shows a representative sample of the total tonnages moved from some of the former GCR depots during April 1951.

Table 1: Tonnages Moved

Depot	Traffic	Tons
Bridgehouses	Small consignments and perishables	9,512
Park Station	Wagon load traffic	1,534
Blast Lane	Fruit, perishables and coal traffic	2,350
Attercliffe	Mainly timber traffic	8,530
West Tinsley	General traffic	2,110

The main marshalling yard on the GCR line was at Bernard Road and there were also coal depots at Harvest Lane and Nunnery. On the line from Woodburn Junction to Rotherham there were yards at Broughton Lane, Rotherham Road and Ickles.

The LMS yards consisted of Wicker, Queens Road, Heeley, Brightside Wharf, Upwell Street and Pond Street. April 1951 saw Queens Road handle 9,614 tons of sundries traffic, regularly dispatching out to Ancoats (Manchester), Bristol, Derby, Lawley Street Leicester and Somerstown (adjacent to St Pancras in London). There was, however, only one shift a day to deal with loading, cleaning out wagons and re-loading.

Wicker was the busier depot with almost 11,000 tons being dealt with in April 1951, 110 loaded wagons being received daily and 120 wagon loads dispatched, whilst Pond Street dealt with full wagon loads for collection only.

The London & North Western Railway (LNWR) reached Sheffield by exercising running powers over the MS&LR from Manchester. In 1895 they opened a ¾ mile branch from Woodburn Junction to a depot named Sheffield City Goods on Bernard Road. The branch was extended a further ¾ mile in 1903 to a site on the corner of Broad Street and Wharf Street. A new depot here was named City Goods, the earlier depot being renamed Nunnery Goods to avoid confusion. Nunnery Goods then dealt with the heavier items. City Goods was also known as Wharf Street; it was different to the other two companies' depots in that it was built above street level, being three storeys high and covering 94,260 square feet. It was contemporarily described as *"an engineering marvel of its time"*, with wagons being moved around with the aid of hydraulic capstans. It also had two hydraulic wagon lifts capable of lowering 10-ton wagons to street level for the loading of goods on to road vehicles. It had 7 roads, Nos. 2–7 each having wagon turntables, all the tracks being set in cobblestones which it was said *"made maintenance virtually impossible"*.

By the 1950s Queens Road was the parent depot to Wharf Street, the latter depot being staffed only on a daily basis and having sporadic traffic. Wharf Street also had large areas let off as fixed space tenancies along with Wicker and Bridgehouses, these being occupied by companies including Allied Newspapers, Batchelors Foods, Cadburys, Fyffes Bananas, Sheffield & Ecclesall Co-operative Society, Gerrards Industries and Newton Chambers Ltd.

During 1959 the Sheffield Area was responsible for 10% of the country's freight and coal traffic, conveyed in nearly two million wagons. The outlying area had around 55 collieries and including general freight approximately 22.5 million tons of traffic was being forwarded, leading to freight traffic gross receipts amounting to £23.7 million.

After coal the next most important traffic was general merchandise, the bulk of which comprised the raw materials for iron and steel making along with its subsequent products. Below: 3 million tons of merchandise and minerals were forwarded in 1961, receipts being in the region of £4 million, whilst over 6 million tons were received in the same period. This discrepancy was due to roughly four to five tons of raw materials being required for every ton of steel produced. Daily wagon loads were in the region of 9,000; around 5,000

Above: Moving a wagon by use of a capstan. The rope was wrapped around the capstan several times, and then attached to the wagon by a hook. The shunter would then stand on the operating stud which set the capstan turning, thus moving the wagon.
Chris Booth collection

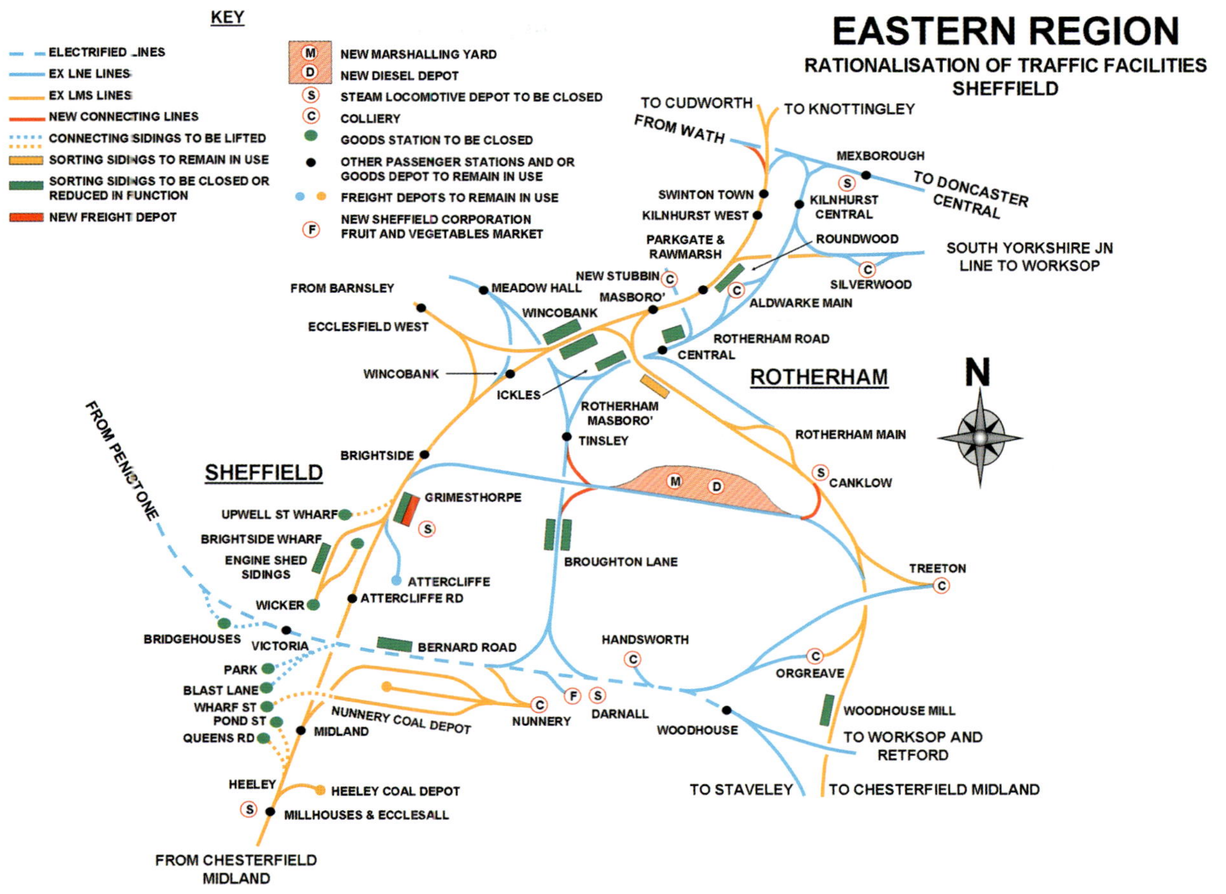

Above: Diagram showing the rationalisation of traffic facilities in the Sheffield Area. Adapted from a British Railways diagram.
Alex Fisher

empty wagons were required in the Sheffield Area for loading each day.

The outlying Sheffield area also had larger marshalling yards at Wath, Mansfield Concentration Sidings, Warsop Junction, Barrow Hill and Seymour Junction which, apart from the inevitable coal, dealt with other goods traffic including cast iron pipes from Staveley, smokeless fuels from Bolsover, pottery & glassware from Chesterfield and Barnsley, foodstuffs from the Batchelors Foods plant at Wadsley Bridge and Bassetts sweets.

Coke was also dispatched from several plants including Orgreave, which had connections to the former MR at Treeton and GCR at Rotherwood (Orgreave Colliery closed in October 1981, the coking ovens in 1990) and the Smithywood Coking Plant on the former GCR Blackburn Valley Line.

Rationalisation

The British Railways Rationalisation Plan for Sheffield's Railways called for the eradication of duplication, and it was dictated that there was to be just one marshalling yard, one goods depot and one traction maintenance depot. Passenger services were to be concentrated on Sheffield Midland thanks to improvement of the Nunnery curve to passenger standards. Sheffield Victoria was to remain, but would only be used for Manchester traffic via the Woodhead Route. In Rotherham, Central station would close and passenger services would be concentrated on Masborough station.

As part of the Eastern Region's 1957 re-organisation, G.F. Fiennes had become LTM for the Great Northern territory of the Eastern Region. Within that territory was the newly amalgamated Sheffield Area and in his book *"I Tried To Run A Railway"* (published by Ian Allan 1967) he describes how he, along with assistant chief civil engineer, Ralph Sadler, and divisional manager, Dick Temple, spent two days looking at where the MR and GCR lines were adjacent with the idea of building a common marshalling yard. They looked at the MR and GCR Chapeltown branches but these were discounted as they did not trap the traffic flow between east and west. They looked at the Roundwood and Aldwarke area, but there was insufficient land available. Killamarsh was looked at next where the MR "Old Road" and the GCR main line ran parallel, but that was considered to be too far south. Beighton was another area where the MR and GCR ran parallel but that was discounted because of its propensity to flooding. Eventually they arrived at Catcliffe where the Sheffield District Railway (SDR) connected to the "Old Road" and also connected at Brightside Junction on the MR "New Road" between Sheffield and Masborough, en route crossing the GCR Woodburn Junction to Rotherham branch near West Tinsley.

In the discussion that followed it was decided that with the construction of some new junctions, trains would be able to serve all points of the compass and the MR and GCR lines would then become one railway. Land was available but as Mr Fiennes pointed out, as the land rose 60 ft above rail level, *"there was a bloody great mountain in the way!"*

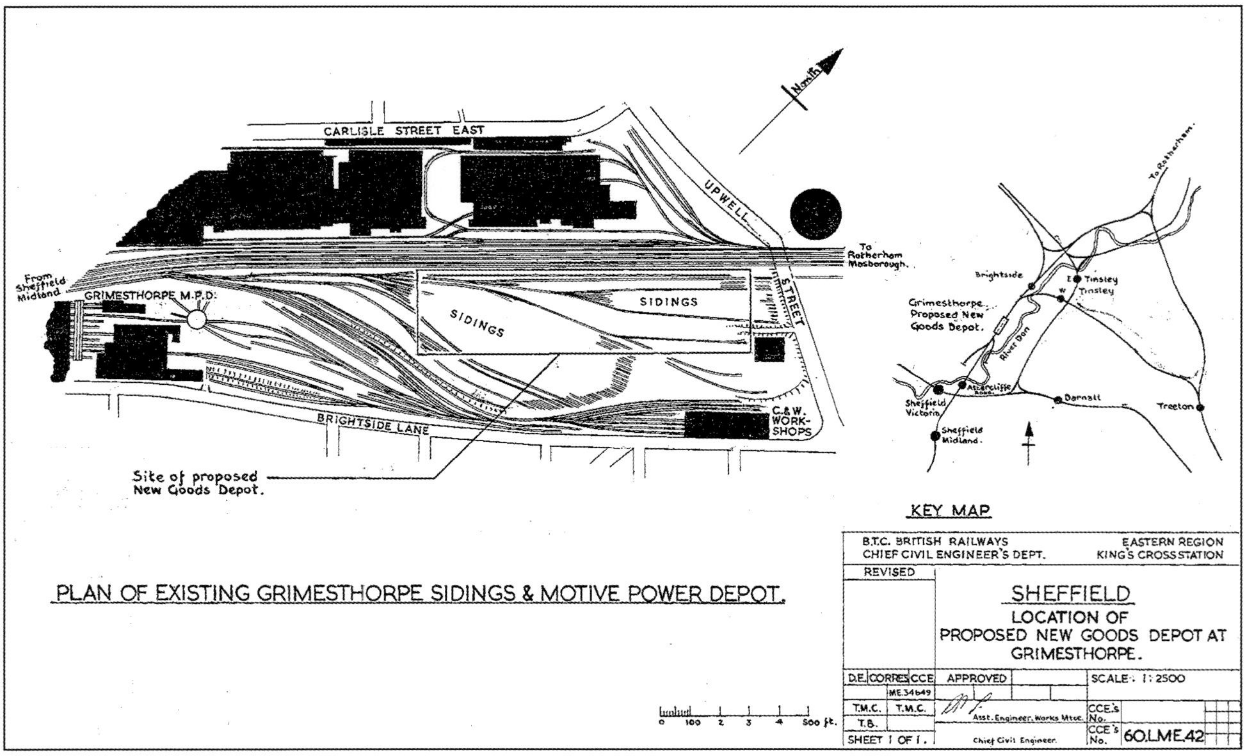

Above: Plan showing the location of Sheffield Freight Terminal in relation to Grimesthorpe steam motive power depot.
Network Rail Corporate Archive

Undaunted, Ralph Sadler, as Mr Fiennes put it *"rolled his head and one eye, waved his hands and said muck shifting is easy nowadays; I can take it away in a matter of months"*. And so the idea of Tinsley Marshalling Yard was taken back to Stuart Ward the operations manager at Great Northern House, for the BR Board to deliberate on.

The Sheffield Area rationalisation proposals were submitted to the British Transport Commission (BTC) in 1960 and after approval by the Ministry of Transport, a BTC Act of 1960 allowed for construction to start on the new yard in August 1961, with the purchase of 115 acres of land from 12 owners. However, as Firth Vickers had developed plans for a new steelworks on part of the land required for the yard, it was necessary to provide them with replacement land. Consideration was also given to the M1 Motorway which was to be built alongside the required land.

Detailed work on the design of the yard was undertaken by the civil engineers' office at King's Cross. Here, a cardboard model of the proposed Tinsley Yard track plan was built, which it is understood from a conversation with a former CCE employee was about the length of a cricket pitch. It was built with cross-sectional profiles,* then had thin card or ply-backed dyeline prints of the track plan stretched over the cross-sections to determine the optimum position for the main control tower. To ensure that controllers had clear sightlines past the Wood Lane overbridge supporting columns, the sightlines were laid out with the aid of cotton. In 1961 there was no computer aided design available!

The new yard was thus constructed to the north side of the SDR, the line being moved from its original position to a new alignment on the north side of the yard. The yard itself was more than two miles long and almost ¼ mile wide, covering an area of 145 acres and containing nearly 60 miles of track.

*According to the Construction Dictionary, cross-sectional profiles are: *"A graphic representation of the sections that are in a linear work, on cutting the axis of the aforementioned work with vertical planes perpendicular to it and defining the trace in elevation. In such profiles, the gradient, the land, the slopes, the levelled areas and the embankments are represented."*

Sheffield Freight Terminal

The Sheffield modernisation plan had called for a single freight terminal and for this the new Sheffield Freight Terminal (SFT) at Grimesthorpe would be built on land vacated by the steam depot and yards, a short distance to the south of Brightside Junction.

The original 1960 authority for the yard and depot work at Tinsley was £6.93 million, with that at Grimesthorpe being £2.323 million, making a total of £9.253 million. By 1963 this had been revised upwards to £8.011 million for Tinsley and down to £1.102 million for SFT, making a total of £9.113 million. Much as today these figures would fluctuate as time went along due to variances in the cost of materials and unforeseen necessary works. The original estimate also included a credit of £1.145 million relating to the value of redundant yards and depots which could not be realised until after the works had been completed. As such an investment allocation of £10.398 million was sought from and approved by the Ministry of Transport. To put the value of this expenditure into context, a capital investment in 1961 of £10 million would be worth in the region of £200 million in 2018.

Concentration of goods services upon Sheffield Freight Terminal improved the average wagon load and eliminated most of the trans-shipments between depots. Handling within the depot was completely mechanised with a 35-ton electric gantry crane which straddled four tracks and a roadway, thus giving access to nearly 50 wagons.

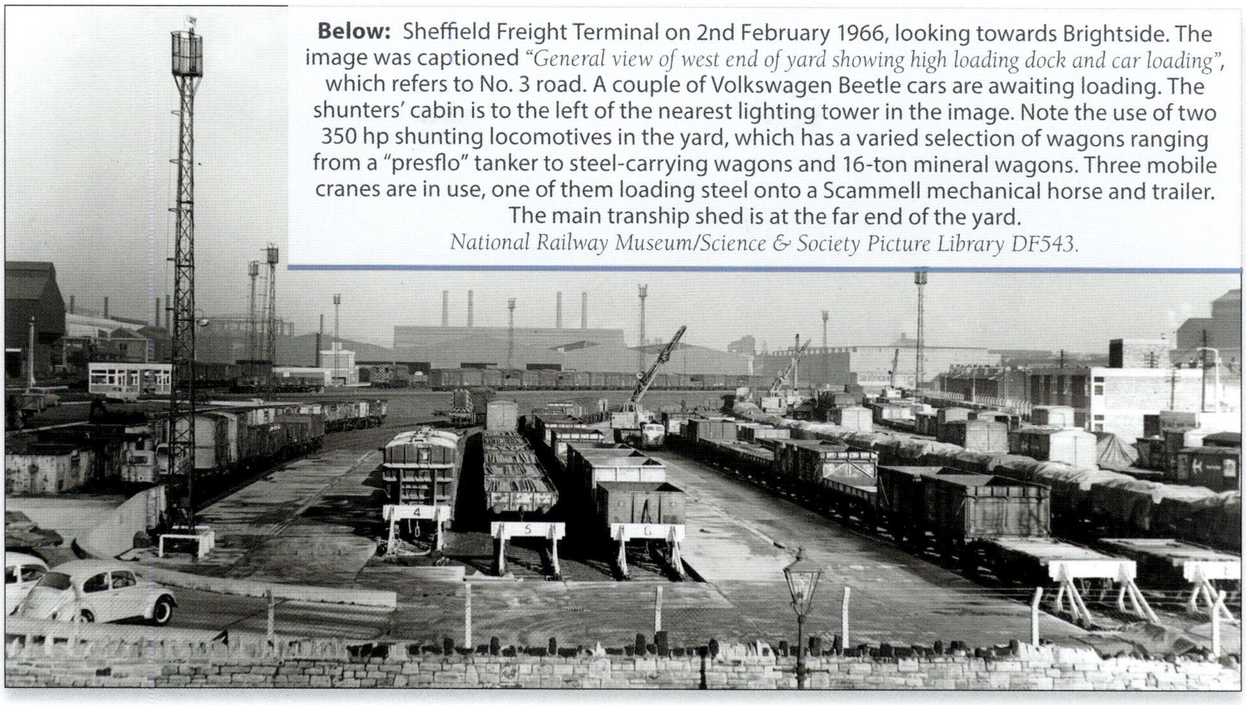

Below: Sheffield Freight Terminal on 2nd February 1966, looking towards Brightside. The image was captioned *"General view of west end of yard showing high loading dock and car loading"*, which refers to No. 3 road. A couple of Volkswagen Beetle cars are awaiting loading. The shunters' cabin is to the left of the nearest lighting tower in the image. Note the use of two 350 hp shunting locomotives in the yard, which has a varied selection of wagons ranging from a "presflo" tanker to steel-carrying wagons and 16-ton mineral wagons. Three mobile cranes are in use, one of them loading steel onto a Scammell mechanical horse and trailer. The main tranship shed is at the far end of the yard.
National Railway Museum/Science & Society Picture Library DF543.

Above: On 4th May 1964, Darnall-allocated Brush Type 2 locomotive No. D5848 departs from Parkgate with a train of specially adapted wagons with cradles for conveying strip coil for Birmingham. 200,000 tons of strip steel were being produced per annum from Parkgate, and these block trains ran five days a week direct to Birmingham – one of the flows not to use Tinsley Yard. The train is about to join the former Midland route and, as this was before the opening of Aldwarke Junction, the area was under the control of Roundwood signal box, which can be seen just above the locomotive. Roundwood was one of the locations considered for a yard, but there was insufficient land available.
National Railway Museum/Science & Society Picture Library TFD96A"

Right: Burnett & Hallamshire (Fuels) Ltd was formed in 1953, when Burnett Brothers and Hallamshire Coal Supplies merged. In 1957 they opened a new coal depot at Nunnery sidings which is seen here. The handling plant had 20 storage bunkers capable of holding at least 60 tons, all high enough to get a 15-foot high lorry beneath. The plant was served by around 60 rail wagons a day, shunted by their own diesel loco. It was expected that the coal concentration scheme at Nunnery would take 100,000, possibly rising to 250,000 tons per annum, to be unloaded via wagon tippler. The site was bought by the British Fuel Company in 1970, however the Sheffield rate rise, British Rail's freight rate rise and the switch to heavier trucks, all helped to make delivering coal to the stockyard uneconomic and the plant was demolished in 1981 (Source www.healeyhero.co.uk). *National Railway Museum/ Science & Society Picture Library*

Above right: Parkway Wholesale Market rail entrance is seen on 20th February 1988. The banana and fish roads were to the right, and fruit and vegetable roads to the left. The market was accessed via a link from the main line close to Woodburn Junction, trains being propelled up to the market as there were no run-round facilities. The gradient from Woodburn Junction was initially 1 in 188, rising to 1 in 34, so a good shove was required to get the trains into the market! The points in the view were worked by the shunter, the signal being controlled from a 2-lever ground frame behind the camera. No. 1 lever worked the gate bolts for the crossing gates (there was a second crossing behind the camera), and No. 2 lever worked the signal. The road in the view led to Darnall steam depot, which was to the left of the image at a lower level. Although Parkway Market still exists, the area has been completely redeveloped with new buildings, and there is nothing to show that a rail link ever existed here. *Andrew K. Overton*

Also available in the yard were four mobile diesel-electric cranes in the 6 to 10-ton range, all being equipped with electro-magnets for movement of the large amounts of steel products which passed through the terminal.

Inside the shed, which was one of the largest in Europe, was a 920-feet long deck with berths for 80 delivery vehicles. It also had separate areas for incoming and outgoing traffic, which was moved around by battery-electric tugs. Heavier boxes were handled by a mobile crane dubbed "*The Iron Fairy*". The new depot dispensed with most of the duplicated cartage rounds and it dealt with around 2,000 tons per day. There were about 300 tons of sundries traffic for delivery and 420 tons for collection daily, traffic being distributed to roughly 160 principal destinations from the new terminal which provided a greatly improved service than from the smaller yards. It was expected that the depot would save about 65,000 wagon journeys per year with the elimination of transhipment between depots.

Closures

Closures began with Pond Street depot closing its gates on 7th October 1961, the site being cleared to make way for a new Sheffield Division headquarters building, to be named Sheaf

Above: Brush Type 2 D5682, new to Darnall on 12th January 1961, waits to depart from the Batchelors Foods Ltd factory at Wadsley Bridge with a train of palletised canned goods traffic on 29th June 1961. The "palvans" were notoriously rough riding if not loaded exactly equally, prone to derailment and subsequently had short lives. This block load traffic was another example not to use Tinsley Yard on opening.
National Railway Museum/Science & Society Picture Library TFD44

House. Queens Road freight depot closed on 11th May 1963, its traffic being diverted to Wicker and Bridgehouses depots; Park Goods succumbed in October 1963. The new Sheffield Freight Terminal (SFT) opened during the summer of 1965, being built at a final cost of £1.102 million. The opening of SFT permitted the closure of the Sheffield goods depots at Blast Lane, West Tinsley, Wharf Street (formerly City LNWR), Bridgehouses and Wicker from 12th July 1965. In addition, SFT would deal with sundries traffic previously handled by the depots at Chesterfield, Barnsley, Sheffield and Rotherham.

The opening of SFT allowed large areas of high-value land within the city of Sheffield and surrounding areas to be released for sale. Maps were made of the areas to be sold, with annotations showing the area of each parcel of land. Examples of this were:

- Plot 5 Engine shed sidings 5.490 acres, 26,570 square yards
- Plot 6 Cardigan sidings 3.520 acres
- Wicker branch 9,220 square yards
- Wicker sidings 41,840 square yards

In Rotherham, Masborough sorting sidings was retained to be rebuilt as a Freightliner depot which opened on 7th August 1967, the first departure from there being a Sheffield to Glasgow Freightliner which was fittingly piped away from the yard by Mr Hamish Ross, traffic manager at Steel, Peech and Tozer.

The beginning of the modernisation of facilities also saw a new Parkway Wholesale Market built (also known as Sheffield Corporation Fruit and Vegetables Market), which covered over 80,000 square feet. Situated behind Darnall MPD it was accessed via a steeply-graded connection at Woodburn Junction. This new market replaced the inner-city fruit and vegetables facilities at Blast Lane

Table 2: Sidings Closed and Land Sold as part of the Sheffield Modernisation Plan

Bernard Road	Closed and site disposed
Broughton Lane "Down" Yard	Closed
Broughton Lane "Up" Yard	Closed as a marshalling yard but kept as a concentration point for private siding traffic between Attercliffe and Meadow Hall which did not provide full train loads of traffic to and from Tinsley
Engine Shed Sidings	Closed and most of site disposed but a portion retained for exchange sidings with Firth Brown and other Wicker branch firms
Grimesthorpe	Closed and site used for Sheffield Freight Terminal
Ickles	Closed but part of site retained as exchange sidings with Steel, Peech & Tozer
Rotherham Road	Closed as a marshalling yard but retained as exchange sidings serving the power station, New Stubbin Colliery and Parkgate Iron & Steel Company
Roundwood	Closed and site disposed
Wincobank "Up" and "Down" Sidings	Closed and site disposed
Cardigan Sidings	Closed and site disposed
Chesterfield Hollis Lane	Closed
Masborough Station Yard	Closed

and Bridgehouses, along with those at Castlefolds Market. Table 2 shows how some of the sidings and land were sold off.

After closure of Bridgehouses and Wicker depots the buildings became a magnet for youngsters to explore, it being common for them to walk down from Bridgehouses through Spital Tunnel, which was nicknamed the *"Firey Jack"*, to the *"ghost town"* at Wicker as they termed it. They used the depot as an adventure park, swinging across the platforms on ropes and it is reported that during the six-week school holidays on 31st July 1966, a fire was started in Bridgehouses and 15 minutes later at Wicker. As the two yards were connected by Spital Tunnel this was suspected to be the route of the arsonist. Wicker was totally destroyed with casualties being incurred by the fire fighters attending, whilst Bridgehouses was seriously damaged. The Wicker site was cleared soon afterwards, but Bridgehouses was not demolished until 1969.

Above: The modernisation of Sheffield's railways included a Freightliner terminal at Rotherham Masborough. Here we see the first departure of the new Sheffield to Glasgow Freightliner service being piped away by Mr Hamish Ross, traffic manager at Steel, Peech and Tozer, on 7th August 1967. Note the coach behind the loco, this probably being used as a brake van for the guard. This terminal did not do as well as expected and had closed by the mid-1970s, the site then being used for movements of steel. It re-opened as a container terminal operated by Newell & Wright in 2015.
Adrian Wynne collection

Goods Yards – Wicker

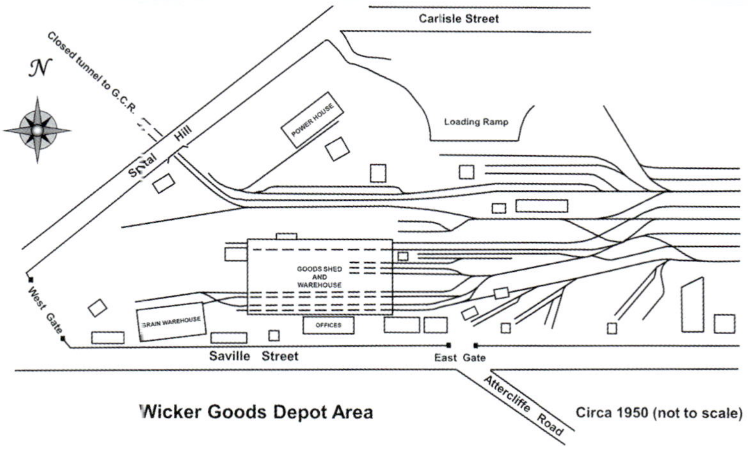

Sectional Appendix Instruction 1st October 1969 Wicker Branch

A stop board is provided on the approach side of the points leading to Nos. 1, 2 and 3 Wicker sidings and the ESC sidings. Guards must satisfy themselves that the points are correctly set and the line is clear for the movement about to be made before authorising drivers to pass the board. A stop board is provided at the exit from Nos. 1, 2 and 3 Wicker sidings and at the exit from the ESC sidings and guards must obtain the authority of the signalman at Brightside Junction, by telephone, before authorising drivers to proceed past either of these boards.

Above: Wicker goods depot was sandwiched between Spital Hill Carlisle Street and Saville Street. Note the closed Spital Tunnel to the GCR Sheffield to Penistone line at Bridgehouses, this being the route arsonists took to get from Bridgehouses to Wicker when they set fire to both depots.

Left: The west gate with a Scammell Scarab and trailer entering the yard. The Scammell Scarab was a 3-wheeled tractor unit produced by the truck manufacturer Scammell between 1948 and 1967. Its name is derived from the notional combination of an Arab horse (replaced by mechanical horse in BR usage) and the word Scammell. Sca-rab = Scarab. The registration RXX 450 dates from August 1955, the last two letters XX denoting a London area registration.
Chris Booth collection

Right: The brick and stone gateposts were still in situ long after the depot had been removed and replaced by the Pentagon Vauxhall car dealership seen here. The image also shows a sculpture commissioned in 1993 by Sheffield Development Corporation and constructed by Mayflower Engineering, entitled "Made in Sheffield". Designed by Amanda King and made in stainless steel, it was placed on the site of the entrance in 1993.

The sculpture and gate post were removed when the Sheffield inner ring road was constructed in 2006 and the cars later made way for a Tesco store which opened on 31st October 2011.
Dave Ball ©Sheffield Hallam University

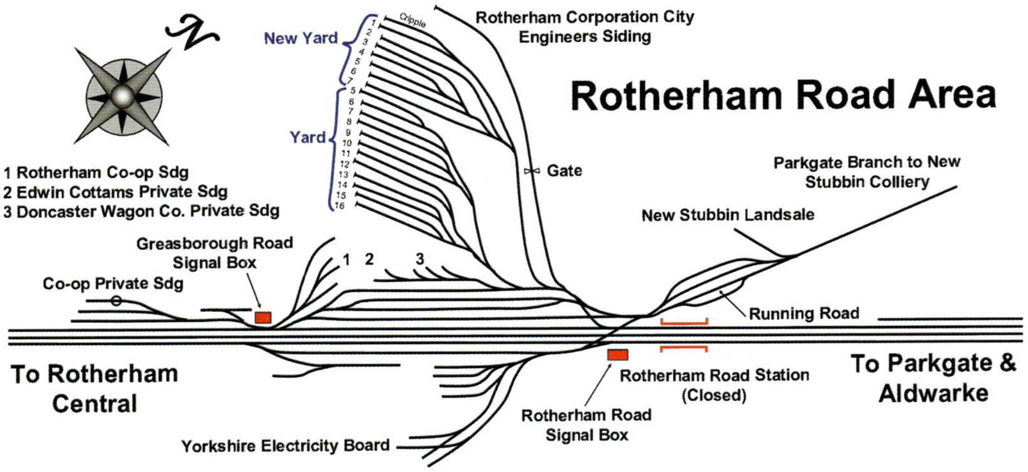

Goods Yards – Rotherham Road

Above: Rotherham Road, between Rotherham Central and Parkgate, was another yard with sidings either side of the line, together with a branch to New Stubbin Colliery and the South Yorkshire Coal & Chemical Company coke works.

Left: Rotherham Road signal box was a non-standard MS&LR design measuring 18 ft x 12 ft, opened in 1909 but second-hand from elsewhere, and latterly with a 32-lever GCR frame. Closed on 12th April 1987, the box was demolished over the weekend of 30th/31st January 1988. This 1981 photo, looking south-west, shows 20223 & 20208 with a loaded steel train, probably bound for Tinsley. 20223 was withdrawn in November 1987 and 20208 in September 1989.

The rusty lines coming in at the bottom right were from the MS&LR line, which ran alongside the Greasborough Canal and beneath the MR to reach the canal head and serve early colliery workings. Opened in August 1873, it also served Earl Fitzwilliam's New Stubbin Colliery, and later Park Gate Steelworks and the South Yorkshire Chemical Works.

The first sod of the colliery was cut on 14th November 1913 by Viscount Milton, son of Earl Fitzwilliam. Sinking was complete by 1915 and its first coal wound in November 1919. In the 1950s the colliery was surfacing half a million tons of coal per annum, and production continued until its remaining seams became unworkable. Production ceased on 6th July 1978 and the colliery closed, having produced 21 million tons of coal. The agreement between Earl Fitzwilliam's colliery company and the MS&LR effectively terminated on 31st December 1978 when the branch was closed. To the left of the picture can be seen the Sheffield & South Yorkshire Navigation. *John Law collection*

Right: The sidings and colliery branch have all gone, and the canal has undergone major changes. In 1983, the canal locks from the New Junction Canal, north of Doncaster, all the way to (just short of) Rotherham were extended to around 230 ft x 20 ft so that large 700-tonne barges and butties could use the locks together. This was part of a push to revitalise the waterway and encourage more traffic. Many new staithes and loading facilities were constructed, along with the *"winding hole"* used to turn barges, seen here behind 66131 as it passes with 6G03, the 13.00 Tinsley to Aldwarke on 27th June 2011.

This scene has changed again in recent times, with the addition of gantries supporting overhead wires for the Sheffield–Rotherham tram-train system.
Chris Booth

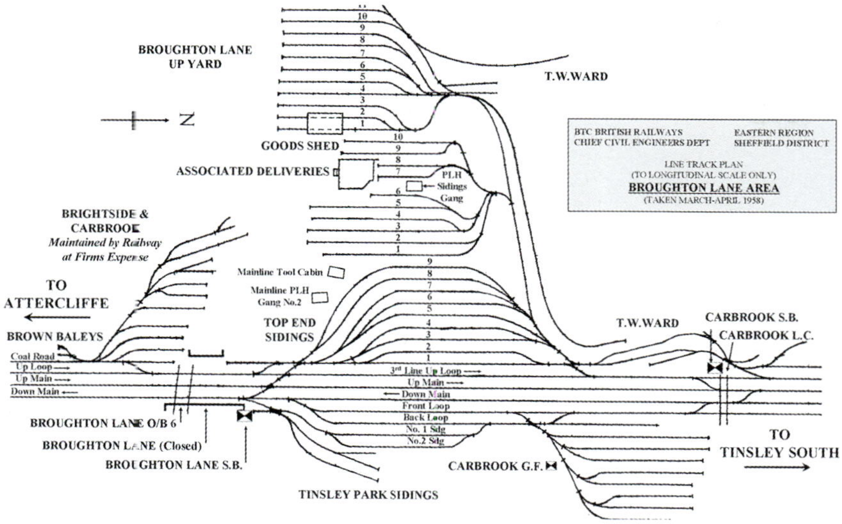

Goods Yards – Broughton Lane

Below: The remains of the former GCR station at Broughton Lane. It is 1981 and the wires are still in place for the connection from Tinsley Yard to the Sheffield–Manchester line over Woodhead. Broughton Lane station was opened in about 1864 and closed in 1956, so had been standing empty for 25 years at the time this picture was taken. The unidentified Class 45 could possibly be en route from Tinsley to Sheffield Midland to take over another express to the capital. Broughton Lane Top End sidings can just be seen beyond the station buildings. *Adrian Wynne*

Sectional Appendix Instruction 1st October 1960 Broughton Lane WORKING TO AND FROM UP SIDINGS

Drivers must not move their engines from the coal stage in the direction of the sidings until instructed by the signalman in Broughton Lane station box who must first obtain the permission of the shunter, nor must engines be turned into the sidings from the down main or up goods lines until permission has been obtained by the signalman from the shunter. Drivers must not move their engines from the up sidings in the direction of the coal stage or the up goods line until instructed to do so by the shunter in charge and the latter, before authorising such movements must obtain the permission of the signalman.

Below: Broughton Lane in the 1960s as an unidentified loco clanks past the signal box with yet another coal train. Behind the train, wagons are stabled in No. 2 siding and between the wagons and the signal box are Tinsley Park sidings, formerly the connection to Tinsley Park Colliery. Note the coaling stage for the yard pilots, between the signals. *Roger Wainwright collection*

Ickles Yard

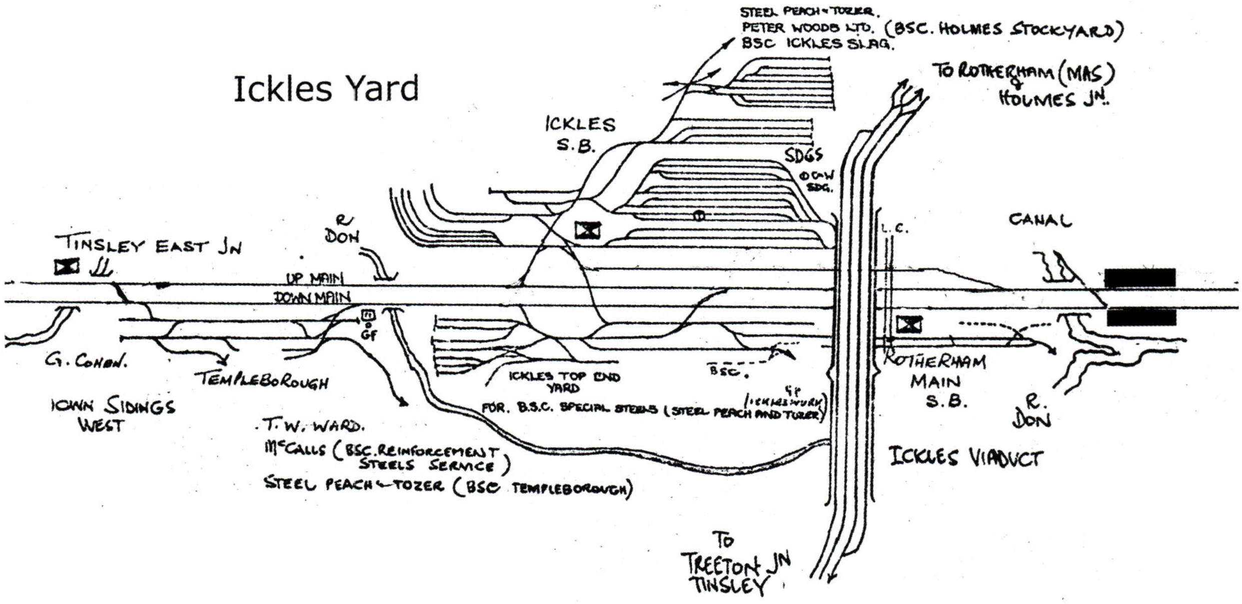

Goods Yards – Ickles

Right: 31214 and an unidentified sister pass Ickles with a train of empty coke wagons for Orgreave sidings. The signals in the "off" position are Ickles section signal No. 61 with Tinsley East distant signal beneath. The small arm below the distant signal is No. 47 Shunt Ahead signal. This is a type of signal which is normally mounted under the signal governing entry to the section ahead, and, as its name implies, allows a train to shunt forward into the section to clear a set of points without pulling off the main starting signal, which could give a driver the false impression that the section ahead is clear. In this case it would have been used to allow the train to move forward and then set back into the "down" sidings east, through No. 66 points, which the train in this view has just passed over. No. 60, the small arm to the right, would have been for access to the "down" sidings west, through No. 50 points.
Roger Wainwright

Right: Ickles Yard taken from the cripple sidings looking towards Rotherham. Steel Peech & Tozer is on the right of the image, whilst the long black building to the left of the image is the premises of Slag Reduction Ltd, the origins of which go back over a century to the days when Stanley Faber-Prest struck a deal to provide a range of steel works services to the Steel Peech & Tozer plant at Ickles. One of the services provided was to remove the slag left behind after the steel making process and dump it in huge tips near the plants. The story goes that one day, while walking across a slag tip at Penistone, Stanley slit open one of his brand new boots on a piece of metal in the slag and when the first magnetic separator was developed a few years later, he launched the Slag Reduction Company, with the aim of breaking up the slag, separating out any metal and selling it back to the steelworks.
Roger Wainwright

Above: Ickles signal box opened in 1929 when the yard was developed in conjunction with the Slag Reduction works. It had a GCR tappet-locked 80-lever frame which was situated at the back of the box. The box closed on 10th August 1984.
Roger Wainwright

Right: Ickles Yard as viewed towards Tinsley East from the Midland Railway overbridge. Steel Peech & Tozer is on the left and the signal is No. 30 which was 416 yards from the signal box. The No. 1 goods line has been lifted.
Roger Wainwright

Goods Yards – Queens Road

Right: Diagram of Queens Road goods yard adapted from a Network Rail diagram. Closed on 13th May 1963, most of the track was quickly lifted with the exception of the three roads closest to the main line, which were latterly used for carriage storage. The 1956 handbook for stations mentioned that Queens Road had a 5-ton capacity crane and was equipped to handle livestock and general merchandise. The closed site was taken over by Arnold Laver and used as a wood yard for many years before being cleared and a B&Q store erected.

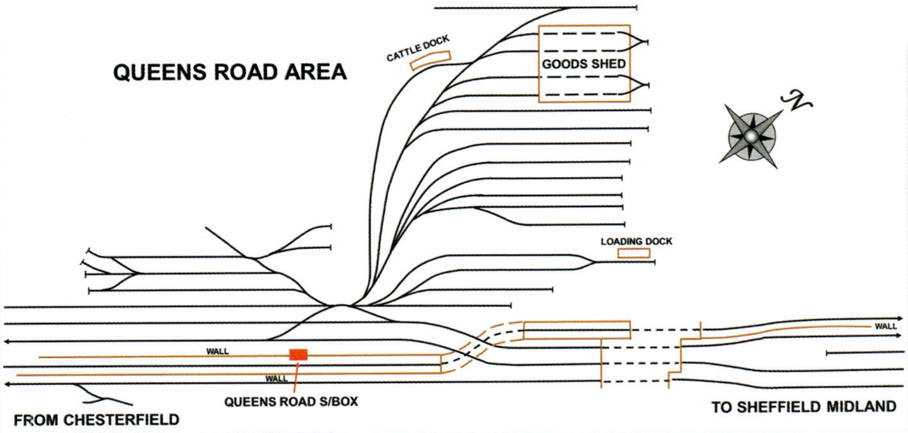

Below: East Midlands Trains 158866 & 158799 work 1L07, the 08.52 Liverpool Lime Street to Norwich past the site of Queens Road goods yard on 25th November 2014. The area is now occupied by a B&Q store but the East Bank Tunnel mouths can be seen. The green area is where the dive-under once was. *Chris Booth*

Below: Queens Road goods depot as seen from Charlotte Road overbridge on 10th April 1962. The two lines to the right of the image are the "up" and "down" Manchester lines from Sheffield Midland station, which is beyond the 80-yard long East Bank Tunnels on the right of the image. Just in view on the right is the dive-under that took the "down" main beneath the Manchester lines to gain access to platform 1.
National Railway Museum/Science & Society Picture Library DF403

Goods Yards – Bridgehouses

Below: Diagram of Bridgehouses depot during the LNER period. The line to Wicker can be seen curving off on the top of the plan, beneath the word "Sheffield". *Diagram courtesy of the Network Rail Corporate Archive*

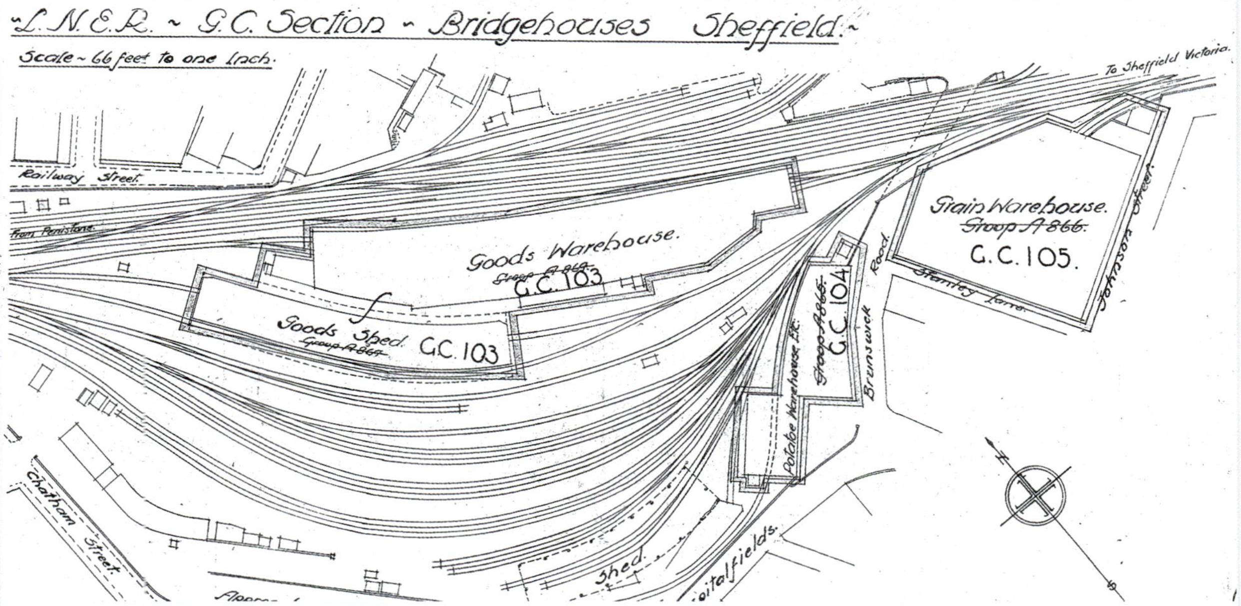

Below: A view of Bridgehouses goods depot between 1891 and 1895, looking towards Railway Street, Sheffield. Sheffield Victoria would be beyond the goods shed (formerly Bridgehouses passenger station) in the distance. In the foreground are a selection of LNWR horse-drawn drays. The signal box on the left is the GCR Sheffield Victoria No. 1, which was superseded in 1908 by a box on the other side of the line.

Bridgehouses was the subject of a BBC TV documentary entitled *"Engines must not enter the potato siding"*, which contained film and memories from former railway workers, made at the time of the rationalisation of goods traffic in the Sheffield Area. The opening of the new Sheffield Freight Terminal. The potato siding was situated high above the road and had a wooden floor, which would take the weight of wagons and vans but not locomotives, hence the sign attached to the building, reminding shunters and drivers that *"Engines must not enter the potato siding"*.
John Law collection

Goods Yards – Park Goods

Above: The former 18-road GCR Park Goods depot is seen in the early 1960s. The depot was used primarily for wood and coal; wagons containing timber baulks can be seen awaiting unloading in the area operated by Fred W. Machin & Sons. To the right of the image, just above the offices of Oxley & Styring Ltd Coal Merchants, are the two-track coal drops with the canal basin to their right. The bridge over the canal can be seen top centre, the line heading towards Bernard Road sidings. Sheffield Victoria station was to the extreme left of the image. *Chris Booth collection*

Below: The LNER knew this as Park Yard and here is a diagram from the 1940s showing the 18-road yard and canal basin. The yard closed in October 1963 and today the site has been redeveloped to include the Hilton hotel and Victoria Quays car park, whilst the arches for the former coal drops are converted to shops and a café adjacent to the redeveloped Victoria Quays canal basin. It is however still possible to discern where the railway bridge that crossed the canal was. *Network Rail Corporate Archive*

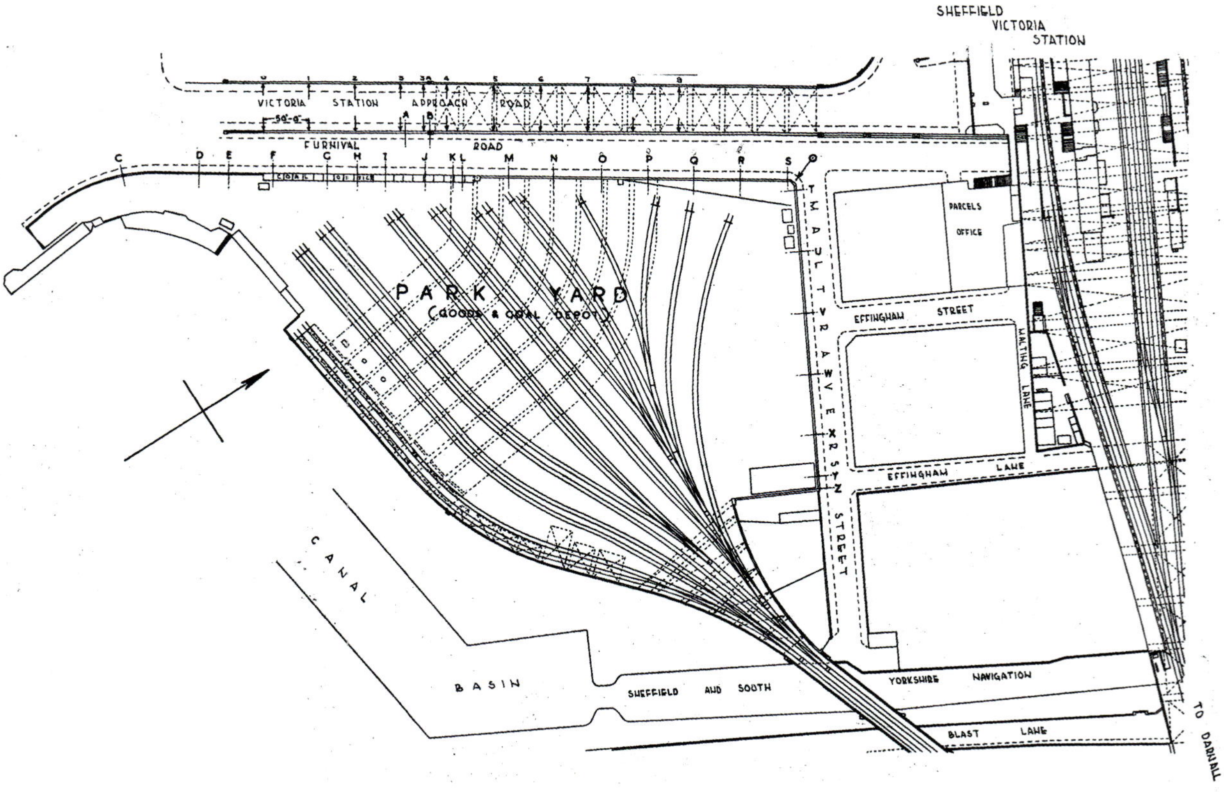

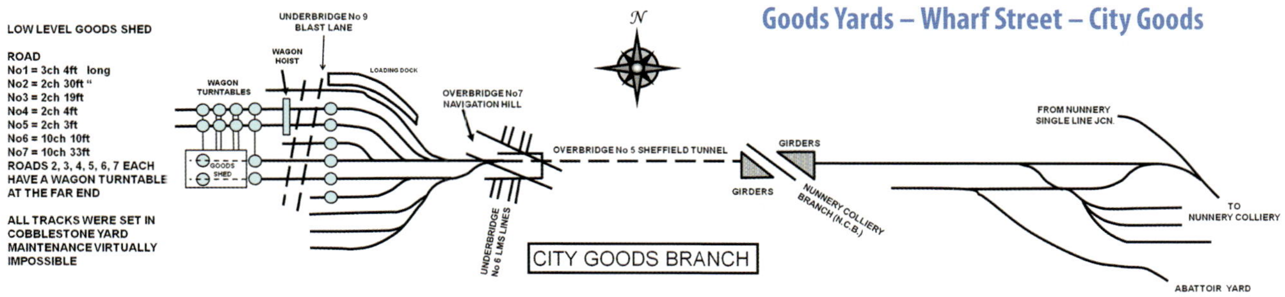

Above: Map of the LNWR City Goods branch. Note the multiple wagon turntables that were used in the depot along with the wagon hoist. *Adapted from a Network Rail map*

Above: The view down Navigation Hill after closure of the depot. Bridge No. 7 crossed the City Goods branch and the Midland Line from Sheffield, as can be seen on the diagram. This area was swept away when the Sheffield Parkway was built and today nothing remains of the depot. However, the Durham Ox public house on Cricket Inn Road, seen top left of the image, remains in situ albeit in derelict condition. This pub opened in 1862 and closed in 1993, latterly seeing use as an office in connection with the building of student flats next door. *John Gabbitas*

Above right: View of the 20-ton hydraulic wagon lifts on Nos. 2 & 3 roads at Wharf Street. These had a 10-ton lifting capacity and were used to lower wagons down to basement level (actually street level) for loading of goods on to road vehicles. *National Railway Museum/Science & Society Picture Library*

Chapter 3

Sheffield Area Dieselisation and Electrification

As has been well documented elsewhere, the electrification of the Manchester, Sheffield and Wath lines was originally planned pre-war and, in conjunction, Darnall would have become a depot for the electric locomotives to be built for this project. Although the electrification was put on hold, in 1943 the LNER built a new steam depot at Darnall and along with this, two pits were constructed in the anticipated location of an electric locomotive depot to service the future Classes EM1 and EM2. Post-war, a two-road brick and asbestos-clad building was constructed over these pits and this opened for business in 1952. Its use by electric locomotives was to be relatively short-lived, as plans changed and a new depot at Reddish opened soon afterwards. However, Darnall did house several EM1 locomotives when new and we know from photographic evidence that at least one EM2 made it there.

The first diesel locomotive in the Sheffield area was 12091, a 350 hp 0-6-0 shunter of what was to become BR Class 11. It arrived at Mexborough depot (36B) in July 1951, remaining there until October. It was to be another two years before any more diesels arrived, 13060–64 (later Class 08s) being transferred to 36B in August 1953, followed by 12103–05 in September. The Class 08s were allocated to Mexborough specifically to replace the Class S1 0-8-4T "Wath Daisy" steam locomotives that were used for shunting at Wath Yard.

May 1956 also brought Grimesthorpe (19A) its first diesel shunters in the shape of 13251–53. December 1956 saw Darnall depot receive its first diesel shunter when 13331 (08261) arrived brand new from Darlington works, followed by 13332–36 in January and February 1957. February also saw 11164 arrive at Darnall, followed in March by 11165. These two locomotives were of the type now known by their TOPS designation as Class 05, despite never officially having been designated as such by BR; both locos left Darnall in June 1957. The previous month had seen 13333 & 13335 allocated to Mexborough. Darnall's footplate staff were sent to Wath to train on the locos, but within weeks the Darnall branch of ASLEF demanded

Above: EM2 27003 at Darnall in 1957, still wearing its black livery with red lining. The loco is standing in front of the original asbestos and brick building, before its reconstruction into a diesel depot. Note the loco is carrying both a red tail lamp disc and an oil tail lamp!

Bestowed with the name "DIANA", the last of the class to receive a name, this loco was withdrawn on 5th October 1968 and sold to Dutch Railways (Nederlandse Spoorwegen), where it became No. 1501. After withdrawal from NS service it was saved for preservation by Werkgroep 1501 and has been used for railtour duties in Holland over the years.
Chris Booth collection

EM1 locomotives on Darnall depot in the 1950s

Left: 26027 and 26036 sit inside Darnall depot on 27th August 1955. Note that the overhead wires continue through the depot; these could be isolated via switch if necessary. *Roy Wilson/Transport Treasury*

Below: 26027 has now moved leaving sister 26036 in full view. Both locos have a red tail lamp shade over one of the electric lights, but also carry an oil tail lamp. *Roy Wilson/Transport Treasury*

that the new locos be double-manned. The response from BR was to move the diesels away from Darnall and reinstate Class N5 steam locomotives for various shunting duties. Grimethorpe depot had been using the diesels without problem since the previous year and it wasn't long before common sense prevailed and the diesels returned.

March 1958 saw 13326 arrive at Darnall, followed in April 1958 by 13127 and 13131. Incidentally 13127, which later became 08092, holds the distinction of having done the rounds so to speak of the Sheffield Area depots. Having arrived at Darnall in April 1958, it went to Mexborough in October 1959, returned to Darnall the following month before making a brief foray to Grimesthorpe in November 1959, returning to Darnall again by December, where it lasted until May 1964 when it was transferred to Tinsley depot.

In 1958 the Eastern Region announced that, in their plans for modernising Sheffield Area traction, Darnall would become the main depot

Below: New Hunslet-built shunter 11164 is seen in the company of its sister 11165, outside Darnall steam depot. The loco was allocated to Darnall from 19th February 1957 until 15th June 1957, when it was transferred to Norwich Thorpe (32A). After renumbering to D2561, it ended its days at Wigan Springs Branch (8F) on 26th August 1967, being scrapped at Duport Steels in Llanelli on 1st October 1972. *Rail-Online ZF-2120-67217-1-002*

Above: Derby "Heavyweight" DTCL (Driving Trailer Composite Lavatory) 56034, attached to power car 50034, emerges from the east end of Darnall depot on 2nd September 1959. This type of unit would be synonymous with the Sheffield to Lincoln passenger service until ousted by newer second-generation units in the 1980s. With the advent of computerisation, this car was renumbered 54034 to avoid confusion with the Class 56 diesel locomotive of the same number. It would be converted to parcels use in 1986, being renumbered 55993 then 54900. *Chris Booth collection*

for the area and the former electric shed would be rebuilt for diesel operation. The original brick and asbestos-built shed was totally demolished and replaced by a standard Eastern Region glass and steel building, which opened in April 1959.

It was initially proposed that Darnall would be allocated 60 sets of new DMUs; due to changes in policy the sets were never actually based at Darnall, but 22 sets did receive maintenance and cleaning here, these being Class 114 Derby Heavyweight units, mainly used on Sheffield to Lincoln services.

The units receiving maintenance were:
From 11/07/1959: 50024/26–31/33, 56024/26–31/33
From 31/10/1959: 50034/35, 56034/35
From 28/11/1959: 50015/22/24/36, 56015/22/24/36

Units also on loan around this time were: 50585/90/632/33 59197/227/31/41, these being returned to the North Eastern Region on 26/12/1959.

Grimesthorpe was to suffer a few transfers away between 1958 and 1960, but they were balanced out by new arrivals and as the depot went into 1960 it still had ten diesel shunters on its books. D4028–30 came to the depot in the summer of 1960 whilst D4031–33 went to Mexborough. From October to December 1960 Grimesthorpe's allocation was further swelled by 13 more diesel shunters, D4035–D4048. However, the depot's allocation of 30 shunters only lasted until September 1961 when they were all re-allocated en-masse to Darnall.

Darnall depot had provided steam locomotives as pilots for shunting Sheffield's freight yards until they were superseded by diesel locomotives. By 1960 the required diesel pilot diagrams were as in Table 1.

Table 1: Required Diesel Pilot Diagrams 1960

Bernard Road	1	Rotherham	4
Park Yard	1	Woodhouse	1
No 6 Down side	1	Tinsley	1
Bridgehouses	2	Wadsley Bridge	1
Harvest Lane	1	Oughty Bridge	1
Broughton Lane	3	Deepcar	1
Grange Lane	1	Penistone	1
Handsworth Colliery	1	Sheffield Victoria	4
Meadow Hall	1		

As the BR Modernisation Plan began to take effect in the Sheffield Area, Darnall prepared to receive main line diesel locomotives to work trains.

Originally given the depot code 39B, this was amended to 41A in October 1955. Under Darnall came nine other sheds from various former districts of the LNER and LMS. As part of the changes, Mexborough became 41F; Wath electric depot was at this point still a sub-shed of Mexborough but with the closure of Sheffield's Millhouses depot on 1st January 1962, the code 41C became vacant and in 1963, almost certainly to coincide with the opening of a new diesel depot in Wath Yard, the code was bestowed upon Wath.

From 3rd October 1960, the BR Eastern Region decreed that the whole of the Sheffield Division would begin diesel training, as a prelude to full dieselisation of the division. December 1960 saw the first main line locomotive arrive at Darnall in the form of English Electric Type 1 (Class 20) D8020. This loco was new to Hornsey depot in October 1959, being subsequently transferred to Finsbury Park in April 1960 and then to Darnall in week ending 3rd December 1960. It stayed at Darnall until September 1962. Following allocations elsewhere in the country, it was allocated to Tinsley in 1966 and again from October 1970, staying there for the following seven years.

The first of many Brush Type 2s (Class 31) to be based at Darnall arrived on 12th November 1960. D5587 was soon followed by D5671 then D5680–93 and D5857–62 which were initially utilised on crew training trip workings to Barnsley. March 1961 saw further Class 20s arrive; D8050–69, were followed by D8021–24 in May 1962, bringing Darnall's allocation of the Type 1s up to 25.

In January 1961 Darnall received its first allocation of a Type 4 locomotive for training purposes when D13 (45001) arrived, although it only stayed until April 1961. June to September 1961 saw D22 (45132) allocated, followed from September 1961 to March 1962 by D101 (45061) and D107 (45120) from March to May 1962. The end of 1961 had also seen a brief allocation of a Class 46 in the shape of D145 (46008) which stayed until May 1962, with the following year seeing D155 (46018) allocated from May to June 1963, these being the only two members of the class allocated to Darnall.

Maintenance and cleaning of Class 114 DMUs at Darnall was to be short-lived as the units were transferred to the new Lincoln depot from 28th April 1962. Cleaning facilities for DMUs were then transferred to Sheffield Victoria to allow space to accommodate the new diesel-electric locomotives.

As more new diesel locomotives were introduced to the Sheffield Area, 33 Brush Type 2s (Class 31) were allocated to Wath, whilst surprisingly, although Canklow never received allocated diesel shunting locomotives, from November 1963 to January 1964 it was home to Type 4 D144 (46007) and from December 1963 to February 1964 Type 2s D5183 (25033) followed by D5184 (25044) from February 1964. Like Darnall it can be reasonably assumed that they were only there for training drivers and secondmen prior to their re-deployment to Tinsley depot.

More Locomotives for Darnall

Further types continued to arrive at the depot. English Electric Type 3s (Class 37) D6676–D6818 were new to Darnall from December 1962 through to March 1963, followed later in August 1963 by the first Brush Type 4s (Class 47) in the batch D1530–1569. The first three had been delivered new to Finsbury Park but from D1534 (47426) they were brand new to Darnall, with D1534–49 being built at

Above: Brush Type 2 locomotives D5840 and D5841 stabled between turns on Canklow depot in June 1965. Allocated to Darnall at the time, they would move to Tinsley in October 1965. With the arrival of TOPS, these would become 31307 and 31308; both locos have now been scrapped. *Paul Braybrook*

Right: D115, later 45067, sits at Darnall depot on 8th March 1964, whilst D5846 (later 31312) is serviced inside the shed. 45067 was withdrawn with collision damage on 31st July 1977 and broken up at Derby in July 1980. *P.Ayton/Chris Booth collection*

Below: Darnall photographed in July 1965. The Brush Type 2s on shed are D5684 and D5841. *Paul Braybrook*

Brush Traction, Loughborough and D1550–69 all being built at Crewe works.

March 1964 saw Mexborough shed lose its allocation of 12 diesel shunters to nearby Wath depot, which by then had lost its allocation of electric locomotives to Reddish; the former electric depot was by then in use as a diesel depot.

Diesel Workings from Darnall

The Type 1s allocated to Darnall were used on local trips to Canklow and Hasland etc. whilst the Type 2s were used on passenger turns to Manchester via the Hope Valley Line. Due to a shortage of diesels at Finsbury Park (34G) during early 1962, several of Darnall's Brush Type 2s were used in the King's Cross area on local services; D5680/1/2/3/5 were among those so used.

A "celebrity" locomotive allocated to Darnall from 7th April 1962, was the experimental D0280 "FALCON" (later taken into capital stock as Class 53, D1200). It was initially used on the "Sheffield Pullman" and "Master Cutler" trains to Kings Cross. During late 1962 and early 1963, it was used on heavy coal trains whilst out-based at Shirebrook West (41J) before being returned to Brush in November 1963. Various other types of locomotive would visit for fuelling, such as "Peaks" and Class 17 "Claytons". From 26th April 1964 all maintenance work was transferred from Darnall to Tinsley; at the same time Tinsley gained the code 41A, with Darnall re-coded 41B.

Decline and Closure of Darnall Depot

Once the freight concentration yard at Tinsley was completed in 1965, it replaced several of the smaller freight yards in the Sheffield Area and at a stroke this reduced the requirement for pilot locomotives. Consequently, as of October 1965 all remaining shunting locomotives were transferred from Darnall to Tinsley, leaving Darnall without a locomotive allocation. Table 2 shows the 1963 allocation, whilst Table 3 shows those locos transferred to Tinsley.

Although the servicing of locomotives came to an end, the depot saw a return to its former use as a DMU servicing point and in the 1970s new Mark 2 coaching stock also began to arrive at Darnall for servicing in connection with their introduction on Sheffield to St Pancras services. Former Class 15 locomotive ADB968002 (D8237) was in use at the depot between April and June 1975 as a source of power for the testing of air-conditioning systems. The Mark 2 coaches would continue to be serviced at Darnall until the arrival of HSTs on the Midland Main Line.

Table 2: 1963 Allocation for 41A Darnall

350 hp Diesel Shunter
D3086 D3127 D3129 D3131 D3251 D3252 D3253 D3254 D3288 D3289 D3293 D3307 D3326 D3330 D3336 D3439 D3443 D3475 D3574 D3575 D3659 D3660 D3661 D3662 D3663 D3664 D3685 D3698 D3702 D3703 D3707 D3727 D4028 D4029 D4030 D4031 D4032 D4033 D4034 D4035 D4036 D4037 D4038 D4039 D4040 D4041 D4042 D4043 D4044 D4045 D4046 D4047 D4048 D4049 D4050 D4051 D4052 D4054 D4055 D4056 D4058 D4059 D4062 D4063 D4064 D4065 D4068 D4070 D4071 D4072 D4073 D4074 D4089 D4090 D4091 D4092 D4093 D4094
English Electric Type 1
D8022 D8023 D8024 D8050 D8051 D8052 D8053 D8054 D8055 D8056 D8057 D8058 D8059 D8060 D8061 D8062 D8063 D8064 D8065 D8066 D8067 D8068 D8069
Brush Type 2
D5682 D5683 D5684 D5685 D5686 D5687 D5688 D5689 D5690 D5691 D5692 D5829 D5830 D5831 D5832 D5833 D5834 D5836 D5837 D5838 D5839 D5840 D5841 D5842 D5843 D5844 D5845 D5846 D5847 D5848 D5849 D5850 D5851 D5852 D5857 D5858 D5859 D5860 D5861 D5862
English Electric Type 3
D6742 D6743 D6744 D6745 D6746 D6747 D6748 D6749 D6750 D6751 D6752 D6753 D6754 D6796 D6797 D6798 D6799 D6800 D6801 D6802 D6803 D6804 D6805 D6806 D6807 D6808 D6809 D6810 D6811 D6812 D6813 D6814 D6815 D6816 D6817 D6818
Brush Type 4
D1500 D1501 D1531 D1532 D1533 D1534 D1535 D1536 D1537 D1538 D1539 D1540 D1541 D1542 D1543 D1544 D1545 D1548 D1549

Table 3: Locomotives Transferred to Tinsley

350 hp Diesel Shunter
D3086 D3127 D3129 D3131 D3251 D3252 D3253 D3254 D3288 D3289 D3293 D3336 D3574 D3575 D3576 D3662 D3698 D3727 D4028 D4029 D4035 D4036 D4037 D4038 D4039 D4040 D4041 D4042 D4043 D4044 D4046 D4048
English Electric Type 1
D8022 D8023 D8024 D8050 D8051 D8052 D8053 D8054 D8055 D8056 D8057 D8058 D8059 D8060 D8061 D8062
Brush Type 2
D5584 D5682 D5683 D5684 D5685 D5686 D5690 D5691 D5692 D5840 D5841 D5842 D5843 D5844 D5845 D5846 D5847 D5848 D5849 D5850 D5851 D5852 D5857 D5858 D5859 D5860 D5861 D5862
English Electric Type 3
D6742 D6743 D6744 D6745 D6746 D6747 D6748 D6749 D6750 D6751 D6752 D6753 D6754 D6796 D6797 D6798 D6799 D6800 D6801 D6802 D6803 D6804 D6805 D6806 D6807 D6808 D6809 D6810 D6811 D6812 D6813 D6814 D6815 D6816 D6817 D6818
Brush Type 4
D1500 D1531 D1545 D1548 D1549 D1550 D1551 D1552 D1553 D1554 D1555 D1556 D1557 D1558 D1559 D1560 D1561 D1562 D1563

During the late 1970s and early 1980s, Darnall was used to store withdrawn Class 123/124 DMUs and a number of locomotives. In 1977 several of the then new Class 56s being imported from Romania were stored here whilst awaiting attention to wiring and bogie problems.

As the use of Class 20 locomotives was being reduced at the end of 1982 and into early 1983, some of the class were to be found in store here, before an upturn in their fortunes saw them being reinstated to traffic. Those reported at the time were, 20003, 20092, 20095, 20106, 20130 and 20211.

Until October 1983, the servicing of DMUs based at Lincoln, Hull Botanic Gardens and Bradford Hammerton Street was carried out by Darnall at night and at weekends; after this date, the facility was withdrawn and DMU servicing transferred elsewhere.

After a period of use by the carriage & wagon department and for servicing of on-track machines, the final closure of all the former steam and diesel buildings came in 1989, leaving them to decay in the hands of the elements and the inevitable vandals; the end came in 1994 with the whole site being cleared.

Today the site is an industrial estate and there is no sign that a depot ever existed here.

Right and below left: Withdrawn Class 123 and Class 124 Trans-Pennine DMUs were stored in the former steam shed for a period in the mid-1980s. These two views show the different designs, Class 123 with the corridor connection and Class 124 without, awaiting their fate in the roofless shed during a snowy January 1985. Condemned coaching stock was also present at this time.
Dave Stones

Below: The former diesel depot was used to store some Class 56 locomotives when new. The first members of the class delivered from Romania to Tinsley had numerous defects which needed rectification before they could enter traffic. 56024 and sister 56023 are seen here receiving attention during April 1977, shortly after delivery to the UK.
Rail Online

Chapter 4

Sheffield District Railway

Had it not been for the location of the Sheffield District Railway, Tinsley Yard could well have been built in a completely different place and could maybe have been a totally different yard, so it is prudent here to take a look at the route that the building of the yard was to destroy.

The Sheffield District Railway was conceived as an independent route by which the Lancashire Derbyshire & East Coast Railway (LD&EC) could reach Sheffield, utilising running powers via the MS&LR line from Beighton into Sheffield Victoria. However, the MS&LR had been one of the opponents to the original LD&EC Bill and, whilst their opposition to this new company did not prevent the LD&EC from obtaining an Act, Parliament did not grant running powers into Sheffield Victoria. Subsequently, several Sheffield businessmen came together to try and garner interest in short connecting lines around the Sheffield area which would be of benefit to both the LD&EC and the MS&LR. At the time of asking in 1894, the LD&EC's finances were in a poor state and they were unable to help, whilst the Great Eastern Railway (GER) also refused the plea; as a result, the plan went no further.

The same businessmen made another plea in 1896, this time to build an independent route from the LD&EC at Spink Hill, through Hackenthorpe, Handsworth and Darnall, with running powers again for passenger traffic via the MS&LR from a spur at Tinsley, to Sheffield Victoria. On 4th August 1896 the SDR was incorporated and, as the LD&EC still had a desire for its trains to reach Sheffield, it was backed by the LD&EC and eventually the GER. Although the MS&LR would still not grant running powers over their line, the GER and LD&EC were to enjoy running powers over the SDR line.

The LD&EC agreed to work the line for 50% of gross receipts, but the SDR would remain in charge of maintenance of the permanent way. However, the lack of running powers into Sheffield Victoria had virtually confined the SDR to being a freight only line. The Attercliffe terminus was well positioned for goods traffic but was too far from the city centre to be of use as a passenger terminal of any significance. In a competitive blow, the Midland Railway

Above: Cutting the first sod of the Sheffield District Railway on 20th November 1896. The Duke of Norfolk is seen here in a photograph from The Sketch newspaper, with his hands on the commemorative spade and surrounded by other top-hatted dignitaries. *Chris Booth collection*

Below: Catcliffe Station looking west towards Brightside when the station was open to passengers, with at least 13 passengers awaiting a train. Catcliffe signal box at the end of the platform was 1135 yards from Treeton Junction. This was a RSCo box with a 29-lever RSCo tappet-locked frame, built in 1900. *Chris Booth collection*

(MR) also promoted a Bill for a line in the following parliamentary session. This railway was to be 3½ miles long from a junction to the north of Sheffield, between the Rotherham–Sheffield line at Brightside and the MR main line at Treeton. This unexpected move led the SDR to begin urgent talks with the MR, as it was expected that obtaining capital would be very difficult and profits on two competing lines was likely to be very sparse.

There was to be a favourable outcome to the talks, with agreement that the SDR would take over the MR scheme which was much shorter than the original plan, calling for 9½ miles. The MR were to grant perpetual running powers to the two companies over the line, and then along the MR line from Brightside into Sheffield, to a point where the spur to the SDR's proposed Attercliffe goods terminal would be formed, with further running powers for passenger trains to continue into Pond Street station.

Harry Willmott, the LD&EC's general manager, liked the new scheme and persuaded his board to support it, thus the way was clear for a further Bill to be put in front of Parliament. The abbreviated route was sanctioned by Parliament in an Act of 12th August 1898, which incorporated the running powers for the LD&EC and the GER over the MR, and permitted the MR running powers over the SDR.

Work Begins

Prior to the MR's proposals, construction of the line began on 20th November 1896 when the Duke of Norfolk cut the first sod at Attercliffe. The main contractor was Messrs C.J. Wills, whilst a further contract for £15,841 was awarded to Messrs Price & Wills to extend the LD&EC's Beighton branch to connect with the MR at Beighton. This connection was south of the MR/MS&LR junction at Beighton, thus through running to Sheffield Victoria could still be accomplished if required.

The route was opened on 20th May 1900 by the Duke of Portland; the opening ceremony included luncheon for 400 guests, amongst whom were the Lord Mayor of Sheffield, the Duke of Portland and the GER chairman, Lord Claud Hamilton. Goods traffic commenced running on 29th May 1900, with passenger traffic the following day.

The SDR Route Described

Starting from the Treeton end on an embankment, the first engineering feature was the nine-arch Catcliffe viaduct across the Rother Valley. The arch over the Rother itself was of 40 ft span while the remaining eight spans were 30 ft, one of which was later bricked up and used as business premises, occupied by a Mr J Russell Boot Repairer along with a Grocer and Fruiterer. Stations were of wooden construction due to the nature of their positions on embankments.

The first station, at Catcliffe, ½ mile from Treeton, was nicknamed "Klondike" by the locals due to its bleak location. Catcliffe was followed by the 80-yard Tinsley Wood tunnel near Brinsworth. Beyond the tunnel was open countryside, then the line rose on another embankment until it crossed the MS&LR Sheffield–Rotherham line on a steel girder bridge. Just west of the MS&LR line was Tinsley Road station, 2m 57ch from Treeton Junction. The remaining distance of around a mile passed through open fields to Brightside Junction on the MR line. Engineering features on this section included Brightside viaduct, a seven-span structure with a lattice girder span of 100 ft over the River Don, and an 80-foot plate girder bridge crossing Meadow Hall Road, beyond which the tracks curved south to meet the MR, a short distance south of Brightside station.

There were two short spurs; the first to a large goods depot built at Attercliffe, temporarily worked by the MR from 20th December 1899, accessed from a branch at Sheffield District Junction, seven chains south of Grimethorpe Junction. There were two girder bridges and a twin span bridge over Sanderson's Mill Race along this short section before Attercliffe goods depot was reached 4m 36ch from Treeton Junction. The 40-acre goods yard required substantial supplementary works during construction, including a diversion of the River Don and raising of the ground level by 12 to 15 ft. The goods shed building itself was two stories high with provision made for possible future extensions. It was provided with hydraulic cranes and a steam-operated overhead crane. Apart from the usual facilities (cattle pens, horse and carriage docks etc.) there was also a platform provided for passengers which, as far as is known, was never used for its intended purpose.

Above: West Tinsley station, looking towards Brightside. First opened on 30th September 1900 as Tinsley Road, it was renamed by the GCR on 1st July 1907. As the station was situated on top of an embankment, it featured wooden platforms each with wooden buildings to reduce weight. It closed to passengers on 11th September 1939. West Tinsley signal box can be seen beyond the far end of the "up" platform. *Chris Booth collection*

There was also a Goods warehouse and yard adjacent to West Tinsley station which, being at road level, was reached by a fairly steep incline. Edgar Allen and Company's Imperial Steel Works was also reached via this route. For such a short line there were a large number of private sidings due to the nature of heavy industry in Sheffield. 1903 saw several other short branches opened to serve local industries including Cooper & Co, Hadfield's Steel Foundry, William Jessop and Sons, Sanderson Brothers & Newbould, and Woodhouse and Rixon.

Above: Tinsley Park Colliery box was 1 mile 397 yards from Catcliffe and had a 34-lever RSCo tappet-locked frame. It is seen here just before work began on the yard which would see its demise. *Chris Booth collection*

Collieries

The only colliery to be served was Tinsley Park which opened on 30th November 1903, with a branch over one mile in length. Tinsley Park Colliery closed in early 1943 and the site was used by the Ministry of Fuel & Power as "West Tinsley Disposal Point" from around 1945 until September 1957. The track was removed in December 1957.

Signalling the SDR

The east end of the Sheffield District route connected with the MR "Old Road" and was controlled by Treeton signal box (renamed Treeton Junction), this being a MR Type 2b box opened on 1st April 1900. Brightside Junction controlled the connection with the Sheffield to Rotherham line at the west end. Signal boxes on the route were provided by the Railway Signal Company (RSCo) at Catcliffe, Tinsley Park Colliery and West Tinsley, with RSCo tappet-locked frames. Attercliffe Goods signal box controlled the confines of Attercliffe goods depot. May 1911 saw a new siding to Hadfield's tip commissioned on the "up" side of the line at the Treeton end of Catcliffe viaduct, with a new ground frame released by Catcliffe signal box.

Passengers End, Freight Continues

The SDR closed to passengers from 11th September 1939 for the duration of WW2 but reopened briefly afterwards from 6th October 1946 until 17th March 1947. Goods traffic continued afterwards; Catcliffe station closed to public traffic in 1955, but Hadfield's tip siding remained in use.

As the years progressed the Sheffield end of the line would remain busy and the 1956 Railway Clearing House "Handbook of Stations" shows the following sidings along the line:

- West Tinsley with private sidings
- Edgar Allen & Company's Imperial Works
- W Cooke's Tinsley Iron & Wire Ropes Co Ltd (also used by Birnam Products Ltd)
- Tinsley Wire Industries Ltd
- Cooper & Turner Ltd
- Vulcan Rivet & Brass Works
- Firth-Vickers Stainless Steels Ltd Shepcote Lane Works (also used by Shepcote Lane Rolling Mills Ltd)
- Firth-Vickers Stainless Steels Ltd Staybrite Works (also used by Firth Derihon Stampings Ltd)
- Hadfield's Ltd East Hecla No.1 Works
- W Jessop & Son's Brightside Works

Tinsley Yard

With the construction of the new yard underway by 1961, the SDR closed as of 30th September 1962 and the track west of Catcliffe viaduct was removed in October 1962 with temporary buffer stops being erected.

Left: The cutting approach to the 80-yard long Tinsley Wood tunnel, viewed from Wood Lane towards Catcliffe on 11th April 1962. Tinsley Park Colliery's "up" distant signal is just beyond the 24-inch gas pipe which feeds the United Coke and Chemical Company. The cutting was filled in with spoil from the yard, Tinsley DMD being built on top of the cutting to the left of the image. Note the dump truck above the tunnel transporting spoil to the tip at Orgreave. *BR DM307 Alan McKenna collection*

Left: The Great Central Railway goods and grain warehouse at Attercliffe, once the LD&EC's property and situated at the end of a short branch off the Midland Railway line at Grimesthorpe. Despite being immediately adjacent to Grimesthorpe (Midland) shed it apparently kept its GCR character until its closure as a BR goods yard in the early 1960s. Shunting was carried out here between 07.00 and 22.00 by a loco from Tuxford, which arrived on a goods working at 06.55 and returned to Tuxford at 22.20. Today the goods yard site is home to a scrap metal business and trains of scrap still use the line occasionally. *Chris Booth collection*

Right: West Tinsley goods shed in use as a pattern store in November 1978. *Ted Evans*

Below: British Railways 1960s Attercliffe goods yard diagram.

Below right: West Tinsley station and goods yard diagram. *Reproduced courtesy of Network Rail Corporate Archive*

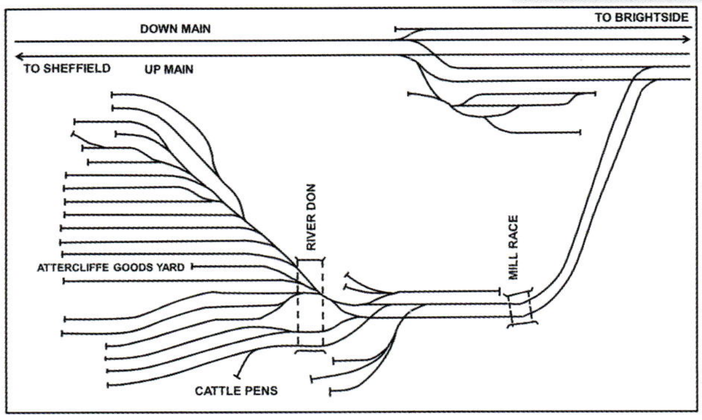

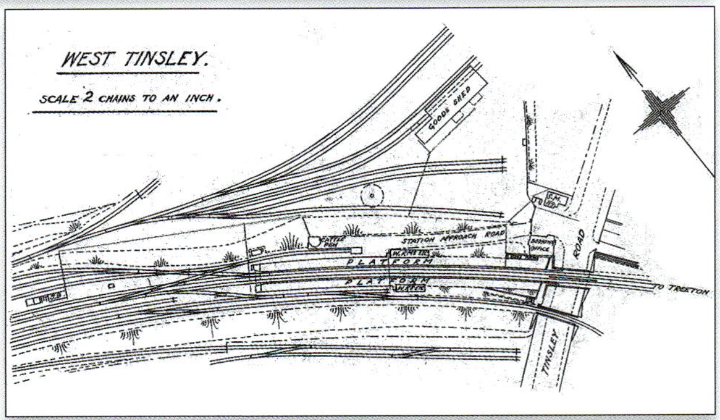

Above: Looking from Parkhouse Lane towards Catcliffe, this is the rail connection to the English Steel Corporation works, which were built on the higher ground to the right of the image. The signals are Tinsley Park Colliery signal box "up" home and "down" starter, photographed on Wednesday 11th April 1962. *BR DM308 Alan McKenna collection*

Above: The view from the top of Tinsley Wood tunnel looking east towards Catcliffe on 11th April 1962. As can be seen, work has begun on the yard as a Ruston-Bucyrus dragline machine is at work on the cutting top. This cutting would be filled with spoil from the workings. *BR DM 3C6 Alan McKenna collection*

Left: The same cutting from the top of the tunnel, this time looking west towards Brightside with Wood Lane bridge in the distance. *Ted Evans*

Below: The 80 yard long Tinsley Wood Tunnel from the Catcliffe end looking west. *Chris Booth collection*

The new north-east departure line at the Catcliffe end of the route was laid between June and September 1962, the construction of this curve requiring a cutting digging to a depth of about 90ft through Hadfield's spoil tip. Once completed this connection remained out of use until September 1963, after which, as the new Tinsley diesel depot had opened, it became the access to the depot from the Rotherham direction. The temporary buffer stops erected at the west end of the viaduct were moved to the bottom of the zig-zag access to the higher level diesel depot, with sufficient standage for two locomotives. All the points into the new reception sidings at the east end of the yard were clamped to prevent access; this trackwork had all been laid between March and June 1963.

The supplementary operating instructions of March 1963 gave notice to delete the line between Treeton and West Tinsley from the appendix, while Hadfield's tip siding at Catcliffe was deleted from the notices by April 1963. However, following closure of the SDR, the new north-east arrival line was still used for inward traffic to Hadfield's tip from September 1963. Outward traffic was via the original south-east line towards Treeton Junction. The tip did not close until circa-June 1967, the ground frame release being taken over by Tinsley Yard PSB in 1965.

Below: The remains of West Tinsley station clung on long after closure and are seen on a frosty day in the early 1960s. West Tinsley signal box can be seen over to the far left and to the right is the Edgar Allen Imperial Steelworks which has since been demolished, the land now forming part of the Meadowhall retail complex. *Steve Ryszka*

Left: The RSCo West Tinsley signal box retained its original GCR colours until its end in 1962. More of the long-gone steelworks can be seen in the background. *Steve Ryszka*

Below: As viewed towards West Tinsley station, the signal box looks cold and unoccupied, although with smoke issuing from the chimney, one of the ancillary buildings looks to be in use. There is a good stock of coal for whoever occupies the buildings! *Steve Ryszka*

Chapter 5

Aldwarke Junction – From Inception to Closure

A feature of the Sheffield rationalisation scheme was a plan for a double track curve to connect the former MR lines with the former GCR lines close to Wath Road Junction near Mexborough. This plan utilised a long-disused trackbed that formed the South Yorkshire curve, a connection originally from the MR, north of Swinton, to Adwick Junction on the GCR Mexborough to Wath route (see map). This connection had opened in 1854 but was rarely used and closed shortly afterwards. It reopened in 1875, only to close again by 1882!

The project to relay the South Yorkshire curve was initially costed at £48,000, but later amended to £53,690. Had the curve been reinstated it would have allowed GCR traffic to gain access to the MR lines and give a second route to the new Tinsley Yard, but due to the GCR line running almost at right angles and beneath the MR lines, it would have suffered a hefty gradient with consequent load restrictions.

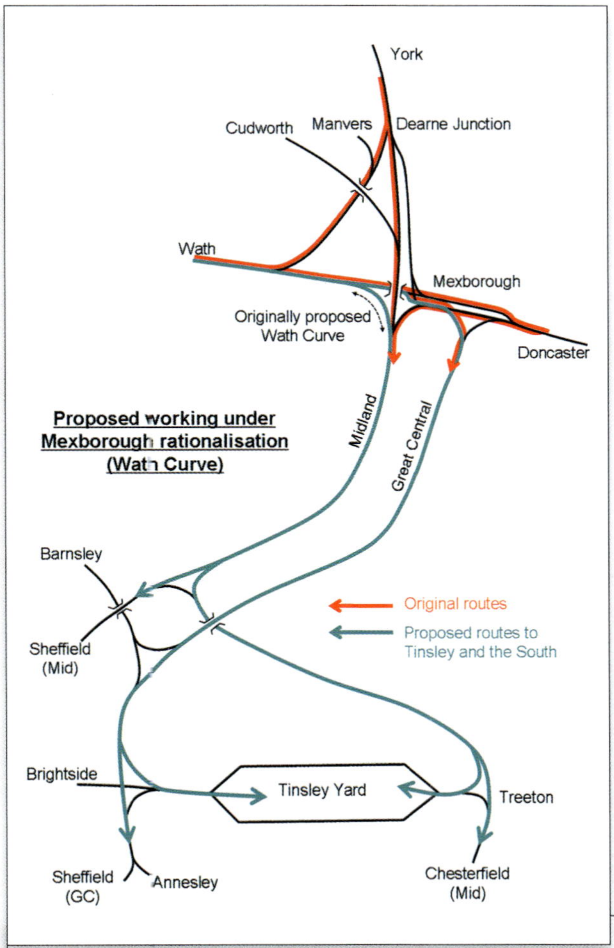

Consideration was given to closure of the former GCR line between Mexborough, Rotherham and Tinsley, but this was discounted as the original South Yorkshire curve project, with its attendant disadvantages, would necessarily have to be re-introduced, and a new chord would be required at Brightside Junction thus making a triangular junction linking the Sheffield District line to the MR. Other civil engineering challenges would have included the expensive bridging of a canal and a road at high level.

There were also 24 private sidings between Mexborough and Tinsley on the GCR Rotherham branch, most of which were connected to the GCR only. The annual traffic handled in those GCR-connected sidings is shown in Table 1 and, as can be seen, closing the GCR route would have resulted in considerable loss of traffic with alternative arrangements having to be made.

Left: Diagram of the proposed Wath connection, showing the original routes of trains from Wath and Mexborough to gain route to the south, which included propelling moves at Dearne Junction, and the proposed route with the addition of the Wath curve. As can be seen, this curve would have cut down on mileage, along with propelling and run-round moves, but its gradients would have been a disadvantage. *Diagram by Alex Fisher*

Table 1: Siding Connections .

Sidings Connected to GCR and MR		
Forwarded	375,000 tons	31,000 wagons
Received	1,225,000 tons	85,000 wagons
GCR-Connected-Only Sidings		
Forwarded	636,000 tons	52,000 wagons
Received	643,000 tons	53,000 wagons
Total GCR Traffic		
Forwarded	1,011,000 tons	83,000 wagons
Received	1,868,000 tons	138,000 wagons

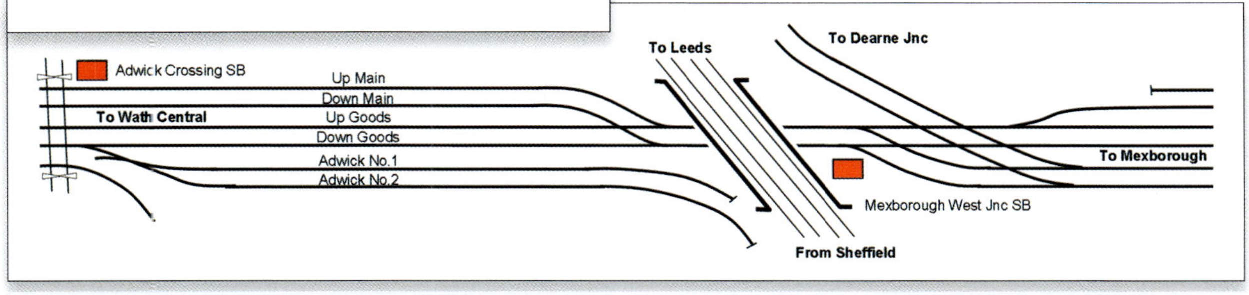

Above: Diagram of lines in the Mexborough West Junction area. The truncated connection from the GCR to the MR lines can be seen, these being annotated Adwick No. 1 and Adwick No. 2.
Reproduced from a Network Rail diagram courtesy of the Network Rail Corporate Archive

To overcome the gradient problem a new scheme was put forward to create a connection at Aldwarke, to the north of Rotherham, where the GCR and MR lines ran parallel. Although this new scheme, at £210,000, was almost four times the cost of the South Yorkshire curve, it would allow much greater flexibility. Most importantly it would also allow trains of load capacity on the GCR or MR routes access to Tinsley, and the option to be routed to arrive at either the east or west end of the yard. Similarly, trains leaving the yard in a northbound direction could leave from either the Shepcote Lane or Treeton ends of the yard.

With the Aldwarke scheme chosen, a small power signal box (PSB) was constructed and commissioned on 7th March 1965. The box was located adjacent to the former MR "up" slow line between Kilnhurst South and Rawmarsh boxes. Here, the double track GCR lines ran almost parallel to the MR quadruple track line and a new junction was created to allow through running between the two routes. New tracks connected the GCR lines to the "slow" lines of the MR in the direction of Leeds and conversely the "slow" lines of the MR to the GCR, in the direction of Mexborough. A flat crossing was installed at the point of intersection of the new tracks. It was not possible to cross between the "slow" and "fast" lines on the MR at this point until 1973 when ladder junctions were installed. At Swinton Town station the double track curve towards Mexborough was closed.

The improvement in operating flexibility contributed to substantial savings in several ways:

- More balanced use was made of the MR and GCR routes.
- The adverse load restriction of the original South Yorkshire curve was avoided.
- Heavy trains were diverted from the severely-graded GC route, thereby eliminating the use of two Type 1 banking locomotives at Tinsley.
- Diversion to the less-steeply-graded MR lines made it possible to run heavier trains, with associated reductions in movement costs.

The estimated annual saving from re-routing services amounted to £45,000, with a further £20,000 saved by eliminating the two banking locomotives at Tinsley, and £2,000 through economies with signalmen's wages with the closure of Aldwarke Main Colliery and Roundwood boxes. After increased annual charges and interest had been taken into account the net annual savings amounted to £49,000 which gave a return of 30% on the additional capital of £162,000 employed (1960s prices).

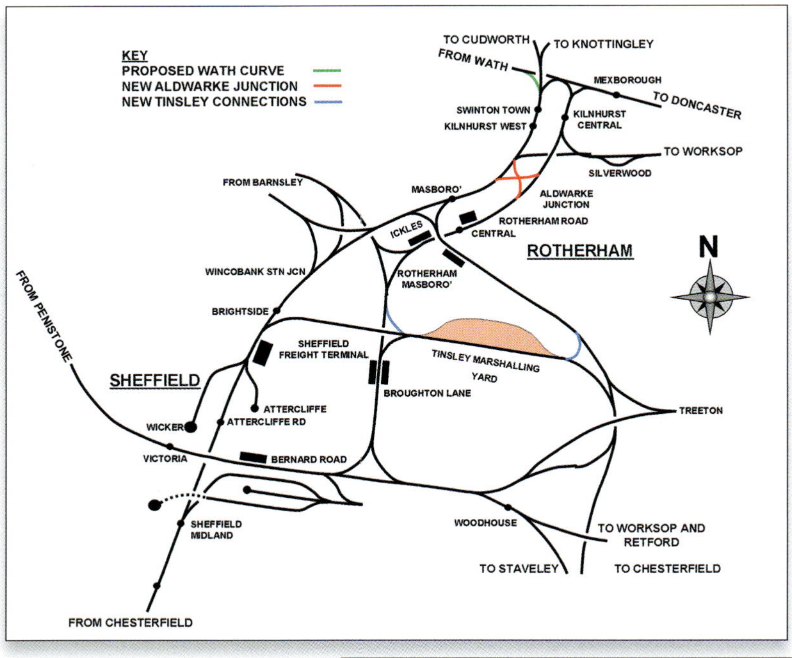

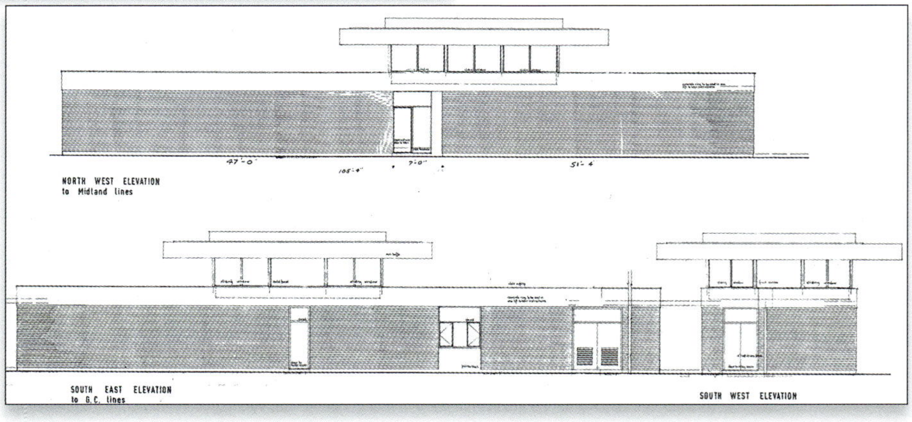

Above: The location of the new Aldwarke Junction showing how access was enabled to both ends of Tinsley Yard by the installation of this junction, along with the new connections at Treeton and Tinsley South.

Right: Diagram of the new Aldwarke PSB.
Reproduced courtesy of the Network Rail Corporate Archive

Around Mexborough, a simplification scheme to remove redundant running lines and reduce signalling was put on hold until the Aldwarke scheme was completed, but it was expected the Aldwarke scheme would lead to a reduction in train delays at Mexborough equivalent to an annual saving of £7,000.

Aldwarke Junction PSB controlled four-aspect colour light signalling on the former MR "fast" lines and three-aspect signalling on the former MR "slow" and GCR lines. After the closure of Aldwarke Main Colliery box, track circuit block (TCB) signalling regulations were introduced and Roundwood box was reduced to ground frame status; the old box was retained to control the connection to the sidings, the single line from Roundwood Sidings No. 1 box and the onward single line connection to Don Bridge East Junction. It was finally replaced by a small ground frame on 5th February 1967.

Closure and Major Changes

Aldwarke Junction PSB closed on 24th February 1980, its functions being taken over by Sheffield PSB. The Aldwarke area would see changes during the late 1980s in connection with the building of a new station at Swinton and the new Swinton curve which would enable through working between Aldwarke Junction and Mexborough via the new Swinton station. From Sunday 10th December 1989, in preparation for the opening of the Swinton curve, the new Swinton Junctions were brought into use along with revised signalling between Aldwarke Junction and Dearne Junction.

Sheffield PSB's control of the Aldwarke interlocking was to be short-lived. The whole area between Aldwarke, Mexborough and Swinton was earmarked for remodelling and, as rationalisation and resignalling continued, Aldwarke Junction PSB re-opened, firstly as a continuously-manned local control point from 25th March 1990, then as a block post from 23rd September 1990. The MR lines between Masborough Station North Junction and Aldwarke Junction were reduced to double track, with the removal of the former main lines. The GCR lines between Aldwarke North Junction and Aldwarke South Junction were also removed and the connection to the GCR in the Mexborough direction was moved to be a single-line ladder junction north of the existing junction, to be named the Roundwood chord. All this removed the complexity of the flat crossing.

The new Swinton station was brought into use from March 1990 and all passenger traffic was thus routed this way, with the former GCR line from Aldwarke to Mexborough being utilised primarily for goods traffic but also available for passenger traffic when necessary. Aldwarke Junction PSB closed for the second time at 00.01 hours on 11th July 1992 and control of the area returned once again to Sheffield PSB, along with control of the new interlocking for the Roundwood chord. The chord was single track as far as Thrybergh Junction, with the existing double-track being retained from there to Mexborough, where the junctions were singled and remodelled.

Control of the area changed once again with the closure of Sheffield PSB on 30th April 2016, its functions being taken over by the York Rail Operating Centre.

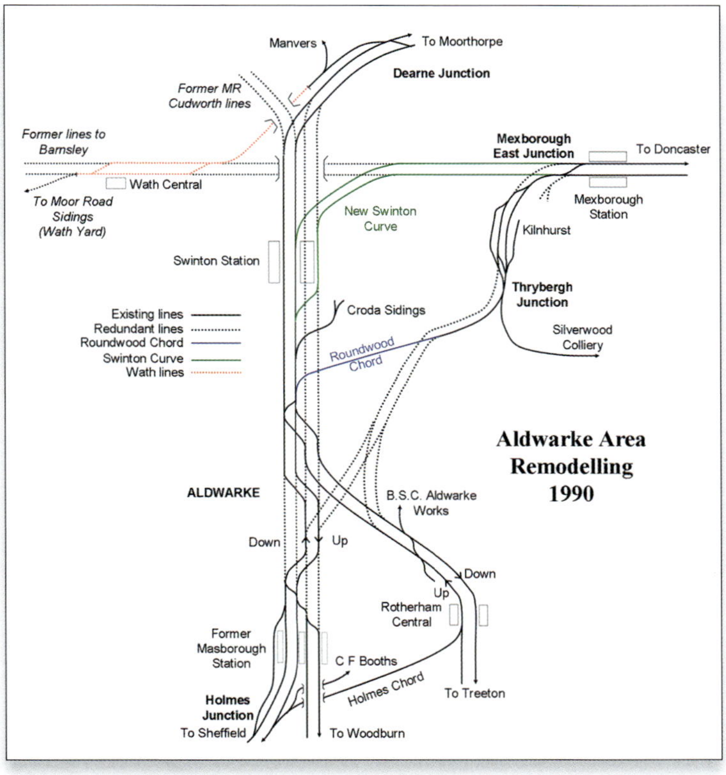

Above: Aldwarke Junction and surrounding areas after remodelling and the construction of the Roundwood chord and Swinton station.
Alex Fisher

Above: On 18th March 1965, signalman J.R. (Ray) Gregory is operating the panel; the train recorder's name is unknown.
Kidderminster Railway Museum

Right: 27th March 1990. Signalman Andrew Overton is seen in charge of the rationalised panel during the brief reopening period.
Andrew K. Overton

Below: On 18th March 1965, Brush Type 4 D1782 crosses from the GCR lines to the MR lines with 8M52, the 13.10 Wath to Toton, conveying 21-ton coke hoppers. The train will have originated from Manvers Main coking plant. A story told by the S&T department was that track circuits caused issues at Aldwarke due to coke falling off trains as they were crossing the junctions. As the coke was porous it soaked up rainwater, which would cause the track circuits to fail and continuously show occupied.

As for D1782, this was new to Tinsley on 14th November 1964, moving to Stratford in 1966 and returning to Tinsley during week ending 9th November 1968. It was transferred to Immingham in week ending 11th July 1970, being withdrawn on 31st January 2003. *BR TFD.115 Chris Booth collection*

Right: A Class 31 passes Aldwarke Junction with a passenger train from the former GCR route to the former MR route bound for Sheffield Midland in August 1986. This view illustrates the junctions perfectly. *Phillip Chrome*

Above: A view of the junctions looking north on 30th March 1990. This was taken after rationalisation had begun and the connection from Aldwarke North Junction (GCR) to Aldwarke South Junction (MR) is out of use. Most of the connection has already been lifted and piles of sleepers and rails are awaiting recovery. A selection of yellow British Rail "chip vans" are in attendance, a type of vehicle not seen nowadays. The area on the far side of the bridge has changed out of all recognition, after open-cast mining was carried out. *Andrew K. Overton*

Chapter 6

Tinsley Marshalling Yard: First Plans

As D.M. Howes, the London Midland Region (LMR) movement officer said in 1962, "There can be few railway subjects on which there is a wider divergence of opinion than marshalling yards". Discussion, he says, "spanned the two extremes from those who believed that marshalling yards should not exist at all to those with the 'playing at trains' mentality who forecast dire consequences from the closing of but a single yard".

There were 966 marshalling yards upon the inception of BR in 1948; nationalisation should have afforded the opportunity to strategically plan the best use of the entire railway system and to eradicate unnecessary duplication. However, in the 10 years following nationalisation only three marshalling yards, designed along modern lines and with up-to-date equipment, were constructed.

The LMR marshalling yard plan provided for a reduction in the number of yards from 111 to 42, the essence being to provide a small number of major marshalling yards, with a rather larger number of what Mr Howes called "satellite yards". The satellite yards were intended to provide feeder services to and from the trunk yards and to serve origin and destination yards in their locality. It was hoped that through this arrangement, intermediate handling of wagons would be vastly reduced and the distances over which wagons moved between shunting and unloading would be substantially increased. With regard to the major yards, two differing concepts were acknowledged. The first type would serve a major industrial area where there was sufficient originating and receiving traffic to locate the yard as near as possible to points of origin and destination. The second type was described as a "through" yard. i.e. not dealing fundamentally with local traffic but rather located at the convergence of a number of routes and thus streaming the traffic for direct working either to destination or destination area.

The plan for modernisation and re-equipment of British Railways was conceived in 1954, re-appraised in mid-1959 and a report on its progress issued in May 1961. New factors were arising in the early 1960s; they may be summarised as follows.

- The final pattern of freight terminals was being determined. This would lead to an increase in the volume of traffic that could be moved on direct terminal-to-terminal services, thus bypassing intermediate marshalling yards.
- Development in road/rail transfer arrangements with the introduction of the demountable body/container.
- The introduction of the MGR principle for the movement of power station coal.
- Development of coal concentration depots for domestic fuel.
- Increase in block working, particularly fuel oil.
- Introduction of new diesel and electric traction allowing heavier loads and faster timings.
- The siting of major new terminals and traction depots alongside new marshalling yards.

Developing a new yard required two stages of consideration. Firstly, a detailed survey of existing traffic flows would be analysed to build a picture of the traffic to be transferred to the new yard. Secondly, the analysis would be revised to include estimates of future traffic levels, taking into account new methods of movement in full train loads, along with the viability and availability of traffic in a market in which BR's share was steadily decreasing. From 1961 to 1962 alone, BR's freight tonnage fell by 5% overall; 13.5% less minerals and 1.4% less coal was carried, whilst a 6.3% loss in merchandise traffic was at least offset by an 11.8% decrease in wagons of merchandise being forwarded – the greater reduction reflecting better usage and, in some cases, the higher carrying capacity of newer wagons.

Whilst the introduction of MGR working and the growth of block loads seems to have been happily embraced, there remained a substantial volume of wagonload traffic that would still need breaking up and re-marshalling. Thus, it was noted that any new yards to be built had to be of the most modern design, be able to achieve rapid throughput but avoid any damage to commodities, with a quick turnaround of locos and wagons.

New yards were also to be sited with consideration for traffic flows, so that maximum length trainloads travelled over the longest possible distances as a standard, not as an exception. Finally, the yard must require a minimum number of shunting staff and locomotives in order to operate as efficiently and economically as possible.

A number of alternative yard layouts were considered when Tinsley was first proposed, differing mainly through their relationship to the main running lines. In terms of both capital and operating costs, a single bi-directional yard dealing with both "up" and "down" traffic over a single hump and through a common siding layout was seen as the optimum design. Experience at the time indicated a peak of more than 4,000 wagons a day would exceed the maximum capacity of such a yard, but 3,500 wagons was an achievable target.

The ideal yard layout would lie between the "up" and "down" running lines to reduce conflicting movements. If it were necessary to build separate yards for "up" and "down" traffic then in order of preference the yards should:

- Lie between the main running lines,
- Sit to one side of the main running lines,
- One yard lie to either side of the main running lines.

Common to all designs, however, was that the wagons should be dealt with in a sequential manner. Thus, the ideal layout should have reception sidings, leading to sorting sidings and finally departure sidings. Additional sidings for secondary sorting, wagon repair and transhipping etc. should be accommodated with the minimum of interference to operations. With the location of Tinsley Yard already identified, an initial diagram showing the proposed layout was drawn up.

By now the initial plans for Tinsley were well advanced, but the method of wagon control to be employed at the yard was still undecided; contemporary press of the time alluded to the possibility of traditional locomotive humping over a secondary yard hump summit. However, the Dowty wagon control system, which enabled continuous movement of wagons over the hump was also under development and following two years of testing at Hull Inward Yard, the decision was taken to adopt this system at Tinsley.

Studies and traffic analysis of the existing yards around Sheffield in February 1964 revealed that Tinsley would have to separate traffic into 111 'sorts' (or sections), of which 71 would be for local yards, companies, private sidings or stations in the Sheffield Division. In addition, segregation of cripples and different types of empty wagons would be necessary.

The draft train plan provided for 57 westbound and 34 eastbound departures from the main yard every 24 hours. There would be 42 departures from the secondary yard and a further 20 from the express freight yard, 13 of which would be destined for onward movement via the Woodhead Route and so would be electric-hauled. Express freight traffic from Grimesthorpe goods depot, private sidings, etc. would be sorted for 17 main destinations. 46 services routed via the main line would run in Class 7* conditions or higher and would require fitted heads.

7* was a BR classification for trains running with braked wagons; basically 7* was a faster running train than a Class 7. A fitted head is a train of wagons that has several vacuum (or air-braked) wagons at the end (or head) coupled to the loco, with the wagon brakes connected to the loco and operational. This increases the brake force of the train and allows it to run at a higher speed. There is a complicated formula that determines the brake force of a train, which includes the type of locomotive, and number and type of wagons.

As many as 70 local trips would leave Tinsley daily; the demands of some stations and private sidings required some locations to be served two or more times each 24 hours.

The final design of Tinsley's main yard incorporated 53 roads, along with a further 24 in the secondary sorting sidings. The three departure roads were expanded into a combined express freight and departure set of sidings, totalling seven roads

With the main and secondary yards defined, the requisite number of reception sidings could now be determined. The initial approach was to assess the anticipated number of arrivals and the time required to prepare a train for humping. This gave the minimum number of sidings required if trains arrived evenly spaced through a 24-hour period; an allowance would subsequently be made for the inevitable bunching that would occur should several trains arrive in a short space of time

Consideration had also to be given to the varying amount of time needed to prepare a train for shunting. Once the loco had been removed from an incoming train, an unfitted train merely required the couplings to be undone between each wagon, whereas partially-fitted and fully-fitted trains also required the vacuum or air hoses to be disconnected. A final variance would come from the number of "cuts", i.e. the individual number of wagons that would be sent one way and another into each siding. The required number of reception sidings could thus be equated to peak arrivals, plus the rate of train preparation, less the rate of shunting. It was calculated that Tinsley would require 11 reception sidings at the east end with capacity to hold between 59 and 74 wagons, as in Table 1.

Table 1: Capacity Allowance of Reception Sidings in SLUs

Siding	Standage	Siding	Standage
1	74	7	64
2	65	8	73
3	65	9	59
4	74	10	59
5	69	11	65
6	63		

SLU = Standard Length Unit, or one 21-foot long wagon

In early yards such as Wath, the reception sidings rose on a continual gradient towards the hump which gave rise to difficulty in humping the trains. More importantly, however, unless the hump locomotive was in position at the rear of the train before the train engine uncoupled, the wagon brakes would have to be pinned down to prevent the train from rolling backwards. In a yard like Tinsley, where every time-saving technique would be employed, this simply wouldn't do. It was therefore decided that the reception sidings should be

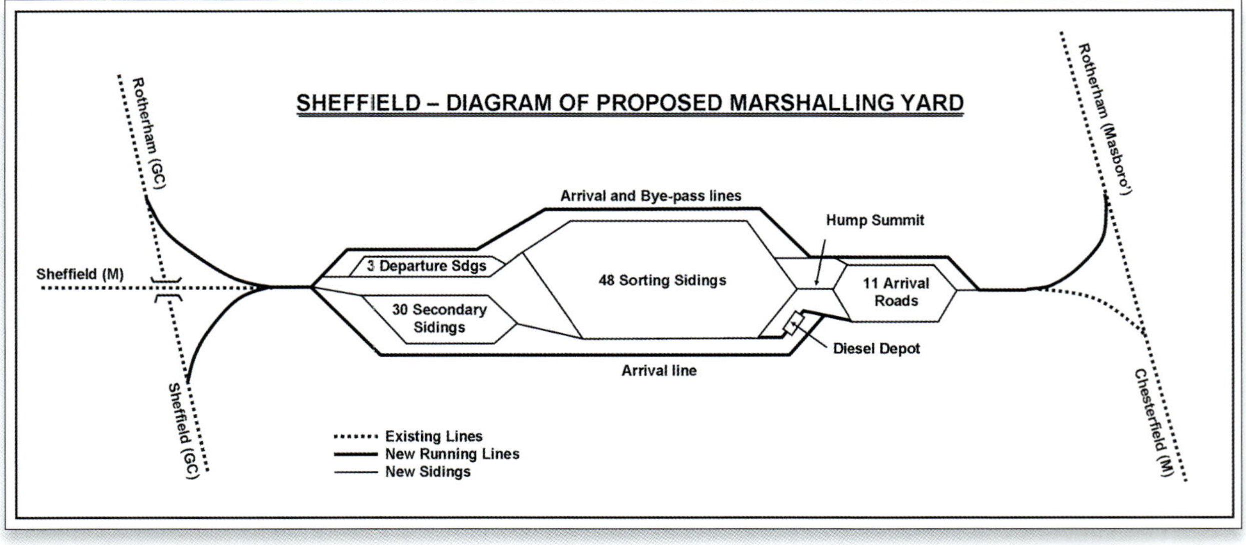

Above: Diagram of the marshalling yard proposed for Tinsley. The eventual yard would be rather different, having a greater number of sidings. *Diagram by Alex Fisher*

laid out in a basin principle with a shallow downward grade at the end furthest from the hump, a flat centre section and a slightly rising gradient to the hump. This would allow wagons to be left without having to pin down any brakes. Ultimately the reception sidings had a 640 ft long rising gradient of 1 in 40 on the approach to the main hump, followed by a short falling gradient of 1 in 20 to obtain quick separation of cuts approaching the first set of points, known as the 'king' points.

Of particular note is that some of the trains into Tinsley Yard would arrive from the Manchester, Sheffield and Wath route behind electric traction and thus a proportion of the reception sidings would have to be electrified.

A final consideration of particular importance to Tinsley was the direction from which trains would arrive. Trains arriving from the west at Treeton Junction, via the north-east and south-east curves, could run directly into any of the reception sidings; access from the west being gained by north and south arrival lines into numbers 1 to 7 and 6 to 11 reception sidings respectively. Trains from Brightside Junction or the south-west curve could reach the reception sidings by north or south arrival lines, but those from the north-west curve could only use the former.

As the plans for the yard layout were nearing completion, a lot of paperwork relating to costs was sent between relevant parties in the British Railways Board (BRB). It is interesting at this point to look at one of these memos to see how the costs had changed. It was sent on 8th December 1964 and is set out here.

B R I T I S H   R A I L W A Y S
EASTERN REGION

T I N S L E Y
MARSHALLING YARD

SHEFFIELD: RATIONALISATION OF FREIGHT FACILITIES

INTRODUCTION

1. Organisational changes in 1957 gave the opportunity to re-plan the railway freight operations in the Sheffield Area on a rational and modern basis. A comprehensive proposal was duly authorised by the British Transport Commission on 15th September 1960 (Min.13/354) at an estimated outlay of £9,253,000. A figure for interest on outlay during construction was not included as this was not the practice at the time
2. The project had three major elements - a new marshalling yard at Tinsley; diesel locomotive maintenance and servicing depots at Tinsley; and a new freight terminal at Grimesthorpe. Technical developments and changes in traffic requirements since planning started have called for major variations which are now explained.

NEW MARSHALLING YARD AT TINSLEY

3. Closure of 9 of the 11 existing yards in the area is still forecast, but the estimated net cost of the new yard has risen from £6,323,000 to £7,357,000 - an increase of £1,034,000.
4. This is due to increased labour and material cost and the following variations:-
   (a) Extension of electrification as compared with traction changeover at Bernard Road (Approved by B.R.B. Min.64/207(a) £241,000
   (b) Adoption of the Dowty system of wagon control. £460,000
   (c) Modification to the design of the yard and its approaches £200,000
   (d) Provision of new connecting lines at Aldwarke £162,000
5. The alterations have caused increased signalling costs but have led to compensating reductions such as earthworks. They also have made possible additional savings which are reflected in the improved return on outlay.

TINSLEY DIESEL LOCOMOTIVE MAINTENANCE AND SERVICING FACILITIES

6. The provision of the diesel locomotive facilities at Tinsley will still achieve the closure of 5 steam motive power depots.
7. The estimated net cost of the new facilities has risen from £607,000 to £654,000, an increase of £47,000 due to the provision of mechanical ventilation and additional plant.
8. This capital expenditure has been included in the appropriate diesel traction schemes which show a separate return.

GRIMESTHORPE FREIGHT CONCENTRATION DEPOT

9. The work of 16 goods stations was to be concentrated at 6 stations, of which only the new freight terminal would deal with sundries traffic. This intention remains.
10. The estimated net cost of the freight depot has fallen from £2,323,000 to £1,102, 000, a decrease of £1,221,000.
11. This reduction stems from:-
    (i) Abandonment of the two level design when BRB opted out;
    (ii) A cheaper and simpler form of construction;
    (iii) Favourable contract prices.

LAND

12. Land transactions are not included in the net outlay but the value of the lands for sale not immediately realised has risen from £1,145.000 to £1,586,000 - an increase of £441,000. An amount of £86,000 has also been obtained from private parties for the exchange of lands.

NATIONAL PLANS

13. The yard proposals were reviewed on reshaping to take into account the future pattern of train operation and the introduction of Liner Trains. Grimesthorpe Freight Depot was similarly reviewed and the design is related to the changes envisaged, including the National Sundries Plan. Earlier this year the diesel locomotive maintenance facilities were considered in relation to the National Traction Plan.

FINANCIAL EFFECT

14. The revised net outlay for the entire project is therefore:-

Original authority	£9,253,000
Reduced costs	140,000
Interest on outlay during construction	725,000
	£9,838,000

15. The annual savings for the yard and freight depot only give an improvement from £935,000 by £179,000 to £1,114,000, which in turn improves the return on net outlay from 10.8% to 12.2%.
16. A summary is shown in Appendix A.

APPENDIX A: Sheffield Rationalisation Scheme Financial Summary

Outlay	Original Authority (1960) £000	Revised Estimate (1963) £000	Variation in Expenditure
Marshalling Yard at Tinsley	6,323	7,357	+ 1,034
Diesel facilities at Tinsley	607	654	+ 47
Freight concentration depot at Grimesthorpe	2,323	1,102	- 1,221
	9,253	9,113	- 140
Interest on outlay during construction (Sept. 1961-May 1965)	-	725	+ 725
Net Outlay	9,253	9,838	+ 585

CONTRACT WORK

17. The value of contract work is £8.04m. All the major contracts have now been let.

PROGRESS OF WORK

18. The marshalling yard formation is complete and the permanent way is laid. The signalling is being progressively commissioned and Dowty hydraulic units are being fitted to the plain track. The extension of electrification is in progress.
19. The diesel maintenance depot is operational. The diesel servicing depot is ready but cannot be brought into use until the yard is commissioned.
20. Construction of the freight depot is well advanced.

PROFITABILITY OF TRAFFIC

21. Almost 40% of the total freight receipts of the Eastern Region originates in the Sheffield District. By far the most important of these traffics come from private sidings and a measure of their profitability is given in the following table.

Commodity	%Contribution over direct Costs
Coal	100
Iron and Steel	50
Petroleum Products	100
Earths and Stone	35
Chemical and Allied Products	80

RECOMMENDATION

22. It is RECOMMENDED that approval be given to the variations described; that the undernoted authorities be CANCELLED:-

Grimesthorpe Freight Depot		Reg. No.
Works & Equipment Committee	£1,521,020	W.E.1147

Min.2345/3 of 9th October 1962 (Cancelling previous authority for £2,323,000 included in B.T.C. Min.13/354 of 15th September 1960, for a total of £9,253,000

Tinsley Marshalling Yard and Diesel Depot

The balance of authority for £6,930,000 contained in B.T.C. Min.13/354 of 15th September 1960

Marshalling Yard	£6,323,000	W.E.1145
Diesel Depot	607,000	W.E.1146

And SUBSTITUTED by authority for a revised net outlay of £9,838,000, viz:-

Marshalling Yard	£7,897,000
Diesel Depot	744,000
Freight Depot	£1,197,000

Estimated effect on Annual Net Revenue: Marshalling Yard and Freight Concentration Depot Only

Net Outlay	Original Authority	Revised Estimate
£8,646,000	£9,094,000	
(Including £635,000 interest on outlay during construction)		
Savings in Annual Charges	£935,000	£1,114,000
Return on Net Outlay	10.8%	12.2%

Above: George Wimpey & Co Ltd and Hoveringham Engineering Co Ltd were responsible for the construction of the diesel depot, and here we see the workforce posing for the official photographer in 1963. The image is from the collection of Owen Owens, who was site engineer. Owen can be seen seated on the front row, third in from the left next to the man in the white coat. *Owen Owens collection*

Below: The start of construction of the yard, as viewed from the Sheffield District Line. The signal is the "up" distant for Tinsley Park Colliery signal box, which is beyond Parkhouse Lane bridge in the distance. Note the single lever in the centre of the image, formerly used by a fogman to place detonators during foggy weather. If the distant signal was at "caution", he would retain the detonators on the track. If a train ran over and exploded the detonators, the driver would know the signal was at "caution" even if he could not see it. The detonators would be removed if the signal was in the "off" position. All the track in this view would be removed and the south-west arrival line would eventually occupy some of the route. The secondary yard would then be built on the left of the image.
National Railway Museum/Science & Society Picture Library

Chapter 7

Construction of Tinsley Marshalling Yard

Before construction of Tinsley Yard could begin in earnest, the BRB had to secure the site and draw up detailed plans for the construction process, not least the massive earthworks that would be needed to excavate the site. As Ralph Sadler had said, "muck shifting is easy nowadays; I can take it away in a matter of months"; he was not wrong!

The contract for the main earthworks, drainage, fencing and services was awarded to Lehane, MacKenzie & Shand Ltd, but before commencing the work it was first necessary to carry out the diversion of certain gas and electricity utilities which crossed the site of the new yard in the vicinity of Wood Lane.

These were Yorkshire Electricity Board's Tinsley to High Hazel and Templeborough to Orgreave 11 kV supply cables, and four gas mains of 24-inch, 18-inch, 16-inch and 6-inch belonging to the United Coke and Chemical Company (UCC). The Templeborough to Orgreave 11 kV cable and the larger UCC gas mains were re-routed and carried across the yard on the superstructure of the new Wood Lane bridge, whilst the 6-inch gas main was diverted along the southern perimeter of the new yard.

Coal Measures

Prior to construction it was necessary to protect the area from possible subsidence from coal workings. It was known that coal seams existed beneath the site as there were collieries in the vicinity – Tinsley Park and Orgreave collieries were working several seams in the area, including High Hazel, Haigh Moor, Barnsley and Parkgate. The National Coal Board (NCB) had announced their intention to work the seam beneath the proposed yard area, which it was envisaged could cause a rapid settlement of some 3 ft 9 in. This sort of movement could not be tolerated as the gradients within the yard were critical; any ground movement would render it unworkable. The electrified lines, north arrival sidings and Catcliffe viaduct were all areas that also required protection.

It was expected that the coal workings would reach the yard area by about 1970 and be worked out by about 1974. To prevent this coal being worked, BR would be expected to pay for the coal over this period. Figures quoted showed the British Railways Board would be required to pay £148,600 to reserve this coal, but payment would only fall due as each part of the seam was reached. The fear was that non-payment of this money would result in the yard closing due to subsidence, the financial consequences of which would far outweigh the costs involved in the reservation.

In the event, some of this cost was recouped when more than 500 tons of coal per week was found in two and a half foot thick seams during the excavation work. Approximately

Above: There was a bloody great mountain in the way" and here it is in the process of being removed! This is the view along the Sheffield District Railway at Catcliffe looking towards Tinsley Wood tunnel and Brightside on 11th April 1962. A Ruston Bucyrus 54RB dragline machine can be seen on top of the mountain in the process of levelling the land. The signals are worked by Catcliffe Station signal box. *National Railway Museum/Science & Society Picture Library DM308*

Above: Lehane, MacKenzie & Shand Ltd won the contract to move the huge amount of earth, and here we see the beginnings of that operation on 11th April 1962. A Thames Trader tipper truck is being filled with earth in the foreground whilst in the background a Ruston Bucyrus mechanical bucket fills a larger dumper truck. The size of the undertaking is apparent! *BR DM312 Alan McKenna collection*

Left: A new bridge to cross Shepcote Lane was constructed by Wellerman Bros Ltd. This is the view of the new bridge under construction on 11th April 1962, with the original Sheffield District Railway bridge just visible behind.
BR DM315 Alan McKenna collection

Below: A 350 hp shunter hauls a motley collection of wagons containing construction materials over the Sheffield District Railway bridge, crossing the Sheffield & South Yorkshire Navigation canal. *BR DM314 Alan McKenna collection*

39,000 tons of coal was eventually recovered, being taken to Neepsend Power Station for burning.

Also on the site of the yard was a former pumping engine house. The supply pond for this was popular with local fishermen but had to be drained to allow construction; when the water level fell it revealed 1,400 fish, which had to be rescued and re-homed in another lake in Rotherham!

Further Costs

A report in the Financial Times newspaper on 3rd October 1967 stated that the BRB had been ordered by the Lands Tribunal to pay £581,000 compensation to Lord Fitzwilliam's Wentworth Estates Company, for the compulsory acquisition of a 93-acre site at Tinsley, to be included in the construction of the yard. This figure was made up of £551,000 for the value of the land, and £29,660 for severance and injurious affection (essentially, if the claimant's retained land is made less valuable as a result of the loss of land acquired, 'severance' compensation is payable. Similarly, if the retained land is made less valuable by the construction and use of the acquiring authority's scheme. 'injurious affection' compensation is payable).

The company was also awarded surveyors' fees, legal charges and costs against the BRB. The Tribunal made an alternative finding of £529,660, including £29,660 for injurious affection, in case it was held they were "wrong in law" on a view they had expressed. After the decision it was disclosed that the Board had offered the company a total of £440,000 compensation.

Moving the Muck

John Pavitt, formerly employed in the CCE office at King's Cross, recalls that one of the first engineering operations was to provide a new road to replace Wood Lane, which ran over Mr Fiennes "bloody great mountain". These works also involved the closure of part of Parkhouse Lane and the demolition of two bridges which carried this road over the old SDR line. Green Lane was also diverted at its junction with Wood Lane and, as Wood Lane could not be closed to traffic, a temporary diversion was laid pending the completion of the new Wood Lane bridge.

John recalls that "This new bridge required the excavation of a large hole so that the engineers could build the bridge over the hole. This appeared to have no rhyme or reason and convinced local people that all concerned were mad! The Sheffield Evening Telegraph enjoyed its jibes at BR; one memorable 1962 headline being 'BR Builds New Bridge over Tinsley Field' with a picture of Wood Lane diverted along the new bridge deck the ingenious Polish engineers had designed, before the underlying structure began to be unearthed from the surrounding landscape".

To build the bridge the land was excavated down to track level, then concrete supports were constructed, and finally the road decks lain on top. Only then could the final clearance of earth take place with the bridge in situ, and after completion the old Wood Lane bridge over the SDR line was demolished.

The contours of the yard demanded the excavation of some 3,700,000 cubic yd of sandstone and shale, the major part of which had to be conveyed to a tip. Of this, some 700,000 cubic yd were tipped on site to form embankments. Some of the material removed from the site was also used to fill in cuttings and the 80-yard Tinsley Wood tunnel on the old SDR alignment. The north-eastern portion of the main yard and the entire reception sidings were in a cutting which reached a depth of 60 ft below original ground level.

Below: BR Builds New Bridge over Tinsley Field!" This photo dated 27th August 1962 shows the method of construction of Wood Lane bridge, by digging a large hole and building up the piers first, then laying the decks on top. Afterwards the land either side would be dug out to form the massive cutting required for the yard to be built. *National Railway Museum/Science & Society Picture Library DM317*

Above: By 11th April 1962, levelling of the west end of the yard was continuing apace. The two buildings on the left of the image are not railway related, but are part of the long-closed Tinsley Park No. 8 pit. The square building is a Newcomen pumping engine house which was used for draining the pit, whilst beside it is a water tower which is still in situ today, providing water to the adjacent steelworks. The pumping engine was originally built in 1851 for Wheal Seton tin mine near Camborne in Cornwall. It had a 70 in diameter cylinder of 11 ft piston stroke in the shaft and was sold to Tinsley Park Colliery in 1874 for the sum of £1,800, plus £120 for dismantling. It was transported from Camborne station by rail to the Tinsley area; however, as this predated the Sheffield District Railway (SDR) it is probable that it was transported from the railhead to the colliery site at Park House Lane by horses. To the right of the image are two bridges over the SDR. The nearer bridge once carried a narrow-gauge tramway from Tinsley Park Colliery to the engine house, transporting coal to feed the boiler. The farther bridge carried Park House Lane; both bridges were demolished.
National Railway Museum/Science & Society Picture Library DM311

When the earthworks commenced in 1961, no firm decision had been made on the type of wagon retardation system to be used. Traditional wheel clasp retarders such as the Frölich type in Whitemoor Yard required a hump height of 21 ft, whereas the Dowty system only required approximately 11 ft. Therefore, the contractor was instructed to work to interim levels pending a decision on the system to be adopted. This decision, given in mid-1962, resulted in a saving of approximately 200,000 cubic yd of excavation.

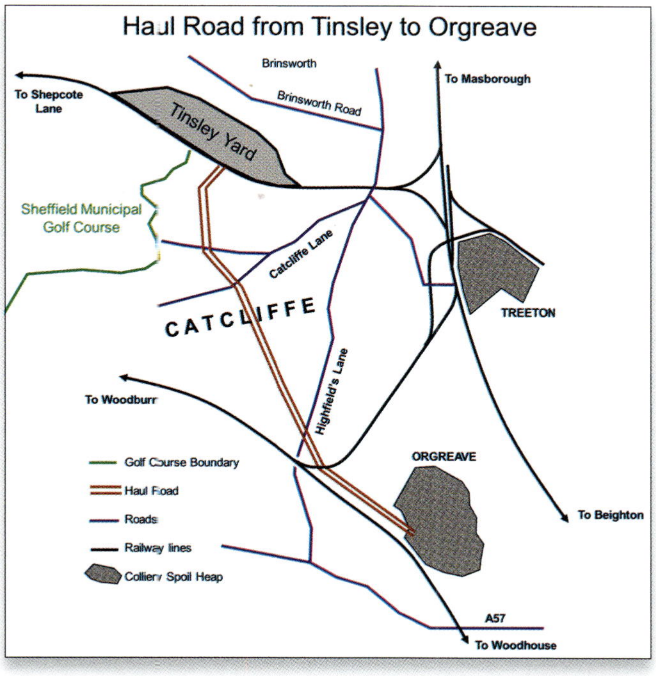

A tip site for the disposal of the surplus excavated material was found on railway land at Orgreave alongside Rotherwood sidings on the Sheffield to Worksop line, a distance of some two miles from the working site. This was later augmented by a second site on NCB land at High Hazel, about half way along the route to Orgreave, each site taking approximately 1½ million cubic yd of spoil.

The problem of transport of this quantity of material over public roads was overcome by the construction of a haul road, some two miles long, across open fields between Tinsley and Orgreave. This 24-ft wide tarmac road allowed trucks to travel at 40 mph, crossing a colliery railway en route via a 100 ft twin Bailey bridge. Vehicle-actuated traffic lights were installed where the road crossed the public highway in two places, and on the bridges which were only strong enough to support one loaded vehicle at a time.

All of this temporary work was carried out by the contractor, with the roads removed and the site restored on completion of the earth works. The main excavation was carried out in planned stages commencing in August 1961, being virtually completed by November 1963.

Above: A diagram of the haul road constructed to move material from the site at Tinsley to a tip at Orgreave.
Diagram adapted from that in the Contractors' Record and Municipal Engineering magazine for 16th May 1962

Above: Dated 4th June 1964, this image taken from the site of the administration block shows the service shed and main control tower under construction. Note that track is laid in the sidings but the point work and Dowty retarders have yet to be laid on the lighter-coloured ballasted area. Track awaiting laying is stacked in panels just beyond the service shed. *National Railway Museum/Science & Society Picture Library DM321*

Above: On 11th April 1962 work was ongoing to remove part of Hadfield's tip at Catcliffe, to allow construction of the new north-east curve from Tinsley Yard towards Rotherham. This image shows earth-moving machines digging out the cutting. The tracks in the foreground are the "down" main of the former Midland Railway "Old Road" (nearest) and Treeton Colliery "down" siding. *National Railway Museum/Science & Society Picture Library DM309*

Construction Machinery

Sandstone was found in layers between softer types of shale; large excavators were used to remove these layers. For the record, the following types of machinery were used in construction: The contractors used Ruston-Bucyrus (RB) mechanical buckets; a 71RB with a 3½ cubic yd bucket along with a 54RB dragline machine and a 37RB bucket.

The excavated material was loaded into dump trucks then transported to the tip at Orgreave. The contractors used six AEC 3673M trucks with 10 cubic yd capacity and six Foden FLD/6/30 trucks with 12 cubic yd capacity. These were fitted with special aluminium bodies, built by G.E. Neville of Mansfield, giving extra payload/capacity without reducing strength.

The dump trucks averaged 3½ journeys per hour on the longer haul roads to the main tip, a round trip of 4.3 miles, whilst they managed five trips per hour on the shorter 1.9 mile trip to the second tip. The 71RB mechanical bucket moved 491,000 cubic yd in a 20-week double-shift operation at an average of 302 cubic yd per hour, whilst the 54RB moved 282,000 cubic yd in 19 weeks of double-shift working, at an average rate of some 200 cubic yd per hour.

The main earth-moving plant was assisted by two Aveling Barford Super MG graders, an International BTD20 bulldozer, a Caterpillar type 68A power shift D8 winch-operated dozer, a 977 Traxcavator and a Euclid C6 dozer. To move the earth to embankments inside the site, there were three Caterpillar DW21 and two Euclid/Terex S21 scrapers.

Trackwork

In early photographs of the yard it will be noted that there are lighter areas of ballast around the pointwork and the hump areas. This was because the ballast in the yard was a mixture of limestone, steelworks slag and ash, with 260,000 cubic yd of material being laid throughout the yard.

Limestone (the lighter material) or slag was used in all areas that were track circuited, the ash being used as bottom ballast up to 3 in below the sleepers, then limestone or slag was laid on top before being compacted by vibrating plates. Some of the ash was obtained from Blackburn Meadows Power Station.

Some of the track and pointwork used was second-hand bullhead and flat-bottom rail, around 70,000 yards of plain line and 600 yards of crossing rail being used. There were also 7,700 linear yards of new 95 lb plain line bullhead rail, and 7,900 yards of switches and crossings, along with 1,700 linear yards of new 110 lb flat-bottomed plain line rail and 16,000 yards of switches and crossings, which all equated to approximately 60 miles of track being laid.

The main track-laying works were carried out in stages. The express freight yard was begun in June 1962 and completed by the end of 1963, with the other yards completed by November 1964. Associated works including rail welding, jointing and fettling up were completed by June 1965.

New connections were laid from the secondary yard to the existing and proposed Firth-Vickers works; on the south side of the yard a connection was made to the new English Steel Corporation works.

Provision was also made for a new roadway giving access from Shepcote Lane to the secondary yard control tower area along the southern perimeter of the yard. This was mainly for the use of staff and maintenance vehicles, or for use by the emergency services.

Above: New bridge works take shape at Broughton Lane during 1962. In the foreground is the Sheffield District Railway line. *BR DM316 Alan McKenna Collection*

Drainage

Such a vast area in so deep a cutting required adequate drainage to avoid flooding.

As the summit of the yard was in the vicinity of the Wood Lane bridge, surface water drainage was designed to flow in an easterly direction into the River Rother and westerly into Sheffield Corporation's drainage system in Shepcote Lane.

Track drainage in the reception sidings east of Wood Lane, and all surface water collected from the diesel maintenance depot, was taken in 9-inch to 21-inch diameter pipes to the east end of the reception siding cutting, from where it flowed through a 24-inch diameter steel pipe into the River Rother at Catcliffe.

West of Wood Lane, surface water from the diesel servicing depot and track drainage in the low cutting at the east end of the south arrival line gravitated through 9-inch diameter pipes into the engine line cutting under the secondary hump bridge, from where it was pumped into the track drainage at the secondary yard. It was not unknown for the pumps to fail in heavy rain with resultant flooding of these low-lying areas.

Track drainage through 9-inch diameter pipes was collected from the main and secondary yards and the express freight sidings, and was finally discharged through a 33-inch diameter concrete pipe into Sheffield Corporation's drainage system in Shepcote Lane.

All soil drainage from buildings throughout the yard, including industrial effluent from the diesel maintenance and diesel servicing depot, was collected into a 6-inch diameter pipe which was situated chiefly under the southern perimeter roadway, and also discharged into the Sheffield Corporation system in Shepcote Lane.

New Connections

Work to build new connections towards both Sheffield and Rotherham (on the former GCR line) at the Shepcote Lane end of the yard commenced in August 1961.

This involved the diversion of the Sheffield & South Yorkshire Navigation canal, the construction of a new double-lift canal lock, the enlargement of a canal basin and the construction of two steel underbridges, the work being carried out by W. & C. French Ltd.

The new bridge spans, one of 70 ft and the other of 164 ft span, were supplied by the Butterley Company; the construction of a bridge over the canal basin, a pedestrian subway and retaining wall and earthworks for the tracks, along with an extension to the canal basin, were carried out by Henry Boot & Sons (Railway Engineering) Ltd.

At the Catcliffe (east) end of the yard a new north-east curve was built towards Rotherham Masborough on the former Midland "Old Road". This supplemented the existing south-east connection from the SDR to the Midland line.

Several sub-contractors were involved in the construction of the yard, all mentioned for completeness:

Site investigations were carried out by Turriff Corporation Ltd, trial Boreholes were made at Catcliffe Tip by Holmpress Piles Ltd and the excavation of trial holes, construction of the bridge over Shepcote Lane and an occupation bridge by Wellerman Bros. Ltd.

Layout of the Yard

The yard was designed for two-way working to avoid the excessive transfers of wagons which would have occurred between separate "up" and "down" yards in an area with such a complex layout as Sheffield.

Above: The results of a drainage pump failure in the secondary yard engine line dive-under! The track is under about 10 feet of water after a massive amount of rain in a storm sometime in 1978. The secondary yard power pack building is prominent; the secondary yard hump tower was just behind this, the hump summit being on the bridge. The signal on the right is SY 1 & 2 hump-shunting signal, which gave instructions for the hump shunter drivers shunting from the mechanical feed road (siding 50), emergency feed road (51) and sidings 52 & 53. *Roger Wainwright*

Tinsley was designed principally as a local yard with over 70% of traffic originating or destined locally, of which some 80% was private siding traffic. This traffic was unsuited for movement by the then-proposed Freightliners or in block train loads.

Under Stage 1 of the yard commissioning, traffic was absorbed from the ex-Midland yards and through loads made up for Toton, Norton Junction, Washwood Heath, Healey Mills, Chaddesden, Leeds, Stanton Gate, Frodingham, Mottram, St. Pancras, Gowhole, Leicester, Brent and Port Talbot. Additional outlets to Hunslet, Nottingham, Carlisle, York and Glasgow were provided by through freights calling at Tinsley.

From the October 1965 timetable, traffic was absorbed from the ex-GCR yards and through services provided to Lincoln, New England, Hull, Birkenhead, Liverpool Brunswick, Colwick, Doncaster, Whitemoor, King's Cross, Ferme Park and Worksop. It was expected that, by this stage, the yard would be working near to its planned capacity of 4,000 wagons per day.

The operational direction of the yard was from east to west, and consisted of four operational areas – reception sidings, main yard sorting sidings, secondary sorting sidings and express freight and departure sidings. A secondary yard was provided to the south-west of the main yard with a smaller hump and 25 sorting sidings with a capacity varying from 36 to 46 wagons, giving a total capacity of 964 wagons. This yard performed primary sorting of traffic for local destinations and for those destinations requiring only short roads. Wagons were propelled from the main hump to the secondary yard hump by Dowty booster retarders installed in the mechanical feed road.

The main yard consisted of 53 sorting sidings in eight balloons, with a total capacity of 3,690 wagons. Capacities of the sidings varied; wagons shunted into the sorting sidings from the hump end were expected to stop at the top of the 1 in 120 gradient, at the west end of these sidings. The main wagon standage figures were based on this, but in certain sidings there was extra allowance for additional standage beyond that point. Table 1 illustrates the siding capacities, Table 2 shows the acceptance of trains into the reception sidings and Table 3 the disposal of locomotives.

Table 1: Main Yard Siding Capacity and Additional Allowance

Siding	S	A	Siding	S	A	Siding	S	A	Siding	S	A
1	80	10	14	68	2	27	77	22	40	64	6
2	73	10	15	68	2	28	77	18	41	64	6
3	67	0	16	75	0	29	77	13	42	64	11
4	63	0	17	75	0	30	77	8	43	64	11
5	63	0	18	77	7	31	77	3	44	64	1
6	71	5	19	77	12	32	76	0	45	64	6
7	71	9	20	77	17	33	76	0	46	61	0
8	71	9	21	77	22	34	77	1	47	53	5
9	64	8	22	77	3	35	77	1	48	53	10
10	64	3	23	77	3	36	77	6	49	43	8
11	62	0	24	77	7	37	77	11	50	Feed Road	
12	62	0	25	77	12	38	77	16	51	53	7
13	68	7	26	77	17	39	77	20	52	53	3
									53	53	0

S= Standage A= Allowance

Table 2: Acceptance of Trains into the Reception Sidings

Entrance to Yard	Approach to Reception Area	Reception Siding
North-East Curve (Treeton)	Direct	All Sidings 1–11
South-East Curve (Treeton)	Direct	All Sidings 1–11
North-West Curve (Tinsley East)	Via North-West Arrival and North Hump Approach	Sidings 1–8
North-West Curve (Tinsley East)	Via North-West Arrival and Reversing Spur	Sidings 1–3
South-West Curve (Broughton Lane)	Via North-West Arrival and North Hump Approach	Sidings 1–8
South-West Curve (Broughton Lane)	Via North-West Arrival and Reversing Spur	Sidings 1–3
South-West Curve (Broughton Lane)	Via South-West Arrival	Sidings 6–11
Brightside Junction	Via North-West Arrival and North Hump Approach	Sidings 1–8
Brightside Junction	Via North-West Arrival and Reversing Spur	Sidings 1–3
Brightside Junction	Via South-West Arrival	Sidings 6–11

Table 3: Disposal of Locomotives from Reception Sidings

Trains Arriving From	Into Reception Siding	Train Locomotive Released on Arrival into Reception Lines
East (Treeton)	Sidings 1–3	Via Hump Engine Release Spur, Hump engine Return Line, Pilot Spur, Reception Sidings, Train Engine Spur, Engine Line to West End, Servicing or Maintenance Depot (Eastern Access)
East (Treeton)	Sidings 4–11	Via Engine Release Line, Engine Line to West End, Servicing or Maintenance Depot

Trains Arriving From	Into Reception Siding	Train Locomotive Released on Arrival into Reception Lines
West (North-West Arrival & Reversing Spur)	Sidings 1–3	Via Train Engine Spur, Engine Line to West End, Servicing or Maintenance Depot
West (North-West Arrival & North Hump Approach)	Sidings 1–8	Via Train Engine Spur, Engine Line to West End, Servicing Depot, Maintenance Depot (Eastern Access)
West (South-West Arrival)	Sidings 6–7	Via Train Engine Spur, Engine Line to West End, Servicing Depot, Maintenance Depot (Eastern Access)
West (South-West Arrival)	Sidings 8–11	(a) Diesel Locos Via Engine By-Pass, Treeton South Arrival, Engine Line to West End, Servicing Depot, Maintenance Depot (Eastern Access) (b) Electric Locos Via Engine Loop, reverse into Engine Line to West End

To the north-west of the main sorting sidings was the express freight yard, with five sidings each holding approximately 76 wagons and a ladder about halfway along. This yard was designed to deal with through express freight trains booked to attach and detach merchandised wagons at Tinsley. It was used for sorting fitted wagons originating at Tinsley Yard, and also as departure sidings for the main yard when not being used by express freight trains.

Eastbound departure lines from sidings 1 to 15 in the main yard, and from the express freight yard, connected into the down by-pass line which skirted the north side of the yard. The eastbound departure line from the secondary yard burrowed under the main hump, to avoid conflicting with other traffic, before connecting into the main yard east departure line. Engine lines to allow loco movement were provided between the diesel maintenance and servicing depots and the receptions, main, secondary and express freight yards. The engine line to the west end burrowed under the secondary yard hump to avoid conflicting movements.

Shunting staff were located at both western and eastern ends of the reception sidings and, by the use of a plunger, they indicated the siding number on which a train had arrived and the destination of the train to be released. The indicated destination was illuminated against the siding concerned on the panel of Tinsley Yard signal box, which extinguished when the appropriate signal was cleared for the move.

Once the locomotive had been removed, humping could begin. Trains in reception sidings 1–5 were humped via the north hump approach and those in sidings 6–11 via the south hump approach. After pushing a train over the hump, the humping loco then followed through to the hump engine release spur and returned to the east end of the reception area via the hump engine return line and hump engine return spur.

Crippled wagon sidings with repair facilities were installed at the west end of the secondary yard, which also had a cattle dock with a fodder store for giving attention to loaded cattle vehicles in transit, although it is uncertain if this facility was ever used.

Maintenance of general equipment in the yard, including lighting, was also the responsibility of the outdoor machinery staff under the supervision of the depot master. The whole yard was illuminated from masts 50 ft high, to give uniform lighting and eliminate shadows. Lamps were of the colour-corrected mercury vapour type.

Carriage and wagon requirements in the yard comprised of examination and oiling of wagons in addition to a small amount of repair work in the two cripple sidings provided; the staff covering this work also came under the supervision of the depot master.

Plans for the new yard were displayed at a town planning exhibition in Sheffield during 1963, showing that the yard was named Tinsley Park Marshalling Yard. It is not known when the word "Park" was dropped from the title.

The Yard Opens

Tinsley Marshalling Yard opened in two stages in June and September 1965, with an official opening ceremony conducted on 29th October 1965. On the same day, an inspection of the new installations was carried out by Lord Beeching (former chairman of the BRB), Mr S.E. Raymond (current chairman of the BRB), Mr

Right: The menu for luncheon held at the Royal Victoria Hotel. In plain English it is shrimps in butter, followed by roast saddle of lamb with mixed fresh vegetables, buttered peas and potato croquettes. Dessert was ice cream frozen in a spherical mould, followed by coffee and a 1959-vintage red wine.

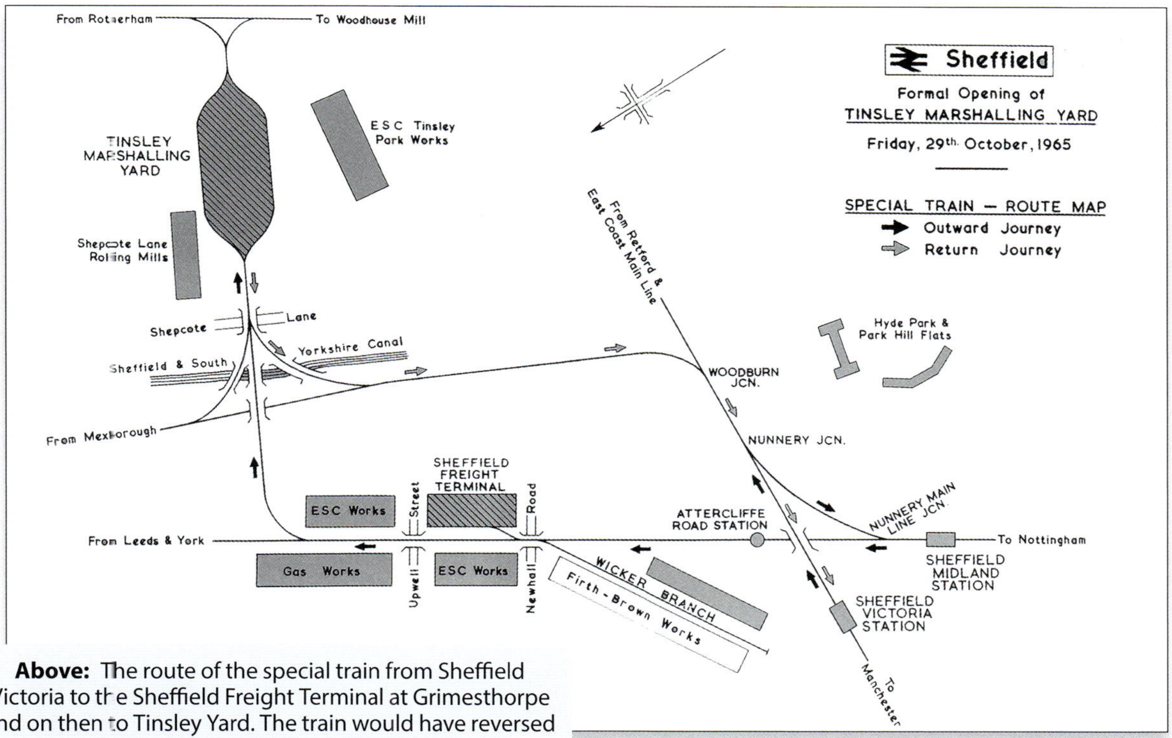

Above: The route of the special train from Sheffield Victoria to the Sheffield Freight Terminal at Grimesthorpe and on then to Tinsley Yard. The train would have reversed at Nunnery Junction and again in Sheffield Midland, before heading to the freight terminal. *From an official BR document*

J.R. Hammond MBE, Mr G.F. Fiennes and Mr H.C. Johnson (the general managers of the Eastern, Western and London Midland Regions respectively), along with a large number of railway officials, industrialists and manufacturers.

To convey the invited guests to the yard a special train was laid on from Sheffield Victoria, departing at 08.53, with an initial stop at Sheffield Freight Terminal where the new depot was inspected. Guests returned to the train, which then departed to Tinsley Marshalling Yard where a temporary platform had been erected on the south west arrival line adjacent to the steps to the administration block. Here, the guests alighted and an inspection of the marshalling yard was undertaken.

After the inspections, Lord Beeching conducted an official opening ceremony and expressed his satisfaction at being able to open such an important scheme. He emphasised this was one of several important features of the reshaping of British Railways plan which would be coming to fruition in the following 18 months. The yard was hailed as the first marshalling yard in the world to be fully equipped with the Dowty wagon control system.

According to contemporary newspaper reports, Lord Beeching told the 250 guests that "British Railways has a highly competent management, but that this was not often recognised" and "they are often obstructed but they are doing

Below left and right: The temporary scaffold platform erected on the south-west arrival line by Mills Scaffolding of Rotherham, to allow guests to alight from the special train laid on for the opening of the yard. *John Byers*

Above: Tinsley Yard in March 1965. D4502, in its original configuration of cab-to-cab, crests the hump after completing a shunt, whilst in the background an 8F 2-8-0 sits adjacent to the servicing shed which is occupied by a Sulzer Type 2 and a Brush Type 2. A further Sulzer Type 2 is outside the breakdown shed. As there were no steam facilities at Tinsley, the steam loco would not have remained there long before heading off, either on another train or to the nearest steam depot, which was Canklow at that time. *Les Nixon*

a good job". He commented that Sheffield had the world's most advanced rail traffic handling system, saying that the inevitable consequence of the reshaping had brought closures "But this has been recognised in that you have reshaped the whole pattern of the railway business" and "this railway management is not numerically strong, but it has developed into a highly competent management team doing a very difficult job". He went on to say, "Now I have left British Railways this is the first time I have felt free to tell you of the difficulties facing this management team so forcibly" and "modernisation involved action which is not always attractive". In a lighter tone he said "I am not best known for opening things" referring of course to his notorious closure programme of previous years. He then unveiled a commemorative plaque on the administration block.

With the opening ceremony completed the guests were transported back to Sheffield Victoria, arriving at 12.50 where, in the adjacent Royal Victoria Hotel, an obligatory luncheon took place at 13.30. After luncheon there were toasts to the Queen by S.E. Raymond, chairman of the British Railways Board, followed by a toast to welcome the guests. This was followed by a presentation to Lord Beeching by the first yard manager, Mr Harry Whitehouse, to commemorate the opening. The presentation was in the form of a silver salver made by Alexander Clark & Co Ltd of Sheffield, complete with an engraving of the yard (this today resides in the NRM and can be viewed on request).

Lord Beeching proposed a toast to British Railways, with a response by S.E. Raymond. The next toast was to the City of Sheffield, proposed by J.R. Hammond MBE, general manager of the Eastern Region, the response being by Alderman J.S. Worrall JP, the Lord Mayor of Sheffield.

The guest list read like a who's who of the time, ranging from Colonel D. McMullen, the chief inspecting officer of railways for the Ministry of Transport, and Mr Geoffrey Freeman Allen, editor of Modern Railways, to Mr L.G. Clugston of Staveley Slag Sales Ltd, the assistant yard manager Mr Goodrich and venerable dignitaries of the Eastern Region of BR, along with representatives from steel works, the NCB, Cravens Ltd, Dowty Mining Equipment Ltd, Ericsson Telephones and so on.

To return the invited guests back south, a special train departed Sheffield Victoria at 15.45 for King's Cross, arriving there at 18.35.

The special train from Sheffield to the yard was used again on Monday 1st November to convey dozens of wives and families of men employed by BR to Tinsley Yard, so that they could have a look at the works involved. They were given a demonstration of hump shunting and the operation of Dowty retarders, before being shown Tinsley diesel depot, where children were allowed to climb into loco cabs and familiarise themselves with the controls, especially the warning horns!

As an aside, the opening day could have gone quite differently as drivers had threatened to walk out on a one-day strike on the day Lord Beeching was to arrive, as they were in dispute over a bus service provided by BR. The idea of this bus service was to give crews regular transport to and from work. Two buses were laid on, travelling in a circular route around Sheffield and Rotherham, one travelling in a clockwise direction, the other anti-clockwise and picking up and dropping off at designated points. The trouble was that the buses were often full and standing and men were regularly left behind; the unions argued with BR that the service was inadequate. After some tense negotiations, the dispute was resolved and the threat of strike action called off.

Above: The yard opened for business in 1965 and is quite busy in this 1966 image. Four coke hoppers are in the process of being humped, whilst the servicing shed is host to a pair of Brush Type 2s and English Electric Type 1, D8066. Just below the main yard hump tower can be seen the Dowty test trolley shed, clad in corrugated plastic. This shed did not fare well and the plastic covering was soon removed. *John Pavitt collection*

Left: Looking along the hump line to the hump summit. Whilst the hump signal is controlled from the main yard hump tower in the background, the small subsidiary signal on the arm is Tinsley Yard signal box TY259 which, when in the "off" position (two white lights), indicated to the hump pilot driver that the route was set to the hump engine return road. This allowed the loco to move forward to the buffer stops, then return back to the reception sidings. This move both allowed the swift return of the loco, and cleared the way for another train to be shunted. *John Pavitt collection*

Above: Tinsley from the air. This view, looking west, shows the diesel depot on the higher ground, with the reception sidings alongside. Wood Lane bridge is prominent in the centre of the image, with the servicing shed, main yard control tower and other buildings just beyond. The curved line in the top left of the image is the English Steel Company access. The bare earth to the left of the diesel depot is the filled-in cutting and tunnel of the Sheffield District Railway.
John Pavitt collection

Below: Tinsley Yard as viewed from the Shepcote Lane end looking towards Treeton. The new south-west arrival and departure curve along with the north-west arrival and departure curve from the former GCR lines at Broughton Lane and Tinsley South can be seen towards the bottom of the image, the line towards Brightside Junction at centre bottom. The Sheffield and South Yorkshire Navigation can also be seen towards the bottom of the image. The A6178 Attercliffe Common is at the bottom, the bridge over Shepcote Lane being dead centre of the image. Shepcote Lane Rolling Mills are centre left, with the English Steel Corporation's works to the right.
English Heritage Aerofilms

Rght: The gradient diagram for all the lines in the immediate area. *Chris Booth collection*

Below, opposite and page 66: The yard diagram as issued in the notices to staff. It was originally on one long sheet and is reproduced over the following three pages. The zig-zag lines on the diagram denote electrified lines. *Chris Booth collection*

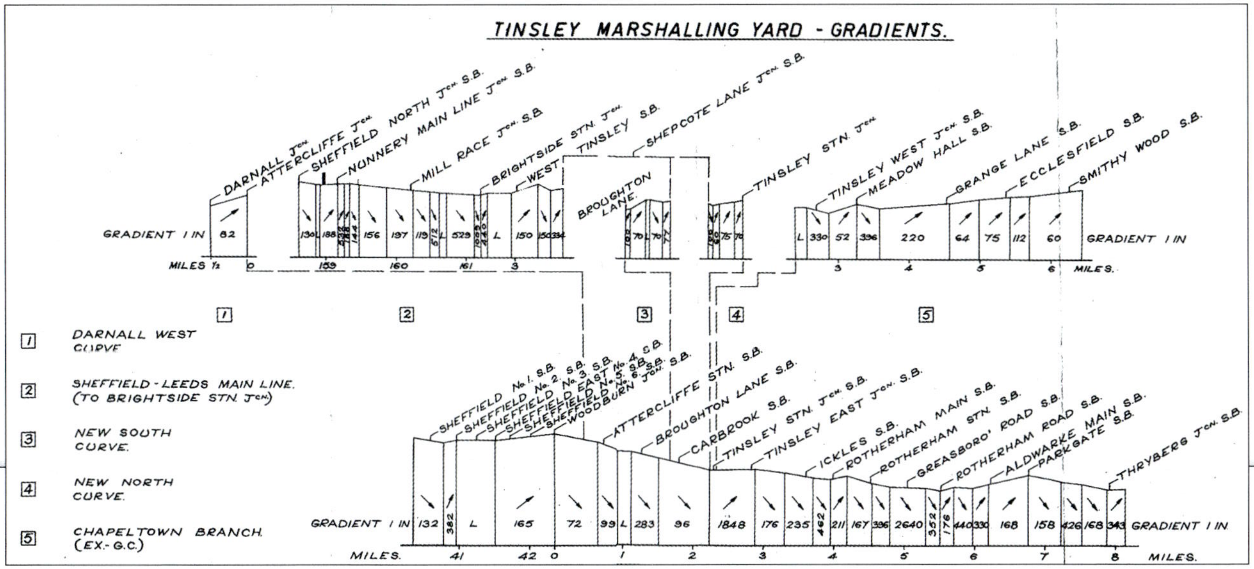

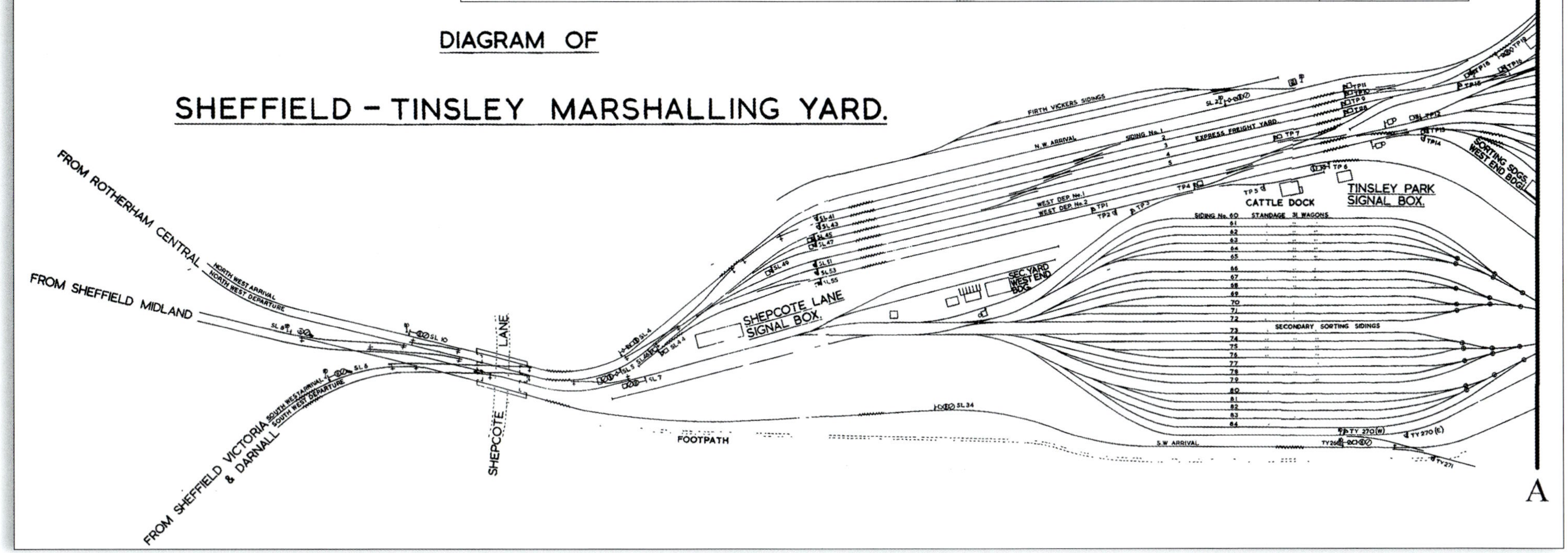

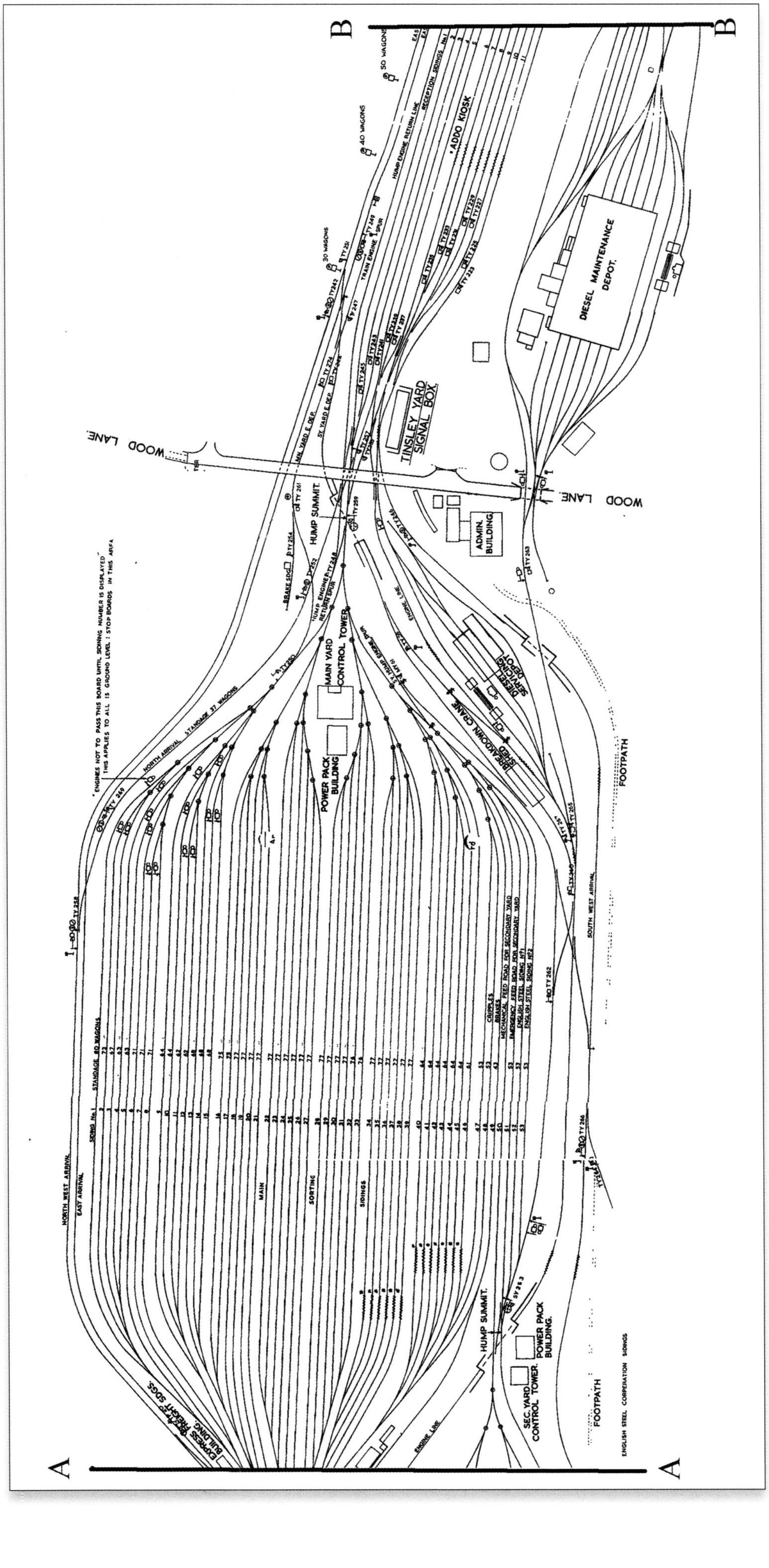
A
B
DIESEL MAINTENANCE DEPOT.
ADDO KIOSK
TINSLEY YARD SIGNAL BOX
WOOD LANE
WOOD LANE
HUMP SUMMIT.
ADMIN. BUILDING
MAIN YARD CONTROL TOWER
POWER PACK BUILDING
DIESEL SERVICING DEPOT
BREAKDOWN CRANE SHED
FOOTPATH
HUMP SUMMIT.
SEC. YARD CONTROL TOWER.
POWER PACK BUILDING.
FOOTPATH
B
A

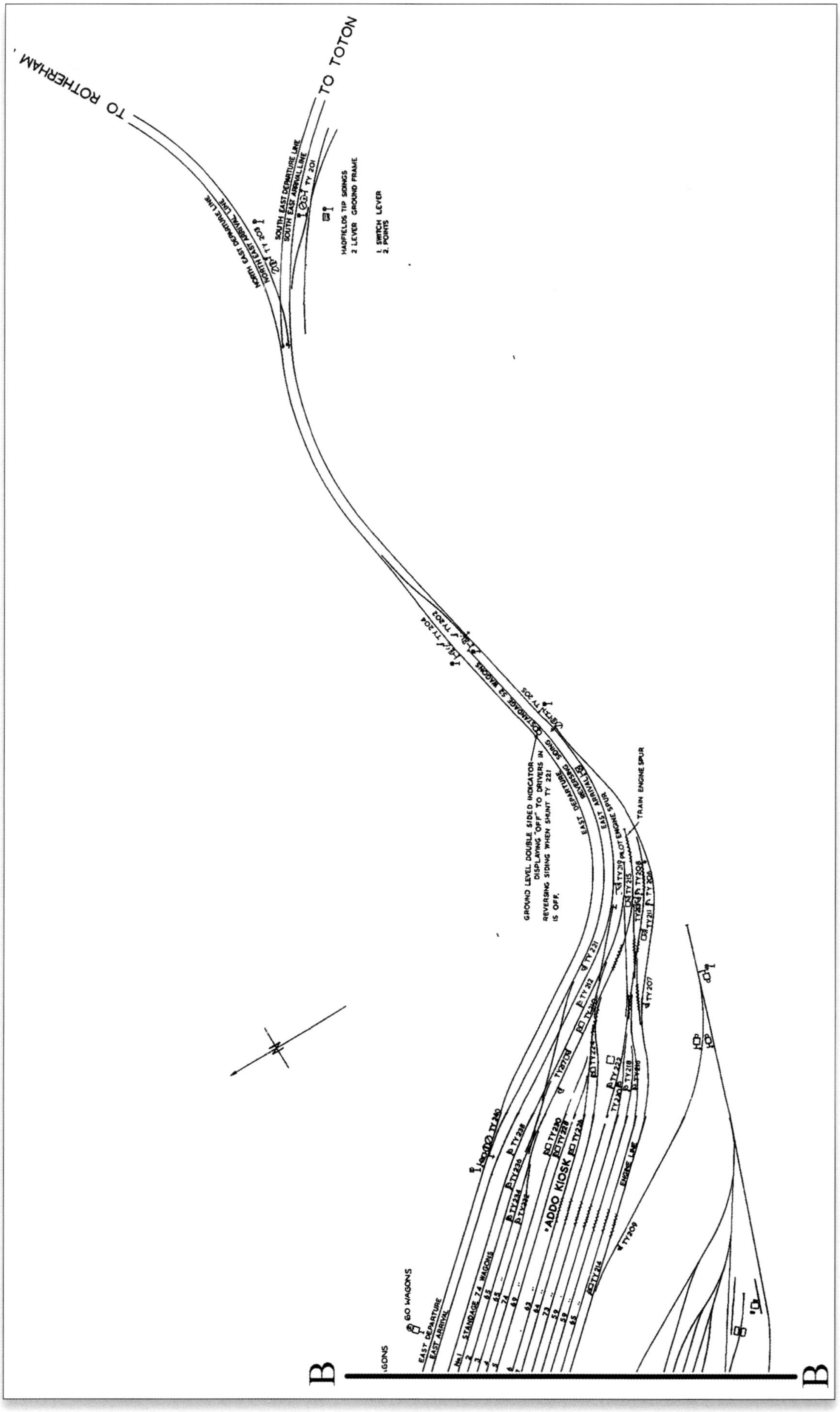
TO ROTHERHAM
TO TOTON
SOUTH EAST DEPARTURE LINE
SOUTH EAST ARRIVAL LINE
NORTH EAST ARRIVAL LINE
NORTH EAST DEPARTURE LINE
HADFIELDS TIP SIDINGS
2 LEVER GROUND FRAME
1. SWITCH LEVER
2. POINTS
GROUND LEVEL DOUBLE SIDED INDICATOR DISPLAYING "OFF" TO DRIVERS IN REVERSING SIDING WHEN SHUNT TY 221 IS OFF.
TRAIN ENGINE SPUR
ADDO KIOSK
ENGINE LINE
60 WAGONS
EAST DEPARTURE
EAST ARRIVAL
B
B

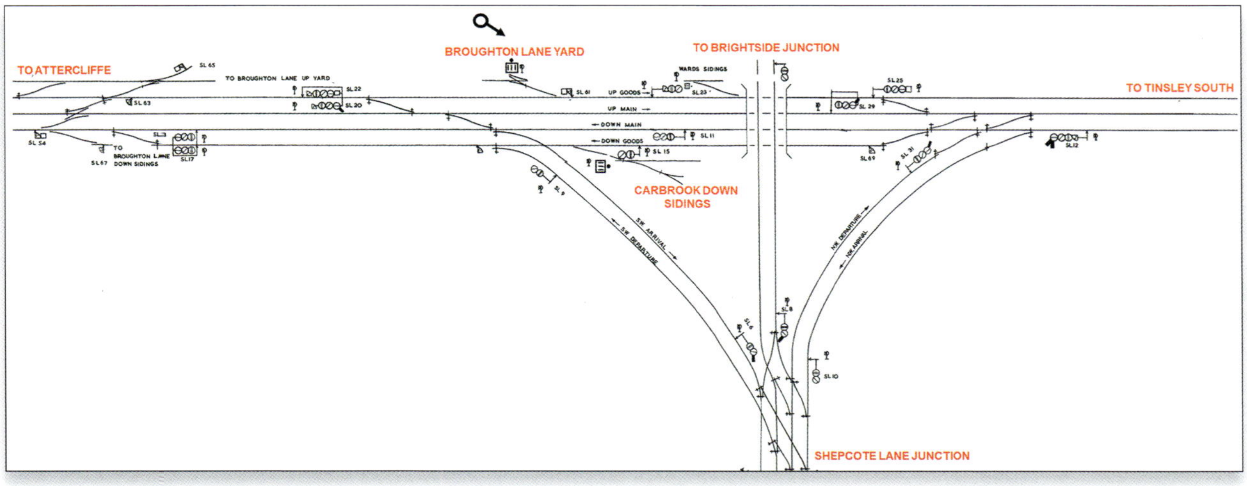

Above: A section of resignalling diagram covering Broughton Lane and Tinsley South to Shepcote Lane, which shows the new south-west arrival and departure curve along with the north-west arrival and departure curve. The line towards Brightside Junction heads straight on across the former GCR Woodburn Junction to Rotherham line. *Chris Booth collection*

Below: Broughton Lane Yard 3-lever ground frame (GF) is in the foreground as 45136 climbs the south west arrival from Broughton Lane Junction to Shepcote Lane Junction on 12th April 1987. The telephone associated with the GF would have rung in Shepcote Lane box. The train is a football special on its return journey from Wadsley Bridge. The arrow on the map above shows the location of the GF and the direction in which the photo is taken. Note that the GF is in the released position. *Ted Evans*

Left: The north-west arrival and departure lines curve away towards Shepcote Lane as a diverted Cardiff to Newcastle HST approaches Tinsley South Junction on 4th April 1988.

The HST is led by 43012, with 43022 at the rear. The steelworks buildings in the background have been replaced by a TGI Fridays restaurant and a hotel – such is the price of progress! *Ted Evans*

Above: A secondman's view of the north west arrival line whilst waiting at Tinsley Yard TY258 signal. A Class 20-hauled train of steel wagons has just moved on to the north arrival line and is heading towards the reception sidings. The guard has a good fire going in his brake van if the smoke from the chimney is anything to go by! *Roger Wainwright*

Below: A view of the reception sidings from Wood Lane bridge. The sidings have a good selection of wagons but there is no sign of any activity; perhaps the Class 13 is already pushing a set of wagons over the hump?
Allan Trotter Eastbank Railway Club

Chapter 8

Tinsley Diesel Maintenance Depot

In the plan for dieselisation of the Sheffield Area it was envisaged the new Tinsley Diesel Maintenance Depot (DMD) would eliminate five steam depots and cater for the maintenance of around 190 main line and 80 shunting diesel locomotives in the Sheffield Division. Initially, those allocated to Wath were to receive maintenance there, but that decision was later reversed.

Development of the Building

The depot was built on a higher level to the south of the reception sidings, and slightly north of the location of the in-filled 80-yard Tinsley Wood tunnel. Access to the depot could be gained from either end of the yard although from the east end was via a dog-leg set of lines. The foundations, boiler-houses and earthworks for the depot building were contracted to George Wimpey & Co Ltd, with steelwork by Hoveringham Engineering Co Ltd.

According to John Pavitt, former senior assistant architect in the research & development (R&D) group in the chief civil engineer's office (CCE) in King's Cross, the design of the DMD was a separate job within the R&D group and was completed before the rest of the yard buildings. The architect was Tony Hall, who had already completed Finsbury Park DMD, the Eastern Region being the first region to go entirely diesel and electric.

John states "There was a demarcation between 'sheds' and 'buildings' in the CCE's department. Sheds were designed by engineers and the architects were only allowed to design buildings over smaller spans. This did not make for unified design but at Tinsley DMD the two end-on maintenance sheds were extended to enclose the accommodation for offices and staff accommodation to form one large portal-framed structure that the architects in-filled with ancillary accommodation".

Working under chief architect H.H. Powell and Eastern Region chief civil engineer A.K. Terris, was Roger Walters – the deputy chief architect. He was very much into research-led innovation on BR; the CCE architects' R&D group co-ordinated central research with BR Derby and London Transport into the design of the buildings.

Building Description

Although other regions used variations of the flat or "north-light" roofs for depots, which were lower and cheaper to build, A.K. Terris insisted upon giving the new standard Eastern Region diesel depot buildings traditional pitched roofs with gable ends. His reasoning was that it would be a natural way for the diesel exhaust fumes to disperse.

Therefore, Tinsley was a single-span steel portal construction, 111 ft 6 in wide and 432 ft long. Fume extraction equipment was incorporated, consisting of 48 2 ft diameter electric fans positioned in pairs across the roof, and spaced equally about the centre line of each berth. Pneumatically-controlled air inlet louvres were also built into the walls to allow fresh air to flow through.

Below: Dated 27th August 1962, this image shows the portal frame under construction, the steelwork being erected by Hoveringham Engineering Co Ltd. This is the view from the west end looking towards Catcliffe. Beyond the structure, work to dig out the canyon to hold the yard has barely begun. this dispels the theory that the depot was built on top of the in-filled Tinsley Wood tunnel, as material from the yard was used to fill this tunnel and, as can be seen, very little material has yet been dug out. *National Railway Museum/Science & Society Picture Library LS509*

Earlier ER steel portal depots such as Finsbury Park, March, Stratford and Shirebrook West, all had berths for three locomotives on each track. Through experience this was found to be impractical as the outer locomotive trapped two locomotives, thus necessitating some time-consuming shunting if the inner-most loco needed to be moved. Having only two berths per road cut these movements to a minimum.

Tinsley was more-or-less two sheds back to back with workshops and stores in between, with six roads at each end of the shed. Each road provided accommodation for two main line locomotives, giving a total of 24 berths.

The tracks inside were supported on short steel columns, with concrete platforms 4 ft 6 in above rail level alongside each berth. The floor level was 3 ft below rail level with headroom of 6 ft 8 in below the working platform; pits were provided 4 ft below rail level. This allowed work to be carried out simultaneously on three levels.

Each area of the depot and sidings was designated for specific types of work, the six roads inside the shed being split and designated east and west. The four outer roads each had a 1-ton hand-operated crane, the cranes on No. 6 road being capable of travelling through the workshop area, whilst at the west end of the depot a 5-ton electric crane covered roads 2 to 5. An additional crane over these roads would be installed during the 1980s.

The east end was usually where servicing and minor running repairs were carried out. No. 6 east bay was used for bogie washes, whilst No. 2 east was for locomotive painting.

The west end was for heavier repairs such as work on cylinder heads, pistons and turbochargers, and the exchange or repair of lubricating oil pumps. Engine changes could not be carried out, although the depot could undertake preparatory work for the removal of engines at main works.

On the south side of the shed was an annexe building containing two sets of lifting jacks for use during the changing of wheel-sets or bogie swaps.

At the rear of the depot was a two-stage locomotive washing plant. As the locos were driven through at a set speed, the first clean was designed to cool the bodyside panels and assist the spread of the cleaning fluid of "Exmover" crystals diluted in water. The loco would then stand for a few minutes before passing through the washing section.

Six main stabling sidings were laid to the west end of the depot building, used for locos awaiting or just off repair and for stabling of locos at weekends. On the opposite side of the depot ran the "proto road", the nickname being given following the testing of prototype MGR wagons in 1965.

Staff

The depot was to receive a complement of 200 maintenance staff; two training rooms were provided for their use on the top floor above the stores and offices. There was also a café for refreshments, but due to financial problems the café facility was closed in the late-1960s.

Staff retention at Tinsley became a problem as competition for skilled labour came from the many steelworks around Sheffield, some of which were offering trained fitters a basic rate of as much as £20 per week plus generous overtime rates. This figure equates to £392.66 in 2018. Additionally, there was competition from the building trade during the summer months, when building work was traditionally at its peak. Consequently, there was a rapid turnover of staff.

Liquid Storage

Other equipment provided included overhead runways on selected roads, bulk storage and dispensing facilities for lubricating and fuel oils, along with piping for pumping away waste sump oil, which was sent for use in the heating boilers. There were ten oil storage tanks in the basement, two with capacities of 4,500 gallons, two with 1,500 gallons and six with 1,500 gallons for the smaller user oils.

Diesel fuel storage was in four tanks, each with a capacity of 25,000 gallons, which were replenished by road or rail. They dispensed fuel through six self-contained pumping units, three at each end of the building. The mid-1960s saw approximately 3,800,000 gallons of diesel fuel per annum being used.

Coolant for main line locomotives was treated with a mixture of water, borax, sodium metasilicate, sodium metaphosphate and an anti-foaming agent to prevent corrosion build up within the engine and pipework. This was mixed on the depot and stored in two 1,500-gallon tanks in the basement.

Below: With building chaos all around, the DMD takes shape on 12th April 1963. The four 25,000-gallon fuel storage tanks are prominent on the right. On the horizon several dumper trucks take spoil away from the building of the cutting to the left of the depot.
Owen Owens collection

Below: A view inside the almost-complete depot on 22nd October 1963. This illustrates the rails laid on short steel columns, with concrete platforms 4 ft 6 in above rail level alongside each berth. The floor level was 3 ft below rail level with headroom of 6 ft 8 in below the working platform. There were also pits 4 ft below rail level. *National Railway Museum/Science & Society Picture Library LS559*

Below: Also dated 22nd October 1963, this image shows the fabric of the building in place, with surrounding work ongoing. Track has yet to be laid to the west end approaches to the doors. *National Railway Museum/Science & Society Picture Library LS558*

Depot Completed

The depot was completed in January 1964 at a cost of £654,000 but, due to restricted access whilst the marshalling yard was still under construction, only daily servicing could be carried out. With the closure of Darnall on 26th April 1964, work was transferred to Tinsley and the full facilities were opened.

April 1964 also saw the opening of a new four-berth diesel servicing depot in the main yard, located at the east end of the main yard sorting sidings, to the south-west of the main hump.

This two-road building was provided with facilities for lubricating, examining and sanding, along with a locomotive washing plant. Fuel was piped from the tanks in the main shed to two 50-gallons-per-minute fuel dispensers.

The commissioning of Tinsley saw completion of the Great Northern plan for their principal diesel locomotive maintenance depots. Tinsley's initial allocation comprised 160 locomotives, which included 350 hp shunters (later Classes 08 & 10), EE Type 1 (Class 20), Clayton Type 1 (Class 17), Brush Type 2 (Class 31), EE Type 3 (Class 37) and Brush Type 4 (Classes 47 & 48).

These locos were soon supplemented by additional shunters from Darnall and a batch of BR Sulzer Type 2s (Class 25). The majority of the Class 47s were fitted with Hawker-Siddeley slow speed control for merry-go-round coal workings.

There were no facilities for servicing steam locomotives at Tinsley, Canklow MPD being retained until steam was eliminated and full dieselisation was in place. Canklow closed on 11th October 1965.

In the surrounding areas, Wath depot had 40 Class 31s under its care by 1965, and a further new servicing depot opened for business on 14th June 1965 at Shirebrook West. Some diesel locomotives were stabled at Grimesthorpe steam depot until the new facilities at Tinsley were completed, the locomotives then moving to their new home.

Train crews also transferred from Darnall to Tinsley on 26th December 1964. The staff complement at Darnall was comprised of 198 drivers, 54 passed firemen, 98 firemen, 18 passed cleaners and 84 maintenance staff, not all of whom would transfer to Tinsley.

By the time of the yard's opening in October 1965, Tinsley depot's motive power allocation had swollen to the following numbers: Class 08 (74); Class 13 (3); Class 20 (12); Class 17 (11); Class 25 (11); Class 31 (57); Class 37 (42); Class 47/48 (73) – a total of 283.

Right: In this image dated 20th May 1963, the portal structure is complete and cladding is being fitted. Internal fitting-out is also under way, but track-laying has yet to commence. *Owen Owens collection*

Below: Track-laying has begun on the freshly-laid ballast and the fuelling points are under construction in front of the building. To the right, laying of drains continues and, on the left, the building with the chimney is the boiler house. *Owen Owens collection*

Above: A higher view taken in June 1963 shows the digging out of the reception sidings area is well advanced, but track-laying has yet to commence. The DMD is still a long way from finished. Orgreave Coking Plant can be seen in the distance, in the top right of the image. *Owen Owens collection*

Right: The fuel storage tanks are complete but work continues on the water tower, from where the image above was taken. *Owen Owens collection*

Below: The south west arrival line dive-under is being built in this view, whilst to the right the depot access roads have been laid. In the yard area, dust rises from the machinery on what looks to be a glorious summer's day in 1963. *Owen Owens collection*

Above: It is now 30th September 1963 and track-laying is well advanced. The lighting tower in the foreground, although looking complete, is awaiting fitment of its mercury lamps. The loco on the left is D3698, which had an interesting history. New to Darnall on 7th May 1959, it was transferred to Grimesthorpe on 15th April 1961 before returning to Darnall on 16th September 1961. It became a Tinsley-allocated loco from 2nd May 1964, and in June 1965 lost its cab upon conversion to the "slave" part of Class 13 No. D4500, later 13003. *Owen Owens collection*

Below: Wet weather has come but work continues in this view from 22nd November 1963. There is still no track in the reception sidings area, but there are plenty of puddles! *Owen Owens collection*

Above: The completed building in early 1965. Just look at the eclectic selection of cars the staff have! On the left just poking out from behind the building is the front of a Leyland FG, no doubt a type of crew bus for permanent way use. Then there is a Standard Ten estate, a Ford Consul, Ford Anglia, Ford Poplar, a sleek-looking Jaguar, a very boxy Triumph Renown and finally a Mini. *John Pavitt collection*

Left: On the top floor above the stores and offices was a café for staff use, shown here when new. The windows gave a good view of the work area below. This area was later used as a training room. *John Pavitt collection*

Below: The ramped main entrance to the lower ground floor afforded access to the subsidiary stores situated 4½ ft below rail level. Short ramps then linked the working floors to the stores area. *John Pavitt collection*

Above: Brush Type 4 D1553 goes through the wash at Tinsley DMD while D6813 & D1540 look on. A Class 20 is seen in the background alongside a diesel brake tender on Sunday 18th April 1965. *Dave Wragg*

Below: The completed depot hosts five types of locomotive at the east end in this view taken on 15th October 1965. On view are examples of Classes 08, Clayton Type 1 (Class 17), with a Sulzer Type 2 (Class 25) behind, an EE Type 3 (Class 37) and a Brush Type 4 (Class 47). The Claytons did not reign long in the area, soon being moved up to Scotland.
National Railway Museum/Science & Society Picture Library LS628

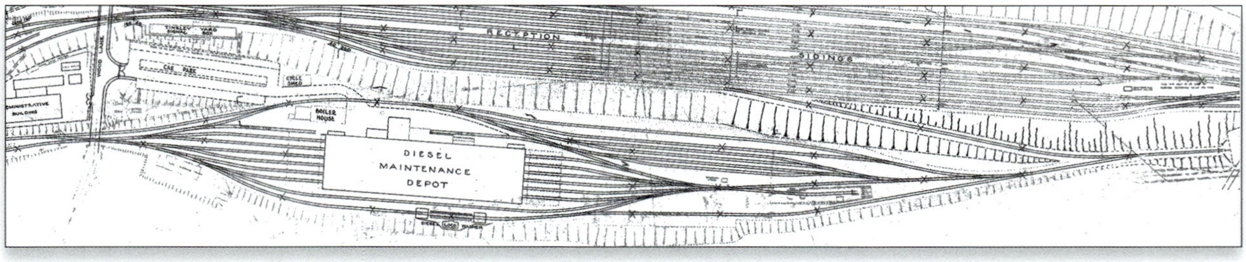

Above: Plan of the completed depot. *Reproduced with permission of Network Rail Corporate Archive*

Chapter 9

Tinsley Yard Buildings

This chapter examines the buildings at Tinsley, giving a description of each type along with details of dimensions and construction methods. By necessity it gets technical in places, but the information is essential when telling the complete story.

Anyone visiting Tinsley Yard, or indeed viewing photographs, will have noticed that most of the buildings looked very similar in design; this is because a new system of construction was used for the first time at Tinsley.

Although modular pre-cast prefabricated construction had been around for many years, for Tinsley a concrete design known as System 'T' was developed by the Eastern Region. The brief for this was "to meet special requirements where several buildings were required, each with a different use".

At Tinsley, buildings were required for the administration block, control tower, signal boxes, crew accommodation, shunters' cabins, carriage & wagon staff and permanent way staff. System 'T' was used for 13 buildings in total.

The division of responsibilities in the chief civil engineer's (CCE) department meant that architects could design only a limited range of smaller buildings; larger sheds were designed by engineers, without architects' input. However, for Tinsley DMD a degree of joint working was developed between architects and engineers, which was reflected in a more integrated design.

System 'T'

The main technical problem for construction of the buildings was the short time available between preparation of the various sites and those sites becoming surrounded by a network of tracks. This ruled out traditional building methods and pointed towards off-site prefabrication of units that could be delivered by rail.

Use of System 'T' meant that the buildings could be erected quite quickly, each building having architectural unity with the other, no matter what they were used for.

To save on construction costs, one of the architects' briefs was to determine the least number of different components (comprising walling, flooring, roof and stair units) required for each building. They also determined that each panel would be of a single thickness so that one side would be

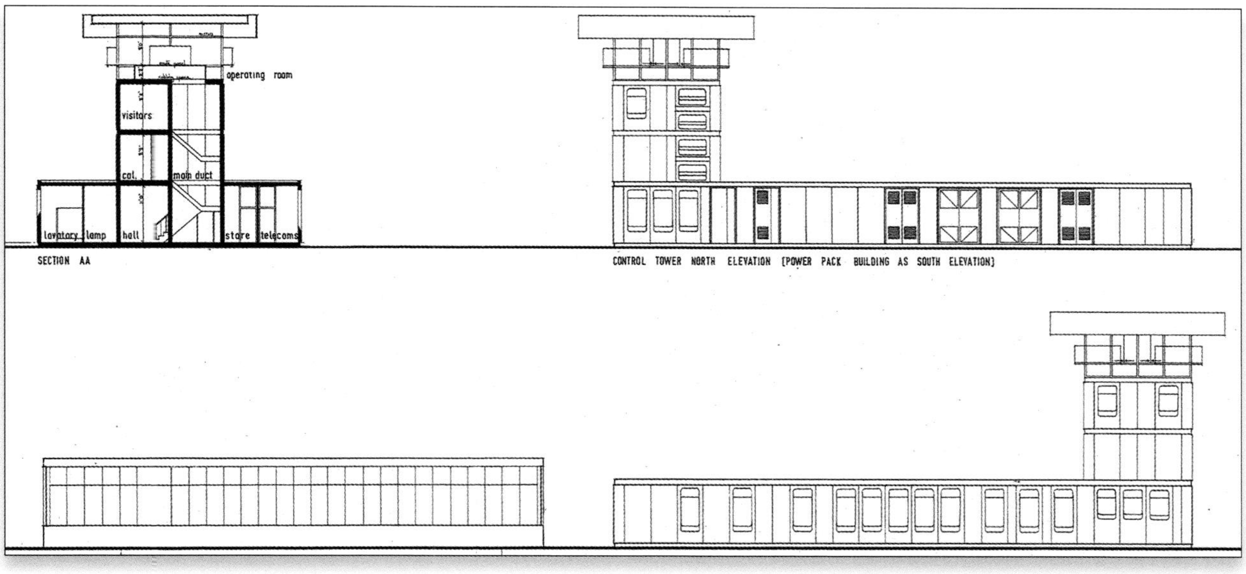

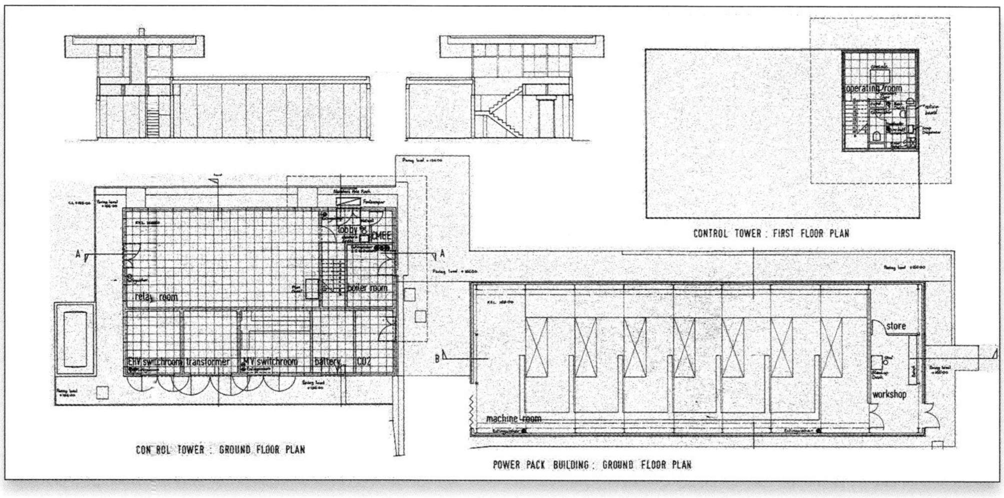

Above: Main control tower diagram.

Left: Secondary yard control tower.

Both reproduced with permission of Network Rail Corporate Archive

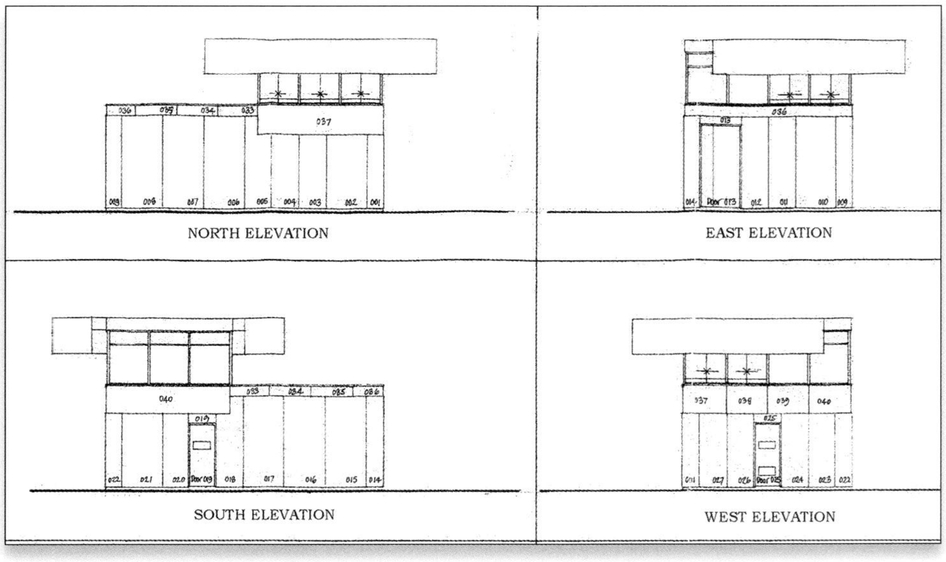

Right: Tinsley Park signal box diagram.

Below: Shepcote Lane signal box diagram.

Bottom: Tinsley Yard signal box diagram.
All reproduced with permission of Network Rail Corporate Archive

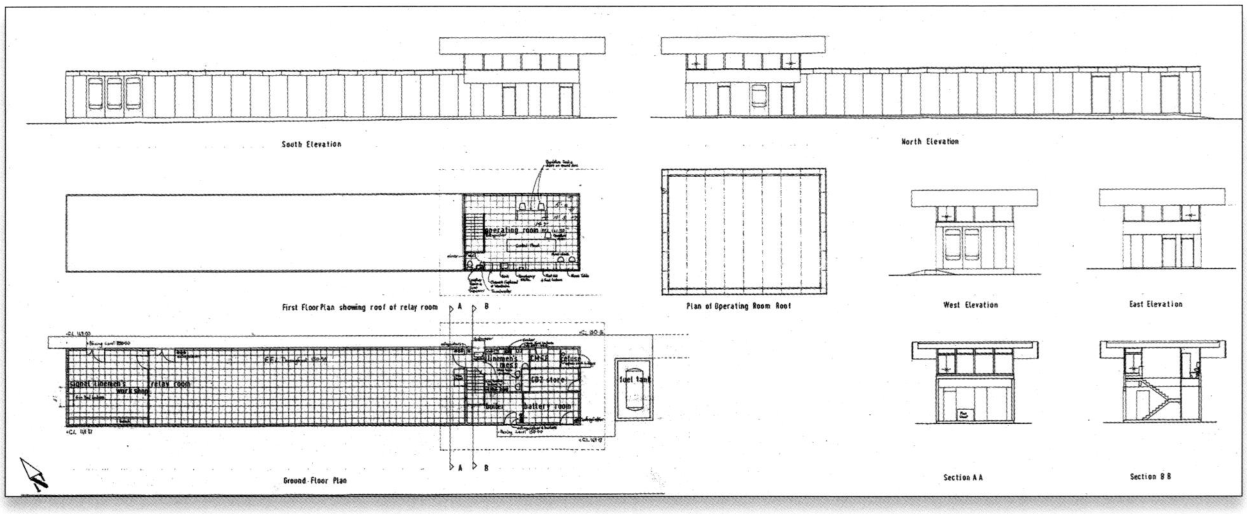

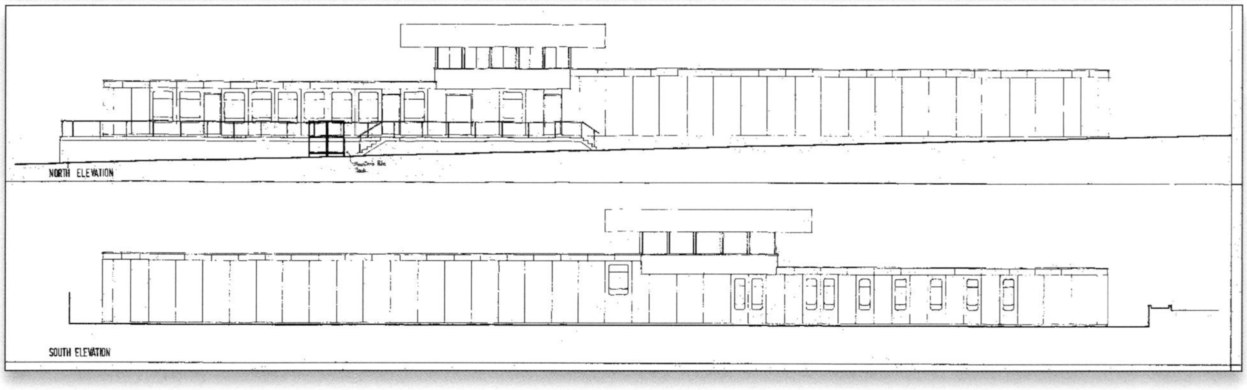

the exterior face and the other the inside finish. Insulation was to be integral through the use of Leca® expanded clay aggregate, this lightweight material providing good thermal insulation.

The exterior facing was of calcined flint aggregate, otherwise known as pebble-dash, with minimum dirt retention being required for good weathering. The design team was fortunate in being able to inspect an array of different exposed aggregate panels that had been subjected to the polluted atmosphere from the many steelworks of Sheffield's east end!

The concrete panels were cast with the external face down and hosed down before the concrete had fully cured, to give a strongly-textured stone pebble-dashed appearance. The panels were 8 in thick in two alternative heights of 8 ft 8 in and 11 ft, and three alternative widths of 1 ft 8 in, 3 ft 4 in and 5 ft. The floor and roof units were generally in two widths and lengths of 8 ft 4 in to 20 ft, in 1 ft 8 in increments. The signal box floor sections were 5 ft wide but especially deep to accommodate wiring ducts.

Panel Fixing Details

In the 1960s there were already concerns about the structural integrity of concrete panel systems; at Tinsley this was addressed by ensuring that all connections were bolted and not reliant on grouting. Steel channels were cast into the edges of the wall units with bolts threaded at both ends to draw the channels together, thus post-tensioning the walls. Tightening the bolts drew adjacent wall panels together, with two strips of foamed butyl rubber acting as a joint seal. The wall units were seated on "Nuralite" that acted as a damp-proof course.

The pre-cast construction could be used for a three-storey walk-up building, with wall tensioning connections in place on each storey just below sill level. The problems of corner joints were overcome by eliminating them; instead T-shaped and L-shaped concrete corner units were connected to the straight walls.

The architectural unity between all the buildings was particularly noticeable in the design of the window apertures. These had rounded corners to emphasis the load-bearing function of the panels and to minimise local stresses. Windows were similar in style to those already used in the transport industry, such as those used on buses etc, with glass set in neoprene gaskets and the upper 16 in an aluminium-framed inward opening hopper. These would be inserted on site

Size Differences

The buildings constructed at Tinsley ranged in floor area from 40 to 14,000 sq ft and fell into three broad categories:-

Small buildings in glass-reinforced plastic (GRP) (shunters' cabins, "Addo" transmission kiosks).

Rolling stock and machine sheds of steel frame construction with corrugated plastics or asbestos cement sheet cladding (locomotive maintenance and servicing depots, power-pack buildings, breakdown crane shed, Dowty test trolley shelter and the boiler house).

Signal boxes, control towers, offices, workshops, stores and staff amenities buildings using System 'T'.

Small Buildings

There was a need for small kiosks to shelter the shunters and their equipment as they entered information on the make-up of trains into the "Addo" keyboards.

The design of these kiosks drew on GRP buildings already designed by Arthur Quarmby and David Appleby at the Railway Technical Centre, Derby in conjunction with Mickleover Transport Ltd, a firm with experience in making GRP milk floats.

The kiosk comprised a simple shelter, 10 ft by 3 ft 4 in, entered through a door with a radiant heating panel moulded into the inner skin of the ceiling. Windows were the same as fitted into buses which allowed the operator to have a view out of the cabin to both sides. For the weary shunter to rest his legs, a tip-up seat was mounted on one side-wall.

GRP cabins for shunters were located in the reception sidings and on the eastern approach to the locomotive depot. They measured 15 ft 9 in x 7 ft 6½ in with worktops and mess facilities provided. Although the drivers had a mess room in the main accommodation block, they also had one of these cabins in the top shed yard for use between turns, which they nicknamed the "yellow submarine" primarily due to its colour! Inside they had added two beds made from coach seats, an unwritten rule when resting being that the driver would always take the left bed and second man the one on the right!

The Dowty test trolley had its own shelter situated close to the main control tower, this being 26 ft long by 13 ft wide and constructed of tubular sections with translucent "Unilux" glass-reinforced polyester cladding sheets.

Power Pack Buildings

The main yard control tower power pack building measured 100 ft x 30 ft and contained seven Dowty hydraulic pumps, two compressors for pneumatic signalling apparatus, stores and a workshop. The steel portal frame rested on a reinforced concrete plinth wall and the sides of the buildings were clad in translucent sheeting. Inside, a traversing overhead crane was installed and folding doors facilitated handling of mechanical components.

In the secondary yard control tower (itself measuring 90 ft x 30 ft) there were seven Dowty hydraulic pumps and a workshop; this was identical in construction to the main yard building.

Control Buildings

The main yard control tower had ground floor accommodation measuring 110 ft x 54 ft to house signalling equipment, electrical switchgear, workshops and amenities for permanent way staff and shunters.

The three-storey tower at the eastern end accommodated fire-fighting and heating apparatus on the first floor, the staff mess room on the second floor and the operating room on

Above left: Artist's impression of a shunters' cabin that was to be placed in the reception sidings. These glass-reinforced plastic (GRP) buildings were supplied by Mickleover Transport Ltd, who were more used to building milk floats.

Above right: Two of the "Addo" data transmission kiosks being delivered on a 13-ton high-sided steel open wagon.
Both John Pavitt collection

the third floor. The operating room had windows on all sides and the roof was projected 6 ft 8 in beyond the windows with a sun baffle to minimise glare. Windows at high and low level opened inwards for easy cleaning from inside, while a light steel catwalk was provided for external cleaning of the fixed panes at eye level.

The operating room route panel and inspector's desk were placed back-to-back on a dais to give an all-round view of the yard; to avoid overcrowding, a waiting area for visitors to the tower was provided on the second floor.

The secondary yard control tower measured 55 ft x 34 ft and housed signalling equipment, electrical switchgear, fire-fighting and heating apparatus; its operating room was at first floor level.

Signal Boxes

David Appleby in the CCE department had researched visual conditions for signalmen in new signal boxes. The problem was trying to allow light into the box, whilst preventing glare from affecting the signalman's view of the indications on the control panel.

This issue was not unique to Tinsley, but unlike other PSBs, signalmen and hump tower operators at Tinsley required a measure of visual contact with the outside railway to efficiently control hump-shunting operations and movements over non-track-circuited sections of line.

The solution was to provide an overhanging canopy to prevent glare, but allow sufficient daylight to enter to reduce excessive contrast on the vertical face of the control panel. The boxes were largely to a standard BR ER design used elsewhere, such as at Aldwarke, but with the specific requirements of the Tinsley environment taken into account.

Tinsley Yard signal box provided accommodation on the ground floor measuring 187 ft x 24 ft for signalling equipment and outdoor operating staff who were working in the middle of the yard. The operating room was on the first floor and measured 25 ft x 20 ft.

Tinsley Park had ground floor accommodation measuring 34 ft x 20 ft which housed signalling equipment; the first floor operating room measured 20 ft x 15 ft.

Finally, Shepcote Lane housed the signalling equipment, lineman's workshop and the mess, heating equipment and electrical switchgear on its 135 ft x 20 ft ground floor. Windows were only installed at the north-east and north-west corners of the operating room but this ensured the signalman had good vision of the panel indications, which were situated on the north wall.

Other System 'T' Buildings

The administration building which adjoined Wood Lane was the largest of those constructed in System 'T', with accommodation on the ground floor totalling approximately 12,000 sq ft. The ground floor housed the train crew accommodation, timekeeper, telephone apparatus room, assistant yard managers' office, train crew inspector, general stores, and heating equipment. The most important office here, as far as the staff were concerned, was the pay office! The first aid room and British Transport Police office were located in a separate block next to the entrance gate.

Accommodation on the first floor measuring 107 ft x 20 ft was for the yard manager and clerical staff. Car parking and cycle storage facilities were provided alongside the building.

The main yard west-end building was single-storey and measured 71 ft x 20 ft. It provided accommodation for shunting and permanent way staff. Shelters for cycles and motorcycles were also provided.

The express freight yard building was also a single-storey building which measured 37 ft x 15 ft and provided

Above left: The main yard control tower receives finishing touches.

Above right: The sun baffles as seen on Tinsley Yard signal box. This photo was captioned "Natural lighting control system in Glass-Reinforced Plastic". *Both John Pavitt collection*

Right: The completed administration buildings, main yard control tower and servicing shed, seen from the water tower. Note the bored looking pigeon! *John Pavitt collection*

Below: The administration block under construction on 13th May 1964. To the left of the image, a Brush Type 2 locomotive is making its way towards the diesel depot, which was fully operational by this time. *National Railway Museum/Science & Society Picture Library DM320*

accommodation for the shunting staff based in this part of the yard.

The secondary yard west-end building measured 74 ft x 20 ft. This single-storey building comprised a small workshop for wagon repairs and an office and store for operating staff, along with staff facilities and the boiler room. Storage bays were provided outside the building, together with a cycle shelter and a gantry for handling wagon components, along with a separate building measuring 37 ft x 12 ft which housed electrical switchgear.

Heating in all the mentioned buildings was from either independent oil-fired low pressure hot water central heating or electric heating.

Building Progress and Cost

The marshalling yard buildings were completed in 17 months at a cost of about £425,000, equivalent to around £9.4 million in 2018. This does not include the cost of the signalling and control equipment, or of Tinsley Diesel Maintenance Depot.

Last Man Standing

The System 'T' buildings appear to have been structurally sound whilst in use although, as the yard shrank over the years, so the buildings became disused and derelict without windows and doors. Vandalism and neglect continued to plague them. The pebble-dashed exteriors weathered well in the main, but the sun baffles to the operating rooms were stripped away over the years.

As the yard was cleared, so the buildings were demolished; all have now disappeared. After Tinsley Yard signal box was burned down it was demolished, leaving the main control tower as the last man standing. Many photos showed it in an increasingly derelict state and after 50 years it was finally put out of its misery by Anthony McHugh Demolition Ltd in April 2015.

Of the smaller GRP buildings there is no trace, but maybe one day one will be found doing duty as a garden shed somewhere in the area!

Above: A night time study of Tinsley Marshalling Yard in January 1977. *Michael Rhodes*

Chapter 10

New Signalling for the Yard

With the construction of such a large yard, new signal boxes would be required to signal both the yard and its new connections to and from the area. The area was resignalled with multiple aspect colour-light signalling; working by track circuit block regulations was introduced. The resignalling, which also provided access between the new marshalling yard and Sheffield Freight Terminal (SFT) at Grimesthorpe, was carried out in three stages.

Brightside Junction

In connection with the resignalling, a new signal box at Brightside Junction was opened on 25th April 1965. This would control the original SDR Brightside Junction and work to Shepcote Lane Junction, but would also control the lines from Sheffield North Junction to Wincobank Station Junction. The new box was a BR (ER) Type 19, replacing the original MR Brightside Station Junction box and also taking over the areas of Upwell Street Junction, Mill Race Junction, Grimesthorpe No. 1 and Grimesthorpe No. 2 boxes.

The drawings for the new box were later given an addendum which reduced the designed length by five feet due to a change in frame requirement; the one installed would be a Westinghouse 37-lever frame to work the points in the immediate area of the box, along with a Westinghouse individual function switch (IFS) panel operating all other points and the signals in a hybrid arrangement. All signalling was either main aspect or ground position light (GPL), numbered B101 onwards. The box controlled 31 electric points, 23 main running signals, 26 GPLs and four ground frame releases.

Treeton Junction

At the other end of the SDR was the MR Treeton Junction box, which was given a new 55-lever Railway Signal Company (RSCo) Great Northern Railway of Ireland (GNoI) frame, to control the new north-east curve from Tinsley Yard towards Rotherham. The type of frame installed was one of those originally destined for the GNoI but not used there, instead being used for installations in Britain.

Broughton Lane Area

The first new connections to the former GCR Woodburn Junction to Rotherham line were the south-west arrival and departure lines, which were

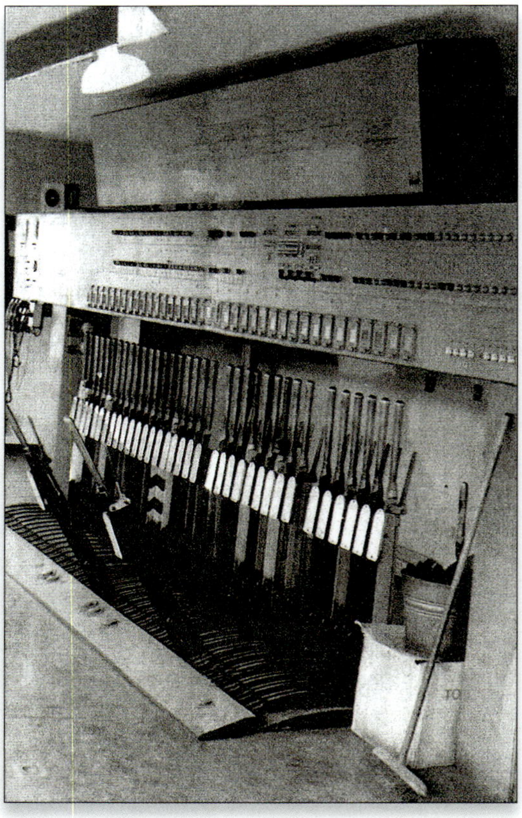

Above: The Westinghouse 37-lever frame with Westinghouse IFS panel above, in Brightside Junction signal box. *Roy Parkes*

Above left: Brightside Junction signal box as viewed from the Midland line in 1979. *Pete Churchman*

Left: Brightside Junction as viewed from a diverted DMU about to take the Sheffield District line towards Tinsley yard in 1978. The signal box can be seen just in front of the train. *John Law*

initially worked as sidings while the yard was still under construction. As access was required for construction traffic, the connection was under the control of Broughton Lane Station signal box.

The south-west arrival line facing point connection was situated on the "up" main, 390 yards north of the signal box, whilst the trailing connection was on the "down" goods line, 410 yards from the box. New semaphore signals were placed to protect the junctions, and a double-sided stop board applying to the south-west arrival line was situated close to the junction with the old SDR line at West Tinsley. This was worded "Stop and Await Instructions" for drivers heading towards the new yard, and "Stop – No Movements Past This Board Without the Permission of the Signalman At Broughton Lane". A telephone was provided at the board for drivers to contact the signal box.

Once sufficient track had been laid to allow access to the new Tinsley depot, which opened before the yard was completed, drivers of locomotives to and from there would also contact this signalman.

Shepcote Lane

As work progressed, and to enable works traffic to continue to access the yard, some form of control was required before the permanent signalling was in place. At the west end of the yard a temporary wooden signal box was erected at Shepcote Lane, this assuming control of access to the new Tinsley depot; drivers of locomotives to and from the depot would then contact this signalman.

Although the signal box received a panel, nothing else is known about its make or type. The opening and closure dates of this box are not confirmed, but it is known that it cannot have opened before 1st May 1964 as notices published at that time showed that Broughton Lane was in control of access to the yard. Conversely, it must have gone by 7th December 1964 when the new Shepcote Lane box opened.

Yard Signal Boxes

In the yard complex itself, three new power signal boxes were to be built.

Shepcote Lane power signal box was fitted with a Westinghouse miniature control desk (MCD) entrance-exit (NX) panel. This box would control the west end, where the line from Brightside Junction and the north-west and south-west lines converged, and additionally signalled trains onto the GCR Rotherham branch between Attercliffe Junction and Tinsley East and West Junctions. When opened, Shepcote Lane brought about the demise of Tinsley Station Junction, Carbrook, Broughton Lane, Brown Baileys and Attercliffe signal boxes.

Tinsley Park box opened on 29th November 1964 with a Westinghouse individual function switch (IFS) panel; it controlled the area at the west end of the main yard and the east end of the express freight yard.

Tinsley Yard box, the largest of the three, opened on 11th April 1965 with a Westinghouse MCD NX panel. It controlled the reception sidings area and the route towards Treeton Junction. These boxes were all BR (ER) Type 18 design.

Train describers were installed in Shepcote Lane and Tinsley Yard boxes where train descriptions, after they were manually inputted, were carried forward on the box diagrams by the movement of the trains; at Tinsley Park trains were signalled by bells.

Above: A driver's eye view of Treeton Junction from the south-east departure line in 1982, during a period of transition as the area was being resignalled to come under Sheffield PSB's control, which it did in October that year. The semaphore signals are Treeton Junction No. 7 (left) leading to the "up" goods line and No. 8 for the "up" main, with Treeton South Junction's distant signals beneath. A two-car Metro-Cammell Class 101 DMU heads along the "Old Road" on a diverted passenger turn. The bridge abutments are the remains of the former GCR access to Treeton Colliery. *Roger Wainwright*

Left: The Great Northern Railway of Ireland (GNoI) frame in Treeton Junction signal box. *Roger Wainwright*

Right: A good view of the signals and box at Treeton Junction looking north. The MR Type 2b box opened on 1st April 1909 to control the junction with the Sheffield District Railway. Treeton Junction was re-framed in 1965 to control the new north-west curve towards Tinsley Yard. It closed on 3rd October 1982 when the area was taken over by the Treeton Junction interlocking of Sheffield PSB. The bridge in the background was the former GCR access to Treeton Colliery. Treeton Junction's home signals on the goods lines (left), being the same height, indicate that the lines towards either Tinsley or Rotherham have equal importance, whilst on the Barrow Hill line, the Rotherham line has greater importance. The small semaphore signal below the full-sized arm on the right-hand bracket is a "calling-on" signal, telling the driver he is entering permissive block and to proceed at caution as there could be another train in front. *Roger Wainwright*

Below: A Driver's eye view of Treeton, heading towards Tinsley Yard with the signal showing "off" for the move. A pair of Class 20s are on the "up" main, whilst in the background a Class 31 waits to come from the Tinsley line on a ballast working. *Roger Wainwright*

Above: To enable works traffic to access the yard, some form of control was required before the permanent signalling was in place. At the west end of the yard a temporary wooden signal box was erected at Shepcote Lane, seen here on the left, just beyond Shepcote Lane PSB. Little is known about this temporary box, but it must have been closed by 7th December 1964 when the new Shepcote Lane box opened. It is seen here on 26th January 1965. *British Railways SB.505/Chris Booth collection*

Control Towers

Tinsley main yard control tower and secondary yard control tower were also BR (ER) Type 18 designs fitted with Westinghouse MCD panels, opening as non-block posts on 11th April 1965 in conjunction with the opening of Tinsley Yard signal box. The control panels in the towers were IFS type being IFS[+] and IFS(Y3)[+] hybrid type. An explanation of these terms follows at the end of this chapter.

Other Boxes Closed

The closure of goods depots and marshalling yards had resulted in the closure of certain signal boxes between 1961 and 1969. Others saw a reduction in use between 1964 and 1969, finally closing that year. These are shown in Table 1.

Table 1: Signal Box Closures

Signal Boxes Closed 1961 to 1969	
Wincobank Sidings	11/04/1961
Firths Sidings	18/07/1965
Elevated Box	18/07/1965
Sheffield Victoria No. 6	03/10/1965
Hollis Lane	12/04/1966
Sheffield Victoria No. 2	15/11/1967
Woodhouse Mill No. 1	31/03/1968
Signal Boxes Reduced in Status in 1964 and Closed in 1969	
Roundwood No. 2	
Masborough Sorting Sidings Centre	
Masborough Sorting Sidings Down Bank	

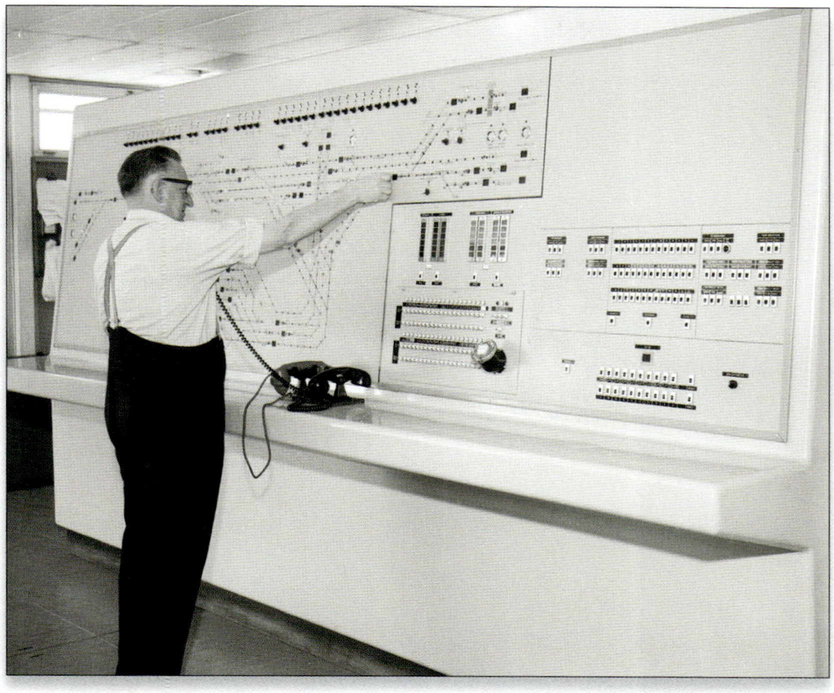

Above: The new Shepcote Lane power signal box was fitted with a Westinghouse miniature control desk (MCD) entrance-exit (NX) panel, which is seen here shortly after opening. *National Railway Museum/Science & Society Picture Library*

Tinsley Yard Signal Box - "New and Old"

Right: Tinsley Yard box as viewed from the reception sidings in October 1965.
British Railways/Chris Booth collection

Below: In this image taken just after the box opened on 11th April 1965, signalman Geoff Holbrow can be seen operating signals in the reception sidings area on the Westinghouse MCD NX panel. Note the long row of black individual point switches on the top of the panel.
British Railways/Chris Booth collection

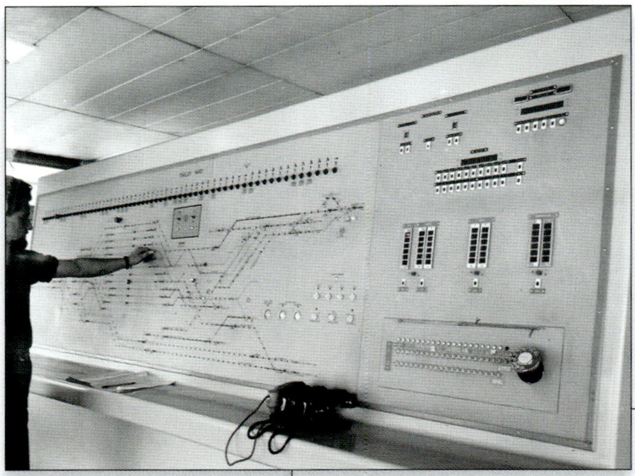

Below: Taken on 23rd May 2008, the signalman's car is parked where the reception sidings and engine line once passed in front of the box. Tinsley depot has gone, replaced by a car dealership and a hotel. *Chris Booth*

Left: This is the same panel but in control of far less track. The vacant space shows where the reception sidings previously were; there are fewer point switches to be seen as all work is now concentrated at the west end of the yard and on the line to SMACC. *Chris Booth*

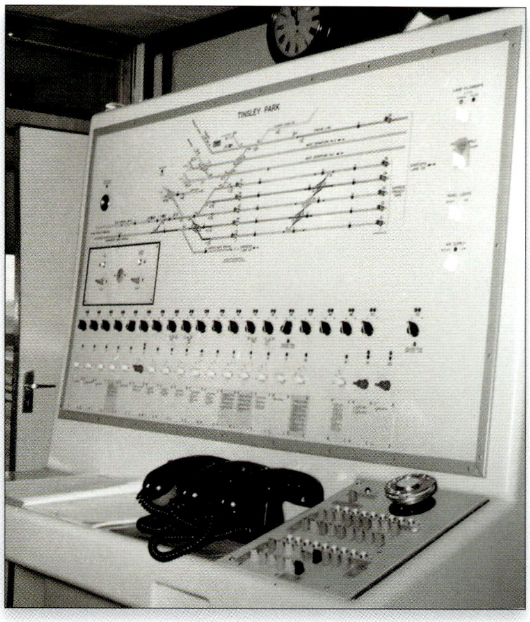

Above left: Tinsley Park signal box on a hot day in 1976. The smallest of the new signal boxes, it opened on 29th November 1964. *Roger Wainwright*

Above right: The Westinghouse MCD IFS panel, which controlled the area at the west end of the main yard and east end of the express freight yard. *British Railways/Chris Booth collection*

Tinsley Park Area Point Motors and Method of Signalling in Fog

Unusually for a power signalling project, many lines close to Tinsley Park signal box were not covered by track circuits (TC). The replacement of signals to "danger" was entirely by signal switch, with no automatic TC replacement being effected and no TC occupation to hold signals at "danger". In the case of shunting movements, there was nothing to stop the signalman flicking the switch to clear the signal for the reversing movement before the first move had fully cleared the GPL; the signal would come straight "off", provided the points were set (there were no TCs to prevent this), so the signal was "off" before the driver got to the other side of it. The signalman, therefore, had to be vigilant and keep a good look out through the windows at all times whilst controlling moves.

Above: Tinsley Park No. 48 points seen on 21st August 1992. This is a Westinghouse electro-pneumatic air-operated style point which could be run through in the wrong direction without damage.
Andrew K. Overton

Two kinds of air points were used at Tinsley – those on the running lines which might possibly be used for diverted passenger trains were conventional and fitted with a facing point lock (FPL) mechanism in the motor. Those purely on siding lines were the yard type with no FPL, and which could

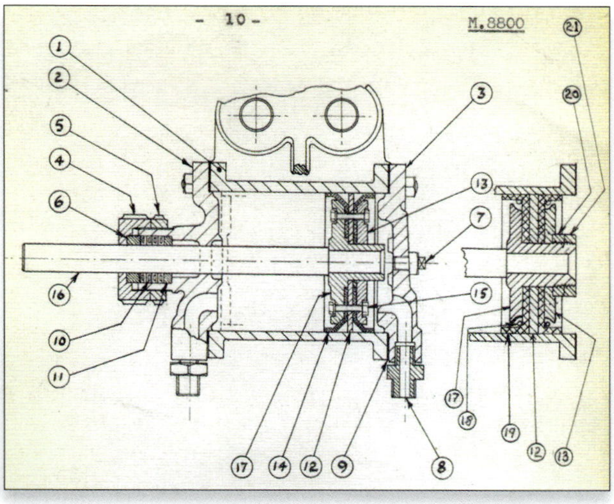

Above: Diagram of an electro-pneumatic point cylinder as issued by Westinghouse for the opening of the yard. The numbers relate to:

1. *Cylinder Body*
2. *Front Head*
3. *Back Head*
4. *Stuffing Box Nut*
5. *Stuffing Box Locknut*
6. *Stuffing Box Gland*
7. *Gas Plug*
8. *Union Nipple*
9. *Gasket for Cylinder*
10. *Packing Rings*
11. *Gland*
12. *Piston Ring*
13. *Piston Follower*
14. *Piston Rubber*
15. *Piston Clamp Bolt*
16. *Piston Rod*
17. *Piston Head*
18. *Piston Expander Ring*
19. *Piston Leather*
20. *Piston Nut*
21. *Piston Locknut*

Left: The main yard control tower as new. *John Pavitt collection*

Middle: Derek Worral sits at the Westinghouse miniature control desk (MCD). This and the secondary yard tower example were opened as non-block posts on 11th April 1965. The control panels in the towers were IFS[+] and IFS(Y3)[+] hybrid type. *Steve Ryszka*

Above: The same panel in the remains of the main yard control tower on 20th May 2009. The fascia plate has been removed at some time and there has been extensive vandalism of the tower. *Chris Booth*

be run through without causing damage. This was not only a cost saving factor but important on yard lines where run-throughs were not unknown. The points at Tinsley Park on the non-TC lines were yard type.

Problems occurred during thick fog when the signalman couldn't see whether a shunt move had completely cleared his points. The radio was useful if the loco was fitted (such as the Class 13) or there was a shunter present, but otherwise it would be a case of waiting for the driver to give a pip on the horn to indicate he was clear of the points. Unfortunately, this was not ideal as other locos in the area could also be sounding their horns. Therefore, the following method of working was adopted:

As the shunt was making its outward move, the signalman would set the points – trailing for now but facing for the return move – to the position required for the return move. The train continued its outward move by running through the points (without damaging them) which motored back to the trailing position as each set of wheels cleared. During the move the signalman would watch the panel and particularly the detection lights* above the point switches on the set of points furthest away from the move. When the detection lights ceased flickering in this manner, the signalman knew the train had stopped and he could therefore operate the return signal switch to "off" for the set-back move.

* Each set of points on a power signal box panel has a switch to manually operate it. Above the switch are three indications – normal, free and reverse – to inform the signalman of the position of the points. Only one light is illuminated depending on whether the points are lying in the normal or reverse position, and the illumination confirms the points are correctly aligned to permit a train over them. This is called detection. If the light is flashing, the correct detection has not been achieved. In the case of Tinsley Park, every time a set of wheels passed through the points and pushed them open the detection was lost and the light would extinguish, immediately illuminating again when the points sprang back and detection was achieved.

The Signalman had to be very careful of moves which were loco-first with the train propelling back into the siding, so that he didn't clear the return signal until he was absolutely sure the outward move had stopped. If he did clear the signal in this circumstance, the driver might be unaware that the whole train had not cleared the pointwork (due to the fog) and would start

Above: The secondary yard control tower seen on 21st August 1992, long closed and gradually being vandalised. *Andrew K. Overton*

to set back with half of the train going back from where it came and the other half taking a different direction, with the inevitable derailment as a result! With no TCs to detect the presence of the train, there was nothing to prevent the signal from clearing to its "off" position with part of the train standing on the wrong side of it. One such incident involved two Class 47s. One was travelling along the engine line and the other heading the other way for the secondary yard. The Tinsley Park signalman operated the points between the bogies of one of the 47s, with the loco trying to take two different routes as a result!

Explanation of Terms

The following gives an explanation of the terms IFS[+] and IFS(Y3)[+].

There are two types of arrangement for where the control buttons/switches are mounted in relation to the illuminated track diagram of the area the signal box controls – combined arrangement or separate arrangement. In a combined arrangement, the control buttons/switches may be physically combined into one module of equipment (the interface) with the diagram, with the controls positioned on the diagram either geographically or separately grouped. Where a combined arrangement is employed, this is denoted by the interface type designation (in this case an individual function switch or IFS) followed by [+]. In a separate arrangement, the control surface may be a physically separate module of equipment from the diagram, and the interface type designation is followed by [-]. Where the arrangement is unknown neither symbol is shown.

Special types of IFS may be provided to control marshalling yards where loose shunting of wagons takes place. There may be no interlocking between hump shunting signals and the yard points and there may be complex additional controls for wagon retardation.

Simple systems have individual point switches for the sorting siding points, which are manually set for each cut. An improved system additionally provides buttons for each destination siding, which will set all the points required from the hump summit – a kind of route-setting but not linked to signal clearance.

In these systems there is mechanical interlocking to prevent more than one button being depressed at a time. The button is released and the next route can be selected once the previous cut has cleared the king points; each set of points will move as required as they are cleared by the preceding wagons.

Point switches have a central position for when the buttons are in use but they can be used at any time to redirect wagons, even when a route has been set through them with the buttons. Other systems are able to store varying numbers of preset siding selections (typically up to 50 in later systems), with only the button which fills the store being held down until one cut is cleared out by passing the king points, whereupon another cut can be added in. This process, with unlimited storage, can be achieved by punched tape.

Therefore, during normal operation the panel works automatically once the cut destinations have been programmed in. It should be noted that in some cases routes which are programmed in at one interface may control points which are actually individually controlled by another interface, which may be situated in another building.

In order to distinguish these yard interfaces and the special circumstances which may appertain to their operation, the definitions shown in Table 2 are used. These definitions are appended to the IFS definition in parentheses, ie. IFS(Y1)[+].

Table 2: Definitions of Yard Interfaces

Name	Abbreviation	Features
Yard Type 1	Y1	Sorting siding points must be set individually.
Yard Type 2	Y2	Sorting siding points may be set either individually, or by the use of siding selection buttons, one cut at a time.
Yard Type 3	Y3	Sorting siding points may be set either individually, or by the use of siding selection buttons by pre-setting multiple cuts.

Above: The secondary yard control tower signalling panel fascia. This was removed after closure of the tower and is now in a private collection. *Steve Ryszka*

Chapter 11

The Dowty System of Automatic Wagon Control

The use of gravity to allow a wagon to freewheel into the sidings was one of the great innovations of marshalling yards; the problem was stopping the wagons exactly where required! Early marshalling yards relied upon wagon chasers – men who literally ran alongside their wagon and pushed down on the wagon's hand brake with a brake stick, in order to bring the wagon to a rest where required. This was both inefficient and very dangerous.

In a desire to reduce damage to wagons and their contents, as well as speeding up the sorting of trains, mechanical wagon retarders were introduced. These devices did the job of the wagon chaser through use of a primary retarder, which squeezed wagon wheels as they passed through, thus reducing the speed of the wagon. Sometimes these were assisted by secondary retarders acting along a similar vein.

These early devices were quite crude in their braking application and due to the large number of variables, such as axle lubrication, wind and general roll-ability of the wagon, it was extremely difficult to accurately predict where a wagon would stop after the application of a braking force. Even when complex electronic equipment was used for prediction, with the wagons free running between the primary and secondary retarders and from the secondary retarders into the sorting sidings, the overall performance left much to be desired.

The ideal arrangement was one which provided continuous automatic control of the wagon's movement, from leaving the hump to buffering up to the last wagon in the sorting siding.

Dowty Description

The Dowty system set out to accomplish wagon speed control by using small hammer-headed hydraulic units bolted at intervals to the inside of the rail. The units exerted vertical pressure on the wagon wheel flange as it passed along the rail. This controlled wagon speeds precisely, irrespective of wagon roll-ability, weight or wind resistance – no independent speed measuring or electronic devices were necessary to help the process. There were two basic types of hydraulic unit – retarders and booster retarders.

Retarder

The retarder was a completely self-contained unit partially filled with oil, which reduced the speed of a wagon travelling

Above: A 350 hp shunter, 13075, shunts Esso tanks in Hull Inward Yard in July 1954. This yard is where experiments were conducted with the Dowty system of wagon control over a two year period from 1961. Two sidings were equipped with the system, which controlled the wagons after they had been pushed over the hump and rolled through conventional Frölich clasp-type retarders. *GEC-Alstom*

Left: Hull Inward Yard as viewed towards the Hump, with the control tower to the left of the image. The white areas opposite the tower are the original Frölich hydraulic retarders. *Mick Nicholson collection*

in excess of 3 mph but had no effect on a wagon travelling at a slower speed. Rapid depression of the piston by the wheel flange of a wagon travelling too quickly lifted a speed control valve, and further pressure opened a relief valve. The effort required to force oil through the release valve had a retarding (slowing) effect on the wagon. If the wagon was travelling too slowly the pressure rise was insufficient to lift the speed control valves, and the oil escaped through it, the unit thus collapsing and having no retarding effect on the wagon.

Booster Retarder

The booster retarder was operated by a hydraulic power-pack, to which it was connected by oil-filled pipelines. As in the case of the retarder, it reduced the speed of wagons travelling at above 3 mph, but for wagons travelling at a lower speed it released energy by thrusting against the wheel flanges, thus increasing the wagon's speed. Booster retarder units used in the sorting sidings had a special non-reversing trip lever in the head, which prevented wagons being propelled back towards the hump should they recoil after buffering up to other wagons in the sidings.

At the hump summit a small amount of arrester units with zero speed settings were placed to ensure wagons separated cleanly when being humped. After this, acceleration was achieved by gravity and the use of 12-foot-per-second booster retarders, which would permit wagons to pass from the hump to beyond the clearance point, where the speed and spacing of the wagons would allow safe operation of the points.

On entering the sorting sidings the wagons then passed through deceleration zones, where the speed was reduced gradually from 12 foot per second to 4 foot per second, at which speed they would close up to the other wagons in the sidings.

Origins

The chief designer of the system was David Bick of the Dowty Mining Equipment Company Limited who, after successfully trialling a new retarder system in their Ashchurch works, offered the new Dowty automatic wagon marshalling equipment system to the North Eastern Region (NER) of BR in November 1959. At the time, the NER was about to embark on the construction of four new concentration yards as part of its freight rationalisation programme. The NER management were impressed with the equipment and from 1961 (and over a two year period) experimented with Dowty equipment on two sidings in Hull Inward Yard. Conventional Frölich clasp-type retarders were fitted at the hump and the Dowty equipment fitted further down the sidings to control the wagons as they rolled through.

On Friday 6th April 1962, a group of interested participants were given a demonstration of the Dowty system in Hull, along with a Dowty automatic wagon coupling system that had been invented by Peter Checkley and Richard Hollis (this coupling system was not perpetuated by British Railways).

The group went on to inspect the Dowty equipment, following which they gathered to observe a demonstration train.

Train No. 1 consisted of 20 empty 16-ton mineral wagons fitted with Dowty automatic couplers. It was shunted over the hump at the normal humping speed of approximately 1.75 mph. The train was to be humped in a sequence of:-

- 6 cuts of single wagons
- 1 cut of 3 wagons
- 6 cuts of single wagons
- 1 cut of 2 wagons
- 3 cuts of single wagons

The group was advised to follow the first six wagons down into the sidings to watch how the retarders worked, followed by how the automatic couplers operated as the wagons buffered up.

Next, a loco was attached to the train and the vacuum brake exhausted to illustrate to the group the speed at which a train of automatically coupled wagons could be prepared for departure.

Demonstration train No. 2 was next, this consisting of a mixture of loaded and empty 16-ton wagons, loaded and empty bolster and double-bogie bolster wagons. Again, the train was shunted at normal speed over the hump and then braked by the Frölich retarders into the two sidings equipped with Dowty units at a speed of 6 mph as in Table 1.

No.of Cuts	No.of wagons	Siding No.
6	1	30
1	4	29
1	2	30
1	1	29
1	3	30
1	2	29
1	2	30
1	2	29
4	1	30
1	2	30
1	1	30

Table 1: Dowty Demonstration at Hull

Siding No. 29 was set to reduce the speed to 2 mph, whilst No. 30 was set for 3 mph; graphs of the wagon rolling statistics were provided to the observers. Details of the make-up of the train, and the speed control of the wagons between certain points entering and leaving the retarding section of siding No. 30 were given over a loudspeaker system.

Although the trials were deemed a success, the Dowty

company was unable to supply enough Dowty units in time for their incorporation into the NER yards under construction, so conventional retarders were installed.

The Regions Show Interest

Although a sum of £425,000 was set aside in 1960 for the installation of Frölich-type retarder equipment in Tinsley Yard, interest in the Dowty equipment was still strong, particularly on the Eastern Region.

As such, a trial of the Dowty equipment was undertaken at Goodmayes "down" yard on the Great Eastern Main Line in Essex. The conditions of the "down" yard were altered to simulate the proposed Tinsley Yard, the hump being re-graded to the proportions planned for Tinsley and unlike Hull, control by Dowty retarders started from the crest of the hump.

On Monday 6th April 1964, the BR technical committee viewed the system at Goodmayes Yard and after another demonstration of its operation, the trials were declared successful. BR ordered its first large-scale application of Dowty equipment for use at Tinsley, along with future plans for its use in the new yard at Scunthorpe. Meanwhile, the London Midland Region authorised the use of 4,300 of the units in Bescot Yard (these were removed in October 1984) and also wanted them for a planned yard in Wigan. The Western Region had also planned to use them in a new yard at Brookthorpe, near Gloucester, which in the event was never built.

To meet the planned requirements for Tinsley and other yards, Dowty Hydraulic Units Ltd were authorised by the BRB to proceed with preliminary planning and tooling-up to a maximum expenditure of £20,000. Under a development contract for the equipment, the Board committed to order the first 20,000 units in any year from Dowty, subject to the price and delivery being reasonable. To ensure the required equipment was ready when construction began on the yard at Tinsley, an interim order was placed with Dowty for the supply of 5,000 booster retarders at a fixed price of £25.10.0d each and 5,000 retarders at £17 each, which would (after the inclusion of royalties to Dowty) equate to £32.1.6d and £21.8.9d respectively.

Above: This image shows a booster retarder with a special non-reversing trip level in the head, which prevented wagons from being propelled back towards the hump should they recoil after impact with other wagons in the sidings.
British Railways/Chris Booth collection

Tinsley's Dowty Equipment

In July 1962, BR accepted a tender from Dowty Hydraulic Units Ltd for the work at Tinsley, totalling £957,282.10.0 which, in 2018 prices, is the equivalent of £21,206,789.72. Table 2 shows a breakdown of some of the costs.

Table 2: Dowty Hydraulic Units Limited Tender for Tinsley Yard

Description	£.	s.	d.
Low Speed Hydraulic Units	357,360	10	0
High Speed Hydraulic Units	210,366	9	0
Hydraulic Track Units	22,465	4	0
Pipework and Other Equipment	247,090	7	0
Installation Costs	31,000	0	0
Part Cost of Additional Tooling for Low Speed Units	5,000	0	0

It was expected that the higher cost of the Dowty system over Frölich retarders would be offset by cost savings, which included civil engineering works at £85,000 and signalling requirements of £5,000. The Dowty system also saved the use of an additional 350 hp shunter that would have been required for pushing wagons together had the Frölich type been used. This would have reduced staff costs and, along with the elimination of wagon damage, was estimated to make an additional saving totalling £76,000. There would thus be a difference in cost of £460,000, but the return on this outlay was expected to be 8.5%.

The main advantages of the Dowty system were as follows:

- Wagon and freight damage due to excessive buffering speeds was greatly reduced, as were delays due to derailments, wrong shunts and stopping short.
- Wagons were at all times under control, irrespective of wind or rolling resistance.
- Operations were unaffected by severe weather conditions.
- Predetermined speed settings could be varied over different sections of the yard as required.
- Distance from hump to clearance points could be reduced owing to the elimination of conventional retarders, weigh rails and speed measuring lengths.
- No electronic equipment was required.
- Wagons close up in the sorting sidings, ready for coupling: no pushing down is necessary.

Left: A Dowty booster retarder with associated hydraulic oil system. The pipework was prone to leakage at the joints if not maintained well, and in this image oil contamination is present. Compare the pipes with those in the previous image, which shows how clean the pipework was when new. Derailments were also a hazard, as the pipes would often be cut, thus allowing oil to gush out and putting that section out of action until repaired. *British Railways/Chris Booth collection*

Dowty in Tinsley Yard

The installation of Dowty equipment in Tinsley Yard required roughly 26 miles of hydraulic pipelines to be installed on two levels. All the main distribution pipes were carried in ducts below sleeper level, with approximately two miles of glass-fibre ducting buried up to 28 in deep. Local distribution pipes rose from the ducts in the four-foot and were then carried along the track at sleeper level with connections to each booster unit, of which there were 23,125 in Tinsley Yard. This is summarised in Table 3.

Table 3: Dowty Booster Retarders in Tinsley Yard

Units	Speed	Type
6461	12 feet/second	Booster Retarders
1273	12 feet/second	Retarders
2680	8 feet/second	Retarders
5337	5 feet/second	Retarders
7344	4 feet/second	Booster/Retarders
16		Arresters
14	Power Packs (tilting head type pumps 120 gallons per minute at 120 hp)	

An additional feature of the Dowty system was that it could keep wagons moving over a short distance. As wagons for local industry required fine sorting in the secondary yard, they could be propelled from the main hump summit along the 865-yard mechanical feed road to the secondary yard without any other form of propulsion other than the Dowty equipment. This precluded the need for a second hump and the use of a shunting locomotive.

The Dowty control equipment and hydraulic pumps were contained in two power-pack houses – one situated at the main control tower and the other in the secondary yard control tower. A small workshop was included as part of the power-pack room at the main control tower, the servicing and repair of equipment being carried out there. Most repair work would be carried out during slack periods, but there were isolation facilities in case of major failure of part of the

Below: Hump shunting under way in the late 1960s as viewed from the main control tower. A train of 16-ton coal wagons is being humped and, although the wagons look the same, they are obviously heading for different sidings and destined for different customers. The wagon on the right is destined for one of roads 40 to 53 whilst the wagon centre is headed for somewhere on roads 1 to 27. The next wagon has just crested the hump summit. Note the LMS-style brake van on the "brake kip" to the left of the image. The Dowty retarders would be doing their work and the clatter of wheels on steel piston heads can be imagined! *Ivan Stewart*

Above: A line of withdrawn Class 45s, headed by 45033, is seen on 27th May 1991. After the end of hump shunting, sections of the yard were used for storage of withdrawn stock and other roads were lifted. The redundant Dowty hydraulic system was gradually removed as the pipework became ever more derelict, as seen here. *Ivan Stewart*

Left: Hump shunting under way. The test trolley can be seen centre left. *Les Nixon*

system. In the event of a pump failure, hydraulic power could be obtained from an alternative pump to ensure sufficient parts remained operational to allow hump shunting to continue.

Testing of Dowty equipment in situ was usually carried out on a Sunday, this being achieved by the use of a specially designed two-wheeled test trolley, propelled by a shunting locomotive. The trolley was fitted with just two wheels to avoid interference with readings from a second set of wheels. The loco would pull the trolley over the retarder units at controlled speeds, just above or just below the designed operating speed of the unit, whilst the recording equipment on the trolley recorded whether the unit retarded, boosted or failed to operate. A continuous tape recording showed which units were defective, allowing swift identification and repair.

Numbered in the departmental series as 040844, the trolley was painted all-over yellow and weighed 7 tons 15 cwt. It was kept in its own shed on a short spur close to the main control tower.

Teething Troubles

As was to be expected in a new installation of this size, maintenance problems began as soon as the yard opened and escalated for about 4 months due to reasons which included:

- Failure of components due to incorrect installation.
- Weaknesses in design and problems not anticipated or unknown from past experience.
- Material defects.
- Inexperience of operating and maintenance staff.

Oil losses occurred in the first few weeks of operation as a number of rubber "O" ring seals were damaged during initial installation, which allowed dirt to enter the hydraulic system. This impaired operation of the Dowty units as it made the pump circuits function erratically and eventually fail. Although filters were installed they proved insufficient to prevent dirt particles entering the system and causing further failures. Another issue was corrosion of the main pilot valves in the servo-control system. This allowed air into the hydraulic system and was difficult to remove.

Although action was immediately taken to rectify any faults as they occurred, they continued and records for November 1965, only two months after opening, showed that failures were occurring at a rate of 200 per week. 90% of these were occurring in the hump area of the main yard and along the mechanical feed road, both areas being the hardest worked.

Over 50% of the failures were attributed to burst or leaking "O" ring seals on the flexible hoses, and a further 25% were failures of the flexible hoses, usually adjacent to the end fittings. Despite the high failure rate, the yard continued to operate and, where failures did occur, small areas could be quickly isolated which resulted in few interruptions to the flow of traffic.

By the end of 1965, loss of oil from the system was running at over 2,000 gallons per week; after modifications to the "O" rings this was reduced slightly to 1,200 gallons. However, it was expected that even after all the modifications had been completed, and an allowance given for losses during unit or hose changing procedures taken into account, a weekly loss of 400–600 gallons was the minimum attainable. This equates to an average of 26,000 gallons lost per annum; it is little wonder the ground was heavily contaminated in these areas!

The booster retarder units were not immune from failures either, the rate of failure rising from 30 per week in April 1966 to 110 per week in October 1966. These failures were attributed to units either failing to perform their correct function, excessive oil leaks, pipe joint failure or failure of the stops on the booster retarders. New hoses and seals abated the small leakages which, along with a programme of replacing hoses in the points area and along the mechanical feed road, resulted in a significant reduction in failures by January 1967.

Nevertheless, problems with oil losses continued, which led to many varieties of seal being tested; there were also issues with fractured joints in pipework. Although the problems were never fully eradicated, it was said in 1969 that, despite minor stoppages whilst faults were identified and isolated, the yard had never been totally shut down because of it.

For a detailed explanation of how the Dowty retarders worked, and memos between BR and Dowty relating to possible future Dowty installations, see Appendix I.

Above: The Dowty test trolley awaits its fate on 24th April 1985, hump shunting having ceased the previous December. Numbered in the departmental series as 040844, the unit was painted in all-over yellow. Originally it was covered by its own shed on this short spur close to the main control tower. *Bob Wallace*

Chapter 12

"Addo" Data Transmission System and Yard Communications

Tinsley Yard's claim to be the most modern yard in the world wasn't just based upon its use of the Dowty wagon control system. It also had a highly efficient method of sorting individual wagons into their required sidings.

In order to hump the maximum number of wagons, it was imperative that the number of wagons in a cut and the sorting siding the cut was destined for, was accurately transmitted from the reception sidings to the main yard control tower without delay.

"Addo"

Across the country, other hump marshalling yards had used a variety of means to transmit information between the ground staff and those in the control towers, such as telephones, pneumatic tubes or radio links. However, these methods could be problematic with uncertainty in spoken communications, and errors occurring when sending the written word to the control tower. Further errors could occur when the information was encoded into punched tape for instructing the automatic electronic controls used to direct the wagons into their nominated siding, i.e. from the reception sidings to the main and secondary yard sorting sidings, under the control of the main yard control tower.

To overcome these problems, Tinsley introduced a new method of working with machines installed by Addo Ltd, a Swedish company based in Malmö.

The equipment consisted of standard "ad-listing" (adding and listing) machines connected to tape punches and tape readers. The keyboard of the "Addo" machine consisted of a simplified symmetrical keyboard with ten digit keys numbered 0–9 and 3 multi (or motor) keys. The keyboard was large enough so that it could be operated even when wearing heavy gloves.

Three "Addo" machines were installed in the yard – two in kiosks between reception sidings 5 & 6 and a third on the ground floor of Tinsley Yard signal box. To interpret the transmitted information, a receiving instrument was provided on the main yard control tower inspector's desk.

Above: The "Addo" machine in the main yard control tower. To the right is the punched tape machine.

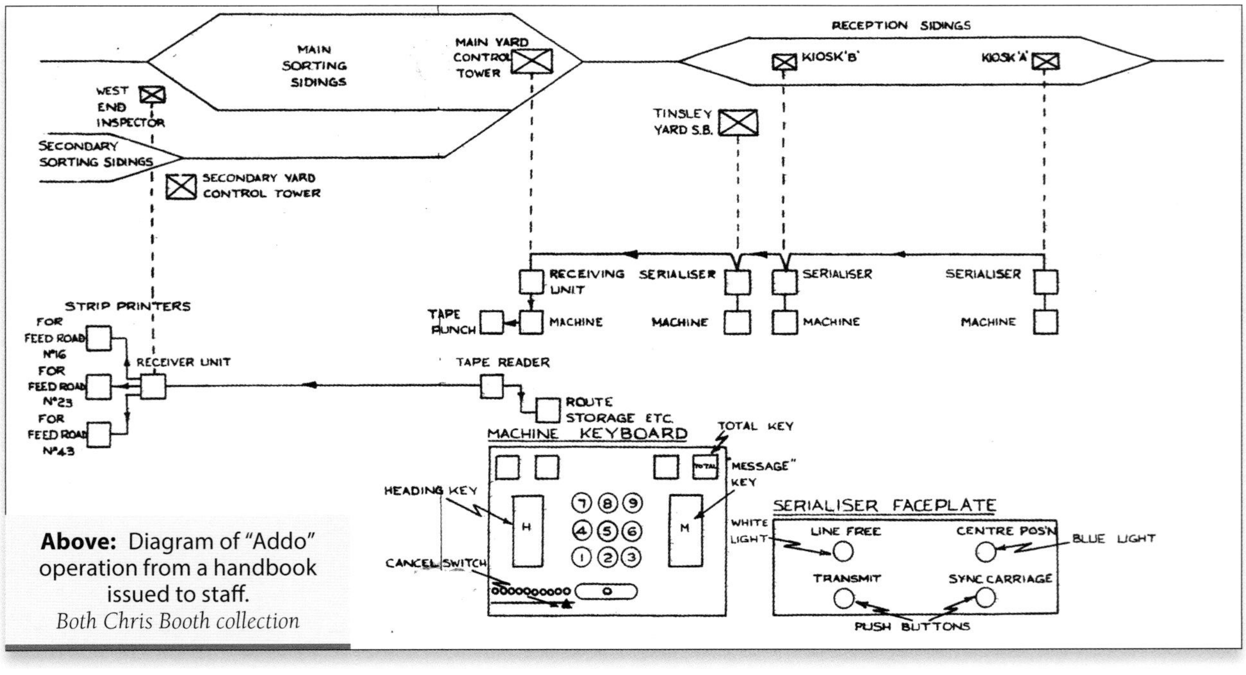

Above: Diagram of "Addo" operation from a handbook issued to staff.
Both Chris Booth collection

Above: The yard inspector tears off the punch tape from the tape machine.

Right: A sample of 5-hole punch tape for automatic setting of route. *Both Chris Booth collection*

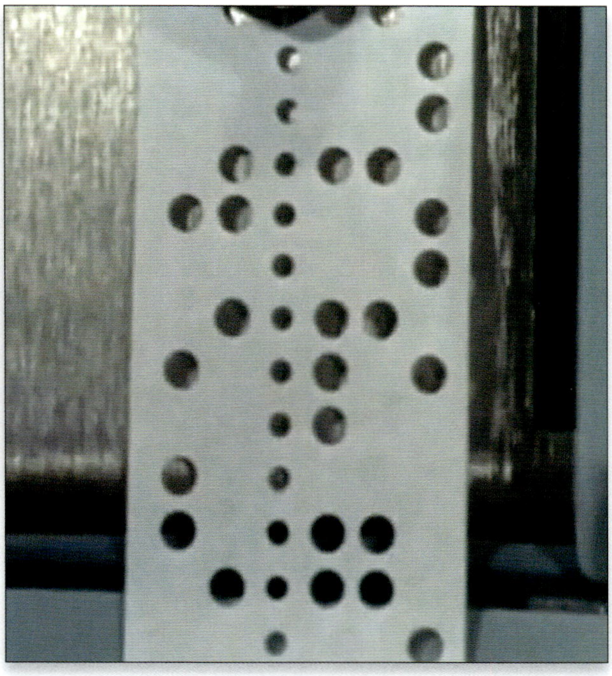

Cutters' Duties

Once a train had arrived in the reception sidings, the process of humping it could begin. After uncoupling the loco, a "cutter" (shunter) would walk back along the train reading the wagon destination labels. Wagons for the same destination would remain coupled; adjacent wagons would be uncoupled if they required a different sorting road in either the main or secondary yard.

As cuts could consist of one or more wagons, so the cutter would uncouple wagons as necessary and make relevant notes for later transmission over the "Addo" machine.

Once the list was complete, he would make his way to the "Addo" kiosk. To ensure that only one cutter could operate the machine at any one time, each transmitter had an indicating lamp which, when illuminated, allowed the machine to be operated, and locked the machine when it was extinguished.

The cutter entered his cut card information, beginning with the train headcode, reception siding number, date and time. This was followed in three columns by destination siding number (which was preceded by a code), the number of wagons in the cut, and any special information such as the presence of a long wagon in the cut.

A wagon destined for a siding in the main yard would usually be destined for a particular trunk route or specific destination. If the wagon or cut of wagons was for a local destination, it would be sent to the secondary sorting yard via siding No. 50, the mechanical feed road.

After each cut had been inputted, a motor key was pressed and the data transmitted to the control tower machine, where it was received both in the form of a typed message and also automatically a tape, into which a code was punched using a combination of up to five holes across the width of the tape and as many as were required along its length.

As the cutter entered the information into the "Addo" machine, it was printed in the same format as on the cut card and thus could be easily verified by the cutter as he was entering it.

Control Tower

Once the information had been received in the main yard control tower, the control tower inspector would tear off the punched tape. The tape machine had a knife edge to ensure that the leading edge of the tape was pointed and the trailing edge fishtailed, to make loading of the tape into the reader easier. The tape would then be stored in one of eleven pigeon holes until the train was ready to be humped.

Once the train was ready, the main hump yard operator would insert the pointed end of the tape into the tape reader mounted on his control panel (it was first necessary to press the master cancellation button to ensure any data from a previous train was purged from the system). After the yard operator had cleared hump signal No. 60, the tape reader machine would set all points automatically into the main and secondary yard sorting sidings.

Failures

Should the data transmitting machine fail, the shunter would take the cut card to the control tower or alternatively telephone the information in, and an inspector would insert the information into his machine in order to produce a punched tape.

In the event of the automatic route setting machine failing, the control operator could manually operate the points in the main yard through the use of individual route setting switches. Although this backup method ensured that the yard continued to operate, it would have been at a drastically reduced rate compared to the fully automated system.

Route Setting Tape

The tape used for route setting was able to indicate a wagon's wheelbase (the distance between wheels). Where a wagon with a wheelbase of 40–45 feet was indicated, a special code of "91" and a white light were shown on the control panel.

Although the automatic route setting operation still selected the correct route, the control operator had to intervene to manually move the points to ensure they did not move beneath the wagons. If any wagons with a wheelbase of greater than 45 feet were detected, they had

Left: The main yard control tower. In the left foreground is the Selectoboard with the "Addo" machine behind. The hump operator is studying a length of punched tape which will be inserted in the tape machine ready for the next train to be humped.
Adrian Wynne collection

clear (or "unoccupied") and the points could potentially move beneath the train.

To prevent this, these particular track circuits used 50 volts AC 50 Hz in the rails, operating a DC relay through a transformer rectifier unit. The increased voltage was especially important along the mechanical feed road, where the Dowty retarders had a slight lifting effect on the wagons meaning the system might momentarily "lose" the wagon.

the code "92" and again a white light illuminated. These wagons posed a particular problem as their large wheelbase could allow the automatic route initiating track circuit to clear between the wheels and thus allow the points to move under the wagon; the route for a cut with these wagons had to be set manually.

Route Storage System

With the information regarding the individual cuts in a rake of wagons now sent to the control tower and printed out in the form of a punched tape, the act of humping the train could begin.

As each train was pushed up to the hump, the tape was fed into the machine and its information electronically stored in the pre-stores route storage system. If the tape machine was inoperable, routes set by the use of the route setting switches were also stored.

Information in the pre-stores system was first indicated on four-way digital route indicators on the operators' panel (which described the special code, route destination and cut length), then translated electronically by the pre-stores system to automatically set the points leading away from the hump in a prescribed sequence, to direct each wagon cut into the desired siding. Route information for a cut registered on the punched tape would be stored to the first storage bank via the pre-stores system.

Track Circuits

To ensure reliable operation of the track circuits around each set of points and along the mechanical feed road, a special type of track circuit was developed in conjunction with the Westinghouse Brake & Signal Company.

Normally, a 6-volt current is sent along one of the rails to a joint in the track, from where it is sent across to the other rail, to travel along it to another joint and back across to its originating point to complete the circuit, hence "track circuit". As a wagon travels onto this track circuit it short-circuits the current, which is indicated in the signal box thus enabling the operator to see exactly where the movement is. Track circuits also lock points, thus preventing their movement beneath trains.

Track circuits only work correctly when the wheels have good contact with the rails. Any contamination or a light contact with the rails could cause the track circuit to show

Sidings Fullness Indicators

In order to prevent a siding from receiving more wagons than it could accommodate, an automatic electro-mechanical indicator on the hump inspector's desk counted the wagons as they entered each siding. This gave an indication of how full each road was and, upon the clearing of a siding, the indicator had to be manually reset to zero in readiness for the next cuts.

Transposition Equipment

Also on the hump inspector's desk was a "Selectoboard", and it was through this board that the required siding (as indicated on the tape) was automatically selected.

The Selectoboard was akin to a giant draughts board with 100 holes along each axis, making 10,000 in total. The holes were grouped into a series of horizontally and vertically numbered squares of 100 holes, each 10 holes by 10, and were all numbered accordingly – from left to right along the bottom, and from bottom to top on the left hand side – with each siding having one peg. Under normal operating conditions i.e. when the controller wanted wagons to enter a specific road, the pegs would be placed in the hole at the point of intersection between the vertical and horizontal rows of corresponding siding numbers, i.e. vertical row 27 and horizontal row 27. Thus all the pegs were normally in a diagonal line from the bottom left to the top right hand corner.

For various reasons, such as a siding becoming full or a mishap, a particular siding could be closed off to further wagons; cuts ordinarily bound for a blocked siding would need diverting to another. To achieve this, the peg for the closed siding was moved vertically to the hole corresponding to the new siding number e.g. if siding No. 13 was closed and its cuts were to be rerouted to siding No. 29, the operator moved the peg from the hole intersecting 13 with 13 to the hole intersecting 13 with 29. Cuts which were marked on the punched tape as destined for siding No. 13 were automatically rerouted to siding No. 29 – quite a remarkable system! When the siding was reopened or cleared of traffic, wagons would be routed back to the original siding simply by restoring the peg to its original hole.

Communication

With an operating length of over two miles, adequate communication between all parties working in the yard

Above: Using the radio, the yard inspector gives instructions to the hump loco driver. The conversation was perhaps as follows… "Tinrail control to pilot Alpha, Tinrail control to pilot Alpha. Over" "Pilot Alpha to Tinrail control, receiving. Over" "Tinrail control to pilot Alpha, increase speed to 3mph. Repeat, increase speed to 3mph. Over" "Pilot Alpha to Tinrail control, message received and understood. Over and out". *Chris Booth collection*

was crucial. Maintenance crews working on the Dowty installations used "walkie-talkie" equipment to keep in touch with the relevant power-pack building.

Two-way public address (PA) equipment connected the yard inspector and operator in the main yard control tower to the signalmen in Tinsley Yard signal box. The call sign for Tinsley Yard signal box was "Tinrail box", whilst the main control tower was "Tinrail control".

In the east end of the main yard and in the reception sidings, loudspeakers were provided to relay instructions from the main control tower to the ground staff. Two-way loudspeaker circuits were also provided between Tinsley Park signal box, the west end of the main yard and to the express freight sidings, allowing staff on the ground to be contacted by the control towers. The ground staff could reply by using external microphones which were mounted on key buildings and conveniently placed posts.

For admin and general purposes, an automatic telephone exchange served the yard area. The exchange also allowed access from the internal circuits to the GPO lines, the divisional control office, local steelworks and the switchboard at Sheffield Midland station. There were also special telephone circuits which connected all the signal boxes in the yard, along with a regulating circuit which connected Aldwarke Junction signal box to the yard inspector.

Cab Signalling

On the Class 13 humping locos, cab radio telephone equipment supplied by PYE Telecommunications Ltd was used to provide communication between the main control tower and the driver of each locomotive. The equipment included a radio transmitter/receiver which used then state-of-the-art valves and transistors. The transmitter obtained its 110-volt power from the locomotive supply, whilst an aerial was mounted on the outside of the locomotive's cab near the roof to receive incoming messages.

Above: As viewed from Tinsley Yard signal box, D4502 is shunting a mixed rake of wagons over the hump on 22nd April 1967. The driver will be watching the hump signals, but will also be receiving a repeated message in the cab from the in-cab signalling system. *Chris Booth collection*

Above: The main yard tower inspector seen on 18 March 1983. *Steve Ryszka*

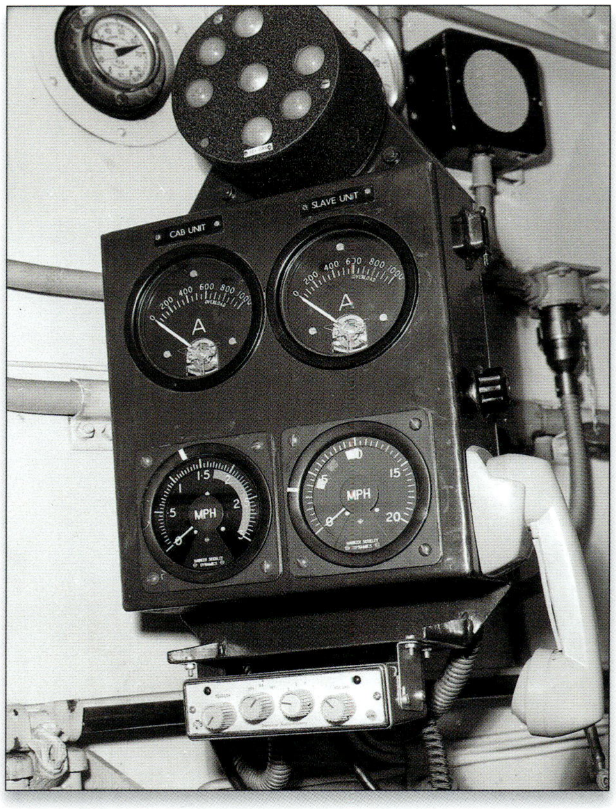

Above: The equipment fitted in a Class 13 cab. From top: Hump signal repeater, two ammeters (one each for the cab and slave units), two speedometers – one for 0–3 mph, the other for 0–20mph – and the radio equipment at the bottom with the handset to the right. *Colin J. Marsden*

The receiver was comprised of a control panel, telephone handset and loudspeaker. The handset was "push-to-talk" and an additional loudspeaker was provided in the cab.

To provide clear and unambiguous communication between the control tower and each driver, each of the hump locos had their own unique radio call sign. D4500 was "Alpha", D4501 "Bravo" and D4502 "Charlie". There was also a fourth call sign "Delta", which was used when an alternative locomotive fitted with a portable system was working in lieu of one of the Class 13s.

Each hump loco also had in-cab signalling for working between the east end of the reception sidings and the hump summit. Information regarding the aspect the in-cab signalling should show was fed from the control tower to the occupied track circuit, which was governed by the condition of the track ahead.

The in-cab signalling unit was attached to the front bulkhead of the locomotive's cab, above the speedometers; it provided a visual repeater of the hump signal ahead of it. The repeater had 7 bulbs which could display three white lights in three permutations to give the driver his instructions (as shown in the diagram).

Indications shown by the hump shunting signals

Three horizontal white lights-
Stop

Three white lights 45°-
Proceed at a humping speed of 2ft/sec (1.36 mph)

Three vertically aligned white lights-
Proceed at 5ft/sec (3.41 mph)

If the hump instructions were changed, an audible beep would sound for three seconds to attract the driver's attention. A vertical set of three white lights would prompt the driver to quickly propel their train towards the hump in order to minimise the time between humped trains. As the front of the train neared the hump, the aspect would to change to three white lights set at 45 degrees, indicating to the driver to slow down to the humping speed of 2ft/sec.

The cab signal did not repeat this signal aspect until the front of the train was at the point where humping speed was required. Should the humping of trains need to be stopped, the repeater would show three white lights in the horizontal position.

Chapter 13

Master & Slave Shunters

As diesel was replaced steam all over the country, the requirement for adequate adhesion weight became a problem. Eight-coupled steam locomotives that had previously been used for shunting (such as the 0-8-4T "Wath Daisies"), weighed around 90 or 100 tons and therefore provided plenty of adhesion. In contrast, a standard 350 hp diesel electric shunter weighs only 48 tonnes.

On dry rails, a tractive effort in excess of 30,000 lbf was adequate for most loads brought to yards by steam traction, but with trailing weights increasing, a single 350 hp diesel locomotive was insufficient. This required the doubling-up of the hump locomotive, with a second shunter seconded from other duties in the yard. This also meant two drivers were required.

Trials began using a Lister Blackstone-engined Class 10 locomotive, with short wheelbase heavy wagons as the test load; trains were successfully moved from the 1 in 260 gradient in the reception sidings, but it could not be certain to restart at the hump signal on a gradient of 1 in 40.

Difficulties were also experienced with holding the train on the hump in certain conditions, as the limit of adhesion was exceeded. The driver only had the brakes of the hump locomotive with which to hold the train as all wagons would at this point be unbraked – it was not possible to use a fitted head for extra brake force as the brake pipes throughout the consist had been disconnected prior to splitting into cuts. As a result, the weight of the train could overpower the hump loco, which could slide backwards with locked wheels!

Twins

Experiments showed that it would be possible to use two slightly modified 350 hp locomotives. A report in the December 1964 edition of Modern Railways magazine stated that the Eastern Region had, for some time, been seeking a higher-powered hump shunter instead of having to resort to pairs of 350 hp shunters, as at Temple Mills Yard in London.

Although construction of a few non-standard diesels to obtain the required adhesion would have been unsatisfactory, the same edition of Modern Railways claimed that Derby works was improvising with a "siamese twin" unit, where two standard 350 or 400 hp 0-6-0s were to be combined as a single unit on a rigid 0-6-6-0 frame.

The article stated that the original cabs were to be removed from each unit, and the bodies joined back-to-back by a new central cab. However, the following month's edition supported a letter from a Mr. K. Marshal, technical assistant marshalling yards, who refuted the rigid-framed proposition and said the resulting loco would be semi-permanently coupled back-to-back, with one of the cabs removed.

Above: In its original configuration of cab-to-cab, D4500 shunts a rake of wagons over the hump in this image from the early years of the yard. Note that the COVHOP wagons cresting the hump are not separating, which could indicate the COVHOPs are one cut, whilst the empty 16-ton mineral wagons will be a second cut. *Rail-On-Line ZF- 6023-08525-1-001*

Right: A pair of Class 08s approach the hump summit at Temple Mills on 11th June 1980. This pair, 08698 + 08546, would require a driver on each locomotive to accomplish the task; a Class 13 only required one driver.
Brian Daniels

Enter the Class 13

The semi-permanently coupled locomotives were formed when six Darlington-built Class 08 locomotives were returned to their birthplace in 1965, for conversion to three master & slave locos (henceforth known as "units" when referred to as a combination), later known as Class 13.

The weight of each loco was increased from 51 to 60 tons, thus giving an axle weight of 20 tons throughout. This was achieved by the addition of deeper and thicker buffer beams and the addition of ballast weights above the intermediate axles. The weights were removable to allow the unit to conform to the restrictions of a normal 350 hp shunter, but special dispensation was given for them to travel to Doncaster works with the weights in position.

This weight increased the maximum tractive effort to 70,000 lbf, which was equal to 26% adhesion; the continuous tractive effort was 32,000 lbf at 3.75 mph. The overall length of the units was now 60 ft 1 in with a total wheelbase of 42 ft 1½ in.

Modifications to the master locos from standard were:

- Auxiliary contacts added to the reverser
- An engine control air-valve fitted
- Generator field resistance steps increased and re-graded
- Auxiliary contacts fitted to the series-parallel switch
- A multiple/single unit selector switch added
- Sanding switches fitted
- Ammeter provided to indicate the slave loco generator current
- Electronic speedometer
- PYE VHF radio telephone
- Cab signalling equipment

The master locomotive was fitted with an electronic speedometer to give accurate readings whilst hump shunting. It incorporated synchronised indicators reading 0–3 mph and 0–20 mph, with provision made for similar equipment to be fitted to the slave loco gear case.

Most notably, whilst the master loco retained its cab, the slave loco had its cab removed to allow the driver to see around the slave loco bonnet. The redundant control equipment desk of the slave loco was plated over and access doors fitted to the front. The former cab then became an open footplate, but to enable the shunter to pass across the loco or ride on it in safety without entering the driving cab, the rear and side framework was extended up to the waistline. The side access steps were also retained, together with the handrails, but doors were not provided. Finally, wide shunter steps with handrails were provided on the four outer corners of the unit.

The intermediate coupling between the locos was by a standard screw coupling, the side buffers being retained. The electrical connection consisted of a single control cable with diesel engine air control pipes on the buffer beam; the unit itself was fitted with train vacuum brakes only.

The slave locomotives had modifications to their control and traction equipment which consisted of:

- Reverser converted to electro-pneumatic operation and auxiliary contacts added
- Engine governor was arranged to be controlled by linkage from an air cylinder
- Generator field-resistance removed and field series connected with that on the master loco generator
- Series-parallel switch removed and replaced by an electro-magnetic contactor
- Master switch cam profile development modified
- Electro-pneumatic sanding valves fitted

Engine starting was via the existing motoring of the generator, but both engines started together, the current being supplied individually from the battery on the unit. The slave loco could be started locally via a slave master switch, lubricating oil pressure switch and water thermostat being fitted to the engine, connected to a fault light in the drivers cab.

The initial coupling arrangement of the two former Class 08s was cab-to-cab, but it soon became apparent that visibility for the driver was poor with this arrangement. There were occasions when drivers misjudged the vehicle length and the locomotives over-ran the reversing spur and collided with the buffer stops.

In 1966 the units were sent to Doncaster works for modification. The master loco was turned around to put the cab at the rear of the unit; further sections of the slave cab were cut away to give better vision. The master & slave units were designed to be fully interchangeable but it is doubtful they were ever swapped around.

Unit Details

D4500, Master Loco D4188 and Slave Loco D3698, Later 13003

D4188 was introduced to service on 11th July 1962 and allocated to Stratford (30A). Rebuilt as a master loco, it was reallocated to Tinsley on 16th January 1965.

D3698 was introduced to service on 7th May 1959 at Sheffield Darnall (41A). It was reallocated to Sheffield Grimesthorpe (41B) on 15th April 1961, before returning to Darnall on 16th September 1961. It was rebuilt as a slave loco and reallocated to Tinsley on 16th January 1965.

Above: 13002 pauses on the hump as a pair of Class 20s bring another train into the reception sidings on 15th November 1981. The horizontal lights on the shunt signal instruct the Class 13 driver to stop hump shunting; the two white lights of the subsidiary signal alongside are lit, indicating that the route is set towards the hump return line, which ends in the buffer stops just ahead. *Les Nixon*

D4501, Master Loco D4190 and Slave Loco D4189, Later 13001

D4190 entered service on 2nd August 1962 and was allocated to Stratford (30A). It was rebuilt as a master loco and reallocated to Tinsley on 16th January 1965

D4189 entered service on 31st July 1962, also at Stratford (30A). Reallocated to Parkeston Quay (30F) on 13th June 1964, it was rebuilt as the slave loco for D4501 and reallocated to Tinsley on 16th January 1965.

D4502, Master Loco D4187 and Slave Loco D3697, Later 13002

D4187 entered service at Stratford on 6th July 1962. Rebuilt as a master loco it was reallocated to Tinsley on 16th January 1965.

Finally, entering service on 1st May 1959, D3697 was allocated to Sheffield Darnall (41A). Reallocated to Mexborough (41F) from 8th August 1959, it was transferred to Wath (41C) on 7th March 1964,

Above: 13001 has just come to a stand on the crest of the hump as its last 16-ton wagons roll down into the yard, whilst a pair of Class 20s approach with another train for the reception sidings. Note the yellow Dowty test trolley in its parking spot close to the main yard control tower. To cover for the failure of more than one Class 13, Class 08 No. D3707 was also equipped with the necessary cab fittings for hump shunting. *Roger Wainwright*

before being rebuilt as a slave loco and reallocated to Tinsley (41A).

When the type was first introduced, Grimesthorpe and Canklow drivers were trained to work the Class 13s. Darnall men, however, were not trained on the locos.

Livery

The units were originally painted in BR green with the inner cab ends green; "wasp" stripes were applied to the outer ends. When the master locos were turned around, the cab end of the master loco also had the "wasp" stripes applied.

The units received TOPS numbers in 1974, 13001 and 13002 whilst in green livery. 13002 remained in this form until it received BR blue livery in September 1977. All three units received BR blue livery, but with minor differences in application to the rear of the slave loco and nose of the master:

13001: Slave loco rear – blue, blue buffer beam. Master nose end – full yellow end.

13002: Slave loco rear – blue, yellow buffer beam. Master nose end – wasp stripes.

13003: Slave loco rear – wasp stripes, yellow buffer beam. Master nose end – wasp stripes.

All three units carried BR double arrows on the master loco; 13001 also carried a pair of arrows on the slave loco.

Daily Work

In the early years of the yard it was essential that at least two locomotives were always available to work, supported by Class 08s in the main yards. The reception sidings were usually full of wagons of coal and scrap steel; as fast as one train was humped another was ready to take its place. It was necessary for crews to have a stand-in relief when they took their meal break, such was the demand to keep trains moving around the clock.

Moments of Glory!

On Saturday 7th October 1978, 13003 was given the unusual job of hauling the BR-run "Sheffield Merrymaker", "Pennine Rambler" railtour from Rotherwood exchange sidings to Tinsley Yard, a journey time of approximately 25 minutes. The tour arrived at Rotherwood behind 37037 and was taken forward from Tinsley over the Woodhead Route by 76010 and 76023.

On 18th February 1984, a second passenger working occurred when 40033 brought Staffordshire Railtours' "The Dore Mat" railtour to Shepcote Lane Junction. Here, 13001 was coupled on and worked from Shepcote Lane Junction (NW arrival line) to Tinsley east (east departure line).

The booked loco change point was Treeton North Junction, but the actual extent of haulage by 13001 was 1 m 20 ch, which took it to just beyond Tinsley PSB, where 40033 took over again and worked the tour to Doncaster (source: *www.sixbellsjunction.co.uk*).

Final Withdrawal

A generator burn-out on the slave loco of 13002 necessitated a visit to BREL Swindon works in 1981, but this was no detriment to the working of the yard as the recession in the

Above: On 18th February 1984, 40033 brought Staffordshire Railtours' "The Dore Mat" railtour to Shepcote Lane Junction, from where 13001 worked to the far end of the yard. Here it is about to pass beneath Wood Lane bridge. *Keith Long*

Below: On 7 October 1978, British Rail ran "The Pennine Rambler" railtour. 13003 is seen here between Broughton Lane and Shepcote Lane en route from Rotherwood to Tinsley. *Chris Booth collection*

Below: On a frosty and murky 19th December 1978, as its driver was changing the points to make a shunt move, 56003 made a bid for freedom and ran away down the zig-zag incline at the east end of Tinsley depot, careering through the buffer stops and ending up on the hard shoulder of the Sheffield Parkway, to the surprise of passing motorists! The loco was rescued by 13002, which dragged it back to the depot. *Roger Wainwright*

Above: The slave unit of 13002 is stripped for re-useable parts before being scrapped at Swindon works on 1st September 1982. *Andy Hoare*

steel industry had seen the requirement for multiple shunting locomotives come to an end.

Traffic in Tinsley Yard had fallen to the extent that there was only sufficient work for two units; many days would see only one in use. It was therefore investigated whether the master loco of 13002 could be rebuilt as an air-braked shunter in its own right. Instead, it was stored from 9th April 1981 and became the first of the class to be withdrawn on 28th June, the whole unit being cut up at Swindon works during October 1982.

13001 and 13003 survived in use until hump shunting ceased at Tinsley in 1984, thus rendering them superfluous. Their unique qualities were not required elsewhere and withdrawal was inevitable.

13001 was withdrawn from service on 20th January 1985, and was moved to Tyseley in a special freight service on 11th February. It was forwarded to BREL Swindon works on 12th February along with 08790 as 9Z54, where the master loco was removed and scrapped first, the slave loco finally being dispensed with in May 1985.

13003 was also withdrawn from service on 20th January 1985, but was sent for scrapping at BREL Doncaster works. Consideration was again given to converting both locos into air-braked shunters (providing speculation that they would be numbered 08959 and 08960) but again this was not proceeded with.

13003 still resided at Doncaster works on 15th September 1985, awaiting scrapping. There was a brief attempt to set up a preservation group to save a class member but, with the scrapping of 13003 later that month, sadly this unique locomotive class was consigned to history.

Chapter 14

Tinsley Yard Electrification

When the Manchester, Sheffield & Wath (MSW) 1500 volts DC overhead electrification scheme was completed in 1954, it originally terminated at Bernard Road sidings (between Sheffield Victoria and Woodburn Junction). A later extension, completed in January 1955, took it further east to Rotherwood exchange sidings.

The original proposals for the yard at Tinsley discussed the idea that MSW electric locomotives would be exchanged for diesels at Bernard Road or Rotherwood, with trains being tripped into Tinsley Yard. Later proposals suggested that electrification to Tinsley should be linked with the Great Northern Main Line electrification project, which had been included in the plan for modernisation and re-equipment of British Railways, issued in 1954.

This proposal was for the electrification of the GN Main Line from King's Cross to Doncaster, Leeds and possibly York at 1500 volts DC which, if the route from Retford via Worksop to Sheffield Victoria was also electrified, would have allowed electric operation from King's Cross to Manchester via Sheffield.

Although the GN plan faltered and was ultimately shelved in 1961, there was perhaps some tangible evidence of the idea that Retford to Sheffield could be electrified, in the form of a level crossing footbridge at Worksop station being lifted by some 8 feet onto brick plinths. Maybe this was in preparation for the project?

Electrification to Tinsley

Electrification of the 4½ miles of former GCR line from Woodburn Junction and Darnall West Junction towards Rotherham into Tinsley would be reconsidered in 1964, when the future of services over Woodhead was being debated by the trans-Pennine routes working party. This body had recommended that the MSW line be retained for freight services only, a case having also been put forward to support the extension of electrification to serve Tinsley. It was suggested that "the workings of the MSW locomotives involved a certain amount of non-productive time which was impractical to absorb in the main line journeys" and that "this spare capacity can be used for working trains between Woodburn Junction and Tinsley Yard without additional locomotives or staff"

Above: Although seen elsewhere, this photo is worthy of inclusion as it was taken just after the yard had opened. Class EM1, E26049 "JASON", departs from the main yard with an electric-hauled Class 8 freight for the Woodhead Route on 28th September 1965. The Class 08 is standing on the engine line adjacent to signal TP6, which is controlled by Tinsley Park signal box. The siding on the right of the image was for the cattle dock, which is just out of view.
British Railways TE253 Chris Booth collection

Left: Multiple working-fitted 76011 + 76025 approach the dive-under on the south-west arrival line, bound for the reception sidings on 23rd October 1979. *Rail-on-Line (ZF-6023-08525-1-002)*

Below: An unidentified pair of Class 76s pass Woodburn Junction and take the line towards Tinsley in this undated shot taken from Woodburn Junction signal box. *Alan Hazell*

The consensus of opinion was that electrifying the route would avoid an increase in transit times for trains between Bernard Road/Rotherwood and Tinsley, and remove the need for two additional Type 3 diesel locomotives to work trip services. It also meant that land valued at £26,000 at Bernard Road could be sold.

Although these factors would increase annual charges by £20,000, savings of £275,000 per annum through a new revised freight train plan (i.e. the concentration of freight over Woodhead) allowed the electrification into Tinsley and the avoidance of delays associated with traction changes in the aforementioned yards.

Work Begins

A tender of £276,413 was placed by British Insulated Callender's Cables (BICC) for the work to extend to Tinsley, and £10,517 for the removal of redundant equipment in Bernard Road yard.

Electrification work began in September 1964, the design and surveying being undertaken by BR staff under the direction of Mr C. Scutt, chief mechanical & electrical engineer of the Eastern Region. The contract for the work was won by BICC but came in at a revised cost of £363,000.

Power for the electrified line was obtained from an existing substation at Woodburn Junction and a total of 16¾ miles of track were electrified. More than 500 foundations for steel masts, plus wall brackets and bridge attachments, were required for the project. These structures were erected with future conversion to 6.25kV AC in mind but, with minor adjustments, they could be converted to 25kV AC, something which of course would never happen with the closure of the Woodhead Route.

In the yard area, electrified lines were:

- South-west arrival and departure lines
- South arrival line and south hump approach line
- Reception sidings 6 to 11
- Engine loop and engine line from east end of reception area to west end of main yard, including engine spur
- Sorting sidings 34 to 45 were also wired at the western end to cater for trains departing for the Woodhead Route
- Express freight and departure sidings 1–5 inclusive

Broughton Lane British Oxygen Co. sidings were also electrified.

As Tinsley drivers were not trained on electric traction, any electrically-hauled trains entering/leaving the yard were driven either by Rotherwood or Guide Bridge crews.

The extension of electrification from Woodburn Junction and Darnall West to Tinsley Marshalling Yard was the last installation of DC overhead electrification undertaken by BR.

Left: Vacuum brake-only 76040 brings a rake of flat wagons along the south-west arrival line towards the dive-under, to access the reception sidings in this shot from 1979. The works in the background is that of the former English Steel Corporation.

Below left: 76022 waits at signal TY246 for access to the reception sidings, with 8E07, Dewsnap to Tinsley. The train consists of 16-ton MCO and MCV wagons containing scrap. Wood Lane bridge straddles the yard and a train is being hump-shunted below. 76022 was the only member of the Class to be painted in blue livery, but retain the original BR crest to the end of its days. *Both Roger Wainwright*

Below: After depositing a train in the yard, 76022 + 76013 pass Shepcote Lane Junction and take the route off left towards Broughton Lane on 23rd September 1980. They will most likely be heading for Rotherwood to collect a train destined over Woodhead. *Michael Rhodes*

Left: Vacuum-only fitted 76038 passes Broughton Lane with a "down" freight of mixed wagons in June 1976. The train consist includes soda ash wagons heading back over the Pennines for Wallerscote.
Arnie Furniss

Below: 76022 + 76013 arrive at Tinsley with a train from Dewsnap sidings on 23rd September 1980.
Michael Rhodes

Above: The view from the brake van as 76040 approaches Shepcote Lane Junction on 6th March 1981 with 8E08, 10.11 TFO Wallerscote to Rockware Sidings soda ash. The loco is about to cross Shepcote Lane bridge and will take the south-west arrival line and head to the reception sidings. Note the lone 16-ton mineral wagon in the consist. *Arnie Furniss*

Below: 76026 and an unidentified sister have just passed Broughton Lane and are now making the climb towards Shepcote Lane, with a rake of empty HTV hoppers on a sunny day in April 1979. *David Ford*

Electric Working Ends

1981 was the final year of electric working. The Woodhead Route closed on Saturday 18th July 1981 once the final trains had run in the early hours of the morning.

Tinsley's final electric-hauled trains are reported to have been in the hands of 76016 + 76010, which arrived with a mixed consist of wagons as 8E73, the 22.26 Dewsnap to Tinsley which, on this final day, was running early. They returned over Woodhead with 23 wagons as 6M10, 23.18 Tinsley to Warrington.

The very last electric-hauled arrival at Tinsley was 8E43, the 20.22 Dee Marsh to Tinsley reception sidings, comprising 30 mineral wagons and hauled by 76028. After depositing his train in the reception sidings, the driver would have taken his loco along the engine line towards Tinsley Park signal box. From here he would have entered either the express freight sidings or one of the electrified roads in the main yard, to collect the very last electric-hauled departure out of the yard – 8M26, 04.00 Tinsley to Warrington.

This final departure would have set off via Shepcote Lane, the south-west departure line and Broughton Lane Junction, to Woodburn Junction, where it would have joined the main line towards Manchester. After passing through the deserted Sheffield Victoria, it passed Dunford Bridge at 03.18 and plunged into Woodhead tunnel, thus bringing to a close the electric period of the Tinsley story.

The Manchester, Sheffield & Wath electric railway story would finish shortly afterwards when 76006 + 76014 worked 6M62, a Parkeston Quay to Edge Hill, from Rotherwood exchange sidings, becoming the final electric train to pass through Woodhead tunnel at 05.08.

With the end of electric working, it was not long before the valuable copper overhead wires were removed from the yard area. However, the electrification gantries remained in situ until June 1982 when removal began, continuing into late 1983. Removal proceeded along the route from Woodburn Junction towards Tinsley, with those between Attercliffe Junction and Darnall West Junction being removed in October 1983. The final electrification stanchions to survive, between Broughton Lane and Woodburn Junction, were removed during November 1983.

Top: Dismantling the overhead gantries at Shepcote Lane in September 1983. The men have attached chains to the lattice-style gantry and cut it off at the base on both sides; it is now in the process of being lowered to the ground to be cut up. The crane in use is a light duty diesel electric crane lettered "diesel rail travelling crane ADRT96307", built in 1958 by Taylor & Hubbard, order No. 1727, and belongs to the CM&EE department. *Steve Ryszka*

Above: After dealing with the gantry in the photo above, the men moved on to the next one. This was a single steel-type gantry; the chains have been attached in readiness for cutting the posts. However, things did not quite go to plan. *Steve Ryszka*

Left: Oooops! After cutting the base on one side, the crane began to take the weight for the second side to be cut. However, the gantry began to swing round and collided with the crane, becoming lodged on the crane roof! Here we see the aftermath; the gantry is still on the roof and the men are looking on in consternation while the gantry is cut up! *Steve Ryszka*

Chapter 15

Yard Pilots

When the new Tinsley depot opened, locos were transferred across from Darnall but continued to be used in the same locations as before. Additionally, several 350 hp shunters were employed in the new marshalling yard.

One loco would be stationed at the west end of the main yard sorting sidings, where it would shunt vacuum-braked wagons to form a "fitted head" for attachment to through trains in the express freight yard. Another loco would perform similar duties at the west end of the secondary yard, and a third was responsible for servicing English Steel Corporation's private siding, adjacent to the secondary yard.

The "fitted head" consisted of vacuum-braked wagons, marshalled at the front of the train immediately behind the locomotive. The rest of the train was usually made up of unbraked loose-coupled wagons; the locomotive's vacuum brake would be connected to the braked wagons to provide greater brake force and thus greater stopping power. The use of braked wagons in the train allowed a higher speed for its journey and therefore a shorter journey time, which in itself made for better train crew, wagon and locomotive utilisation.

The 1970s

The 1970s saw continuing decline of Britain's heavy industry, Sheffield being no exception with many smaller yards in the area closing. This contraction also meant the requirement for shunting locomotives declined, although by early 1979 Tinsley's Class 08 allocation still consisted of 28 locomotives as shown in Table 1.

Although many locomotives were allocated to Tinsley, some were driven by crews from other depots. For instance, Sheffield-based men would work Sheffield station pilots, Nunnery carriage sidings and the Deepcar pilot, whilst Rotherham men would drive the Meadow Hall pilot. Locos were also hired by private companies, including notably Blue Circle Cement at Hope, where the loco would shunt Earles Sidings. This loco was exchanged weekly, being worked to Hope by a crew from Tinsley.

In a Tinsley driver's words "We hoped to have a clear run from Sheffield over the Hope Valley Line but unfortunately, due to the 15 mph maximum speed of the loco, the congested route, and depending on the timing, we would invariably be put into every refuge siding along the way to clear the line for a passenger train. This would make it an extremely long shift taking a replacement loco from Tinsley to Earles Sidings and bringing back the one for servicing". According to one driver, one of the best parts of the journey, if undertaken early on a summer's morning, was the ability to see through Totley Tunnel which was free of engine fumes!

By 1982 the requirements for shunting locos had changed again and Tinsley's Class 08 allocation had dropped to 23, as shown in Table 2.

Table 1: Class 08 Allocation 1979

08008	08115	08223	08335	08507	08862
08024	08131	08240	08386	08539	08878
08033	08186	08244	08447	08729	08880
08075	08208	08260	08459	08813	
08076	08219	08266	08504	08861	

Table 2: Class 08 Allocation 1982

08022	08244	08510	08729	08878
08033	08335	08523	08749	08879
08208	08389	08543	08782	08880
08209	08492	08655	08866	
08219	08507	08678	08877	

Above: An unidentified Class 08 brings a trip working of scrap empties into Tinsley Yard, passing Shepcote Lane signal box. The bright yellow wagon behind the loco is a Railease POA scrap wagon, possibly the prototype of the design, RLS5900. *Roger Wainwright*

08434, 08485 and 08870 were re-allocated from Wath to Tinsley in January 1983, and with the closure of Wath depot a month later, Tinsley-based Class 08s took over the shunting of Wath Yard.

Although many locos could be found stabled on Tinsley depot at weekends, normal practice was to stable shunting locos at the yards in which they worked during the week. Tables 3 to 5 set out the shunters' sphere of operation in 1966, 1979 and 1985 respectively.

Table 3: Work Allocation of Class 08s – 1966

Freight Terminals		Tinsley Yard Shunters	
2	Rotherham Road	1	Tinsley Yard
1	Rotherham Central	5	Tinsley Depot
4	Ickles Yard		
1	Tinsley Gas Works		
1	Rotherham Masborough		
2	Sheffield Brightside Goods	**Stations and Departmental Yards**	
1	Parkgate Junction Midland	1	Sheffield Midland Station Pilot
1	Bridgehouses Goods	1	Sheffield Victoria Station Pilot
1	Langwith Junction	1	Darnall
1	Deepcar	2	Beighton Junction
1	Canklow Yard		
2	Wadsley Bridge		

Table 4: Work Allocation of Class 08s – 1979

Freight Terminals		Tinsley Yard Shunters	
1	Wadsley Bridge	1	Tinsley Express Freight Yard
1	Deepcar	1	Main Yard (West End)
1	Ickles Yard	1	Secondary Yard and Transfer Pilot
1	Rotherham Road NCB Sidings	1 AB	Secondary Yard and Trip to BSC English Steel Corporation
2	Rotherham Area Pilots based in Holmes Yard		
1	Meadow Hall	**Stations and Departmental Yards**	
2	Beighton Junction Permanent Way Yard	3 AB	Sheffield Station Pilots
2 (1 AB)	Brightside NCL Depot	1	Nunnery Carriage Sidings
1 AB	Hope (Earles Cement)	1 AB	Darnall C&W Sidings
AB	Air Braked	Regular loans to NCB collieries at Manton, Treeton and Smithywood. Also to Barrow Hill, Shirebrook and Wath depots	

Table 5: Work Allocation of Class 08s – 1985

Freight Terminals		Tinsley Yard Shunters	
1 AB	Broughton Lane Sidings. Trips to Meadowhall	1 DB	Main Yard Pilot West End
1 DB	Sheffield Freight Terminal	1 AB	Main Yard Pilot West End. Trips to BSC Tinsley Park Works
1 DB	Hope Blue Circle Cement Works Hire Loco	1 DB	Express Freight Yard
1 DB	Brightside Pilot. Trips to BSC River Don Works/Upwell Street (Parkin & Johnson)/Firth Brown Steelworks/Coopers Metals/ Sheffield Freight Terminal/Attercliffe Yard (mornings)	1 DB	Tinsley TMD Pilot, also Trips to Darnall C&W Depot as required.
2 U	Beighton Permanent Way Depot. Trips to Woodhouse Sidings/Holbrook Sidings. A third loco is supplied if traffic demands.	**Stations and Departmental Yards**	
1 U	Deepcar Sidings. Trips to BSC Stocksbridge Exchange Sidings	1 DB	Darnall C&W Repair Shops
		1 DB	Sheffield Station Pilot. Stables at Shrewsbury Road Sidings Loans
AB = Air Braked, DB = Dual Braked U = Unspecified		1 U	Barrow Hill for use in Barrow Hill Yard or Seymour Junction Yard
		1 AB	Shirebrook TMD for use at Worksop
		1 AB	Wath Yard. Trips to Manvers Main Coking Plant

Right: The driver takes a breather as Ickles yard pilot, 08870, shunts a rake of steel slab-laden wagons, most likely from the works of Steel, Peech and Tozer which was adjacent to the sidings. *Roger Wainwright*

The subsequent closure of Wath Yard, allied with a general reduction in the need for shunting and trip working, meant that by 1985, the total requirement for 350 hp shunters allocated to Tinsley had fallen to just 12, which were 08244, 08335, 08386, 08447, 08459, 08504, 08507, 08539, 08729, 08813, 08878 and 08880.

Namings

The practice of bestowing unofficial names on Tinsley's locomotives began in 1987, when fitter Phil Hodgkiss applied names to three Class 20s. The painted names soon spread to other classes including Classes 08 and 09, with several names appearing. A list of these can be found in Table 6.

Table 6: Unofficial Shunter Names

08141	Manvers	08678	Staveley
08210	Tuxford	08691	Escafeld
08308	Langwith	08745	The Tyke
08389	Saxon	08749	Great Central
08434	Midland	08857	Darnall
08436	Beighton	08870	Millhouses
08492	Barnsley	08871	Barrow Hill
08500	Thomas 1	08878	Grimesthorpe
08509	Wath ETD	08879	Earles
08510	Canklow	08880	Mexborough
08514	Hengist	08919	Cadeby
08543	Rotherwood	09008	Aldwarke
08581	Imogen	09013	Shepcote
08661	Mercian		

Some of Tinsley's Class 08s were outstationed at Shirebrook, but by November 1985 Shirebrook's requirement for shunting locomotives was virtually eliminated. Almost all trains in the area were fully air-braked, thus removing the need to shunt vacuum-braked wagons to form a "fitted head".

Tinsley's Class 08 locos were also loaned to other depots, such as Wath (before its closure), Barrow Hill and Worksop.

Change in Policy

From 18th January 1987, the Eastern Region's policy on the allocation of Class 08 shunters was changed, so that just a single depot was responsible to each area mechanical and electrical engineer. This meant that nine depots lost their allocations. Lincoln and Shirebrook's locomotives went to Doncaster, whilst Barrow Hill's became Tinsley's responsibility. Barrow Hill's last Class 08 was 08871 and, when it left, the depot was devoid of allocated engines. Shirebrook also lost its own allocation when 08729 and 08824 went to Doncaster.

Barnburgh Loan Locomotives

A notable use for Tinsley's 08s came in 1987, when the NCB hired a Class 08 at Barnburgh colliery to replace one of their internal shunters which was out of action. Two more followed later, used to move coal in unfitted wagons from Barnburgh colliery to the Manvers complex, where it was discharged and loaded into MGR trains. Two locos would be in use daily with the third acting as a stand-by loco.

Their use at Barnburgh ceased when the last coal was tripped to Manvers on 16th June 1989, the locos being laid up to await collection by BR, before being returned to Tinsley.

Trip Notices

The 1988 BR trip notice listed five pilots, numbered No. 2, No. 3 and No. 5 to No. 7, plus one dual-braked pilot designated T04. This last pilot loco was allocated to work Aldwarke new site, along with Templeborough, C.F. Booths, 11-inch mill, Attercliffe sidings and Sheffield Freight Terminal. Its full working is shown in the accompanying illustration.

No. 2 pilot worked in the main yard and was allocated to shunt Speedlink traffic between 06.00 on Monday and 06.00 on Saturday.

No. 3 pilot worked throughout Tinsley Yard but was allocated to shunt just metals sector traffic between 06.25 on Monday and 05.25 on Saturday.

No. 5 pilot was allocated to Beighton Yard for departmental traffic, shunting between 07.00 and 17.00, Saturdays-excepted. It also worked traffic between Beighton depot and Woodhouse Yard.

No. 6 pilot was also a departmental turn, working at Masborough S&T Yard.

No. 7 pilot worked at Sheffield Midland station, shunting parcels trains around the station area as required.

TOPS Power Report

After the closure of Sheffield control, the area was taken over by Leeds control. One of the Leeds staff members was Kim Collinson, who retained a lot of paperwork formed on a

Below: One loco to receive a Phil Hodgkiss repaint and naming was Sheffield station pilot 08691 "Escafeld". The painted name was first applied on 16th September 1987, along with a Tinsley rose and a repaint into rail blue with light grey roof and yellow side rods. In 1989 the loco was repainted again into green, still with grey roof, but with the addition of a cast nameplate and polished BR arrow. Here it is seen on a snowy 3rd February 1994 between duties. *Chris Booth*

daily basis in the control office, which would normally have been destroyed after use. Kim has been kind enough to allow access to his archives; a selection of information gleaned from them is shown here.

At 13.24 on 20th December 1988, Kim made a TOPS request for a power report on Tinsley's shunting locomotives. This is essentially a request for the location of all Class 08s allocated to Tinsley.

This report showed the following:

Out of Service	
08335	Thomas Hill Works, Kilnhurst
08283	C.F. Booths, Rotherham
08141, 08210	Tinsley Yard
08244, 08678, 08878	Tinsley TMD
Locos on hand OK:	
08308, 08434, 08436	Manvers Main, British Coal
08543	Rotherham Main
08879, 08919	Tinsley Yard
08492, 08870, 08880	Tinsley TMD
08691	Sheffield Midland
08857	Beighton depot

With the advent of the Sectorisation in the 1980s, shunting locomotives were allocated to BR sectors and sub-sectors. A further power report made in July 1989 is shown in Table 7.

Table 7: FSHT Pool (Tinsley Depot Shunting Locomotives) July 1989

Diagram	Loco	Description of work
401	08509	T04 Trip
412	08492	No. 2 Pilot Tinsley Main Yard
413	08543	No. 3 Pilot Tinsley Yards
415	08870	No. 5 Pilot Beighton
416	08691	No. 6 Pilot Masborough Steel Terminal
417	08857	No. 7 Pilot Sheffield Station
421	08308	Hired to Blue Circle (Hope)
422	08434	Hired to Manvers Main
423	08436	Hired to Manvers Main
424	08508	Hired to Manvers Main

Maintenance Cover: 08749, 08879, 08880, 08919

```
                               T04 (METALS/SPEEDLINK)
MIN.POWER REQUIREMENT 08
TURN NO. DRIVER  GUARD                    DRIVER  GUARD
TI.506    MO      MO                        MSX     MSX
         05 20   05 19                     05 01   05 00
         13 29   13 16                     13 09   12 56
          8 09    7 57                      8 08    7 56

         PNB to be taken between 3rd & 5th hours as convenient

                                                              DAYS
           ACTIVITY                          ARR.   DEP.   WTT   RUN
TI.506     LP     Tinsley TMD                       05 45  LD    MO
                  Tinsley MYW                05 54  06 35  9
                  Sheffield FT               06 52
           FS     Service Attercliffe Sidings and Sheffield FT
                  (Guard to assist Shunter as required)
                  until
                  Sheffield FT                      12 09  9
                  Tinsley MYRL               12 31  12 41  LD
                  Tinsley MYW                12 52
           RELD   by TI.507 at 12 52

           LP     Tinsley TMD                       05 26  LD    MSX
                  Tinsley MYW                05 35  06 16  9
                  Sheffield FT               06 33
           FS     Service Attercliffe Sidings and Sheffield FT
                  (Guard to assist Shunter as required)
                  until
                  Sheffield FT                      11 49  9
                  Tinsley MYRL               12 11  12 21  LD
                  Tinsley MYW                12 52
           RELD   by TI.507 at 12 35

           ACTIVITY                          ARR.   DEP.   WTT   RUN
TI.507     REL    TI.506 at 12 52                                MO
                  Tinsley MYW                12 52  13 36  9
                  Sheffield F.T.             13 53
           FS     Service Attercliffe Sidings and Sheffield FT
                  (Guard to assist shunter as required)
                  until:-
                  Sheffield F.T.                    15 30  9
                  Tinsley MYRL               15 52  16 02  LD
                  Sheffield F.T.             16 13
           FS     Service Attercliffe Sidings
                  (Guard to assist shunter as required)
                  until:-
                  S.F.T.                            19 09  9
                  Tinsley MYRL               19 31  19 41  LD
           DISP   Tinsley TMD                20 00

           REL    TI.506 at 12 32                                MSX
                  Tinsley MYW                12 32  13 16  9
                  Sheffield F.T.             13 33
           FS     Service Attercliffe Sidings and Sheffield FT
                  (Guard to assist shunter as required)
                  until:-
                  Sheffield F.T.                    15 30  9
                  Tinsley MYRL               15 52  16 02  LD
                  Sheffield F.T.             16 13
                  Work Trips & FS as required Sheffield
                  Freight Terminal, Attercliffe Sidings and
                  Sheffield Midland Station.
                  (Guard to assist shunter as required)
                  until:-
                  Sheffield F.T.                    19 19  9
                  Tinsley MYRL               19 41  19 51  LD
           DISP   Tinsley TMD                20 10
```

Above: The 1988 trip notice for the dual-braked pilot designated as T04. This loco was allocated to work Aldwarke new site, along with Attercliffe sidings and Sheffield Freight Terminal. Its full working is as shown here.

```
AFA. LOG 24 HOURS ENDED 0600 19/12/85.
::::::::::::::::::::::::::::::::::::::::
1230.  ALDWARKE NOW IN ORDER ALL ROUTES.
::::::::::::::::::::::::::::::::::::::::
1430.  08749 DERAILED 1 PAIR WHEELS WEST END OF 19 ROAD.
TINSLEY MAIN YARD. DERAILED AT DOUBLE SWITCH POINTS DURING
SHUNTING OPERATIONS.  CAUSED DISPUTED.
DRIVER W. A. RICHARDSON.  MAINTENANCE SUP. REQUESTED TO ATTEND.
BDV,S ARRIVED 1810 .  RERAILED 1830.  TRACK IN ORDER 1840.
DEPARTED AT 1840.
..............................................................

AFA TINSLEY LOG TO 0600HRS 12/11/87.
************************************

0740HRS

X MN YD SUPVSR...P02 08880 DERAILED TINSLEY YARD WEST END
SHUNT SPUR 0735HRS 11/11/87.   DRIVER ATKINS S.I.C. PELL.
HAULING 6 WAGONS X WEST END TY YD 14 RD (GROSS WT VEHS 358T)
TOWARDS WEST END SHUNT SPUR UTILISING LOCO B.F. ONLY.
DRIVER STATES UNABLE TO STOP DUE CONDITION OF RAIL.
STOP BLOCK PUSHED BACK APP 6' REAR WHEELS LOCO DERAILED
FRONT WHEELS LIFTED.   BDV'S ORDERED 0745HRS.
LEEDS OC, S&T, PWAY, ELECTRICIANS INFO.
BDV'S ARR 0920 RERAILED 0950 DEP 1000.   SHUNT SPUR OUT OF USE
PENDING PWAY REPAIR.   LOCO NO DAMAGE AND BACK TO TRAFFIC.
S&T/ELECTRICIANS IN ORDER.
```

Above: Kim Collinson kept records of some derailments in Tinsley Yard. The extracts above are from the control log. They record two derailments involving Class 08s from 1985 and 1987 respectively.

Further Allocation Decline

As the workload for shunting locomotives in the Sheffield area continued to decline, Tinsley lost its pair of Class 09s, Nos. 09008 and 09013, to Cardiff Canton in late-September 1993.

By 1996 the allocation was down to just 10 Class 08s – 08413, 08535, 08575, 08694, 08745, 08751, 08879, 08880, 08905 and 08946. One year later, the

Above: Seen on 8th September 1990, 09008 still wears BR blue livery with painted "Aldwarke" name. In front of it is Tinsley's other Class 09, 09013 (then named "Shepcote") and two-tone grey-liveried 08509 "Wath ETD". In the background are (left to right) 20178, 37004, 37070 and 56003. *Chris Booth*

allocation was 08413, 08512, 08535, 08575, 08694, 08751, 08879, 08905 and 08946. In 1998, the final year of Tinsley's existence, official lists showed just five examples left at Tinsley, although of those 08413 and 08751 were shown as unserviceable, and 08905 and 08946 were at the sub-depot of Saltley. This left just 08879.

Final Class 08

On 27th March 1998 the final two Class 47s on depot, 47375 Tinsley Traction Depot and 47145, were posed with the final Class 08 on shed, 08879, before the Class 47s departed for Saltley, bringing closure to Tinsley depot. However, there was still a requirement for yard pilots to work the SMACC and SPACE traffic, so 08879 continued to work in the main yard.

Three weeks after the depot's closure, 08879 required fitters' attention following a small fire in the exhaust system. When the yard shunter rang the depot, he was told by the only man available, Phil Hodgkiss, that the depot had actually closed and there was no one available to help! In Phil's words, "after a pleading and begging session I said that if he could get the pilot up to the top shed, someone would take a look at it", that someone of course being Phil! Fortunately no serious Right:damage had been done and, after running the engine on a fast tick-over for a couple of hours, the loco was fit to return to its yard duties. More of Phil's reminiscences can be found in his book "Life and Times of a Railway Fitter".

The New Century and the Rundown of Shunters

Closure of Tinsley depot notwithstanding, the requirement for pilot locos in the yard continued into the 21st Century.

On 31st March 2003, Immingham depot's 08605 and 08623 were in use, with 08597 in use as pilot at Rotherham steel terminal at Masborough.

Only 08784 was shown to be at Tinsley on a list of shunter locations on 1st January 2004, but it was joined by 08528 from May 2004. That October saw the failure in service of 08528, no doubt partly due to the arduous work of shunting steel wagons up the incline to the SMACC plant. It was replaced by 08824 from Doncaster.

By March 2005, 08784 had returned to Doncaster, with 08953 joining its sister at Tinsley. The next month, 08953 was replaced by 08685, the former moving to Aldwarke steelworks where EWS had assumed responsibility for shunting work. In September, 08953 again replaced 08685 at Tinsley, but was subsequently replaced itself in 2006 by 08994, one of the locos that received a cut-down cab for use on the Burry Port & Gwendraeth Valley (BPGV) line in Wales.

The requirement for two Class 08s in the yard continued to decline, and by April 2006 just 08994 was working in the yard. At the end of May 08994 moved north to Polmadie depot in Glasgow; 08784, 08428 and 08662 all served intermittently at Tinsley thereafter.

Shake-up

In November 2007, EWS became part of the DB Schenker empire – a division of the German rail operator Deutsche Bahn AG. After a shake-up of locomotive pools, 08623 became the resident shunter, relieved by 08784 for a few weeks in early 2008. In April, 08623 was sent to Doncaster Carr depot for attention, thereby leaving Tinsley without a resident shunter for almost a month. Train locomotives

Right: Tinsley's affinity with Sheffield Children's Hospital led to the repaint and naming of 09008 after the hospital at the 1990 open day. Its blue livery also received a grey cab roof and solebar, white buffers, cast metal nameplates, cast metal Yorkshire rose depot diamond logos, and a logo celebrating Tinsley TMD's 25th Anniversary Silver Jubilee 1965–1990. The loco carried the "Sheffield Children's Hospital" name for three years, and received repaint into general grey before transfer away from Tinsley in late September 1993. The nameplates were removed and stored at Tinsley. The loco is seen on 13th October 1992. *Chris Booth*

performed shunting moves for what was probably the first time – but a sign of things to come.

08623 resumed its Tinsley duties in May 2008, continuing until March 2010 when it was sent to Toton for a C-Exam and engine repairs. Instead of being moved by low-loader, it was unusually transferred by rail behind a Class 66. The same Class 66 returned with 08742 in tow, but the latter unfortunately suffered shifted cranks and was returned to Toton in July 2010. Taking its place was another of the ex-BPGV cut-down Class 08, 08993, which had previously been working at Sheerness. On completion of a C-Exam at Doncaster in September, 08405 arrived to replace 08993.

Approaching the End

The building of SIRFT had necessitated the re-routing of trains for SMACC, as the route via the hump underpass was no longer available. As such, a new route was constructed from the west end of the yard to connect with the original SMACC route near the shunt neck, which considerably shortened the distance each train had to travel.

10th March 2011 saw 08804 arrive at Tinsley by rail to replace 08405, which was dispatched to Doncaster for attention. This proved to be the final swap.

The following month, DB Schenker deemed that train locomotives could perform shunting tasks more efficiently and decided to remove the shunter duty at Tinsley. As such, on 12th April 2011, 08804 was photographed sporting a headboard proclaiming "Last Tinsley Pilot 1965-2011" as it worked 8T02 from SMACC to the yard. Thus, the pilot duty ceased from 29th June.

Shortly afterwards the 08 was removed to Aldwarke and all shunting in Tinsley Yard was then performed by main line locomotives, a situation that remains today.

Left: The Ickles yard pilot, 08434, is stabled in the shunt neck at the west end of Ickles yard in 1984. The 4-lever ground frame to the right of the loco was electrically released from Ickles signal box by No. 48 lever. It allowed access into "down" sidings west. Note the missing arms from the bracket signal, indicating that Ickles signal box has closed. Closure was on 10th August 1984, thus dating the photo to late 1984. *Roger Wainwright*

Right: An unidentified Class 08, acting as the Meadow Hall pilot, shunts at the long-closed South Yorkshire Railway, Meadow Hall & Wincobank station. Opened on 11th September 1854, it closed on 7th December 1954. This view is taken from Meadow Hall signal box looking northwards. In the distance is Roe Brothers' scrapyard; the pilot is shunting a rake of condemned wagons into the scrapyard. Today, the station building is still in situ and the trackbed is a pleasant cycleway. *Roger Wainwright*

Left: The daily "wagonload" pickup service to the Allied Steel and Wire Ltd private sidings at Meadow Hall, diagrammed for Tinsley-allocated 08870, has split the points and derailed. It awaits a BR re-railing team on 24th April 1985.

The building behind was originally part of the Yorkshire Engine Company's works, an independent locomotive manufacturer in Sheffield. *Gordon Edgar*

Right: As viewed from the brake van of a departing train, 08424 prepares a short train in the express freight sidings. Note the line of 20-ton coke hoppers in the adjacent siding; these may have come from Orgreave coke works, which was not far away. *John Miller*

Below: Working as No. 6 Pilot at Masborough steel terminal, 08509 shunts the terminal on 28th January 1998, three months before closure of Tinsley depot. The "Old Road" appears at the bottom right of the image and disappears towards Rotherham Masborough. This was the site of the former Sheffield Freightliner terminal, which closed on 5th January 1976 due to dwindling traffic volumes brought on by improvements to the motorway network. It re-opened in 2015 as a container terminal, operated by Newell & Wright. *Chris Booth*

Above: Imagine the scenario – the Class 47 has travelled light engine along the north-west arrival line and has stopped behind TY251 GPL (the tall signal to the left of the Class 37). The driver has changed ends, crossed over and is now heading along the former east departure line into the yard. The Class 08 is standing at TY242, awaiting the road to move towards Catcliffe and stop behind TY251 GPL. The Class 37 is standing at TY249 waiting to enter the yard once the 47 has made its move. Once the 37 has moved, the 08 will set back through the points and propel its train into the yard – a busy little period captured forever! To the right, vegetation is beginning to take over the lifted reception sidings. *Ivan Stewart*

Below: Relief driver Keith Lewis stares into space as his steed for the day, 08749, awaits further work at the west end of the yard on 26th September 1985. A pair of Class 20s trundle towards the camera, and in the distance a second pair are at work. *Michael Rhodes*

Above: Barnburgh Main colliery was one of four pits in a complex that sent coal to Manvers coal preparation plant, the other three being Manvers, Wath and Kilnhurst. The last three transferred coal underground, but Barnburgh sent coal overland to the plant. On-hire Tinsley-allocated 08492, "Barnsley", trundles along the branch from Barnburgh Main colliery to Manvers Main with a mixed rake of NCB internal user wagons, crossing the River Dearne on 15th March 1989. *Andrew K. Overton*

Above: 08870 "Millhouses" shunts at Barnburgh colliery at 10.55 on 8th February 1989. *Andrew K. Overton*

Right: No. 5 Pilot was allocated to work at Beighton permanent way yard, where it also undertook trips to Woodhouse sidings/ Holbrook sidings. Here, Tinsley driver Pete Howard poses with his steed, 08308 "Langwith", on 31st August 1988. *Steven Gay*

The Shunt from SMACC: Due to the yard's layout in its latter days, a convoluted series of shunts had to be undertaken to move a train from SMACC to the yard.

Above: 08844 hauls its train away from the run-round loop at the SMACC works on 8th May 2001 – (1) on the diagram. *Stewart Donohoe*

Left: This view shows the gradient to the shunt neck. 08844 has traversed the track on the right to the buffer stops, changed direction and now heads towards the next change of direction (2). Wood Lane bridge can be seen to the left; the train is almost on a level with the roadway. *Stewart Donohoe*

Below: A diagram of the route taken for a shunt from SMACC to the yard. Numbers indicate the location of the images.

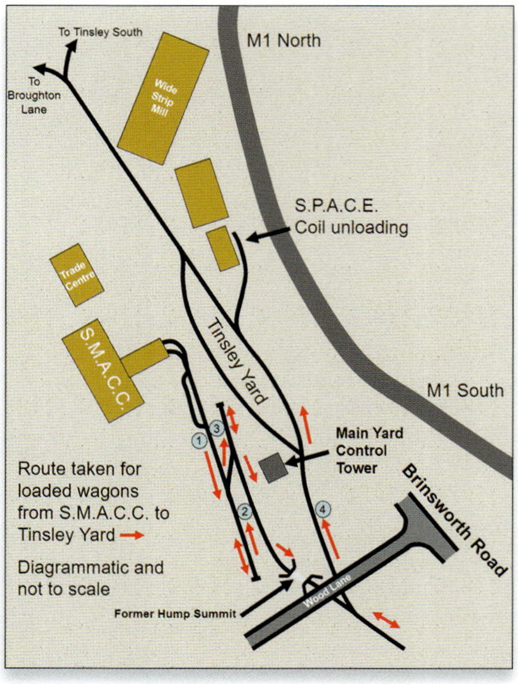

Top left: After changing direction from the previous image, 08844 begins its descent towards another set of buffer stops and another change of direction (3). The train will finish its journey on the far side of the yard, where more steel wagons can be seen. *Stewart Donohoe*

Left: Following another reversal, 08844 would have continued via the former secondary yard east departure line, past the site of the servicing depot and breakdown crane shed, beneath the former hump line (the bridge parapet can be seen to the extreme left), and beneath Wood Lane bridge before coming to a stop. On 24th May 2000, 08428 propels a train of slabs past the site of the hump summit and towards the derelict main yard control tower (4). *Stewart Donohoe*

Below: As well as SMACC, the yard pilots also shunted the Stainless Plate and Coil Expansion (SPACE) works. 08927 is in the coil unloading area on 14th July 2001. Under normally circumstances, the train locomotive would split the train in the yard, shunt half of the wagons into SPACE, and then stable in the yard whilst the 08 shunted the wagons into position beneath the crane. The SPACE works closed in 2005 with the loss of over 600 jobs. *Stewart Donohoe*

Chapter 16

Breakdown Train and Snow Ploughs

At the zenith of Britain's railways, the railway companies owned as many as 125 breakdown cranes between them. They could be seen countrywide attending derailments and mishaps of varying natures, from the derailment of a single wagon or locomotive, to major accidents. As the number of wagons of varying types (some of them of a pretty ropey nature!) dwindled over the years, so did the requirement for cranes to re-rail them.

According to official figures, there were still some 415,832 wagons on Britain's railways in 1970, with 10,182 derailments occurring nationwide in that year. By 1987 the number of wagons had dropped to 74,247, with a corresponding reduction in the number of derailments to 3,171. The majority of those were of a minor nature, probably a result of the deplorable condition of track in colliery yards, or from rough shunting in marshalling yards and sidings.

The development of hydraulic lifting jacks was another factor in the falling requirement for re-railing cranes. It was far easier and cheaper to send for the "packing vans" and breakdown gang, than to source a crane. Bruff road/rail vehicles followed, which radically reduced the need for a breakdown train at all. The Bruff Manufacturing Company was based in Suckley, Worcestershire, and initially manufactured agricultural machines, until the development of this revolutionary recovery vehicle for British Rail. It was basically a four-wheel-drive lorry chassis, capable of travelling on roads, rough terrain and railway tracks, and was therefore able to attend most derailments.

The ongoing withdrawal of unfitted wagons and closure of small yards saw derailments reduce in number, to the degree that there were as few as 125 derailments by 2001.

Tinsley's Steam Breakdown Crane

Tinsley Yard was equipped with a breakdown train which included packing vans and a 45-ton steam crane. Tinsley's breakdown train notionally covered an area from Bradford in the north to Chesterfield in

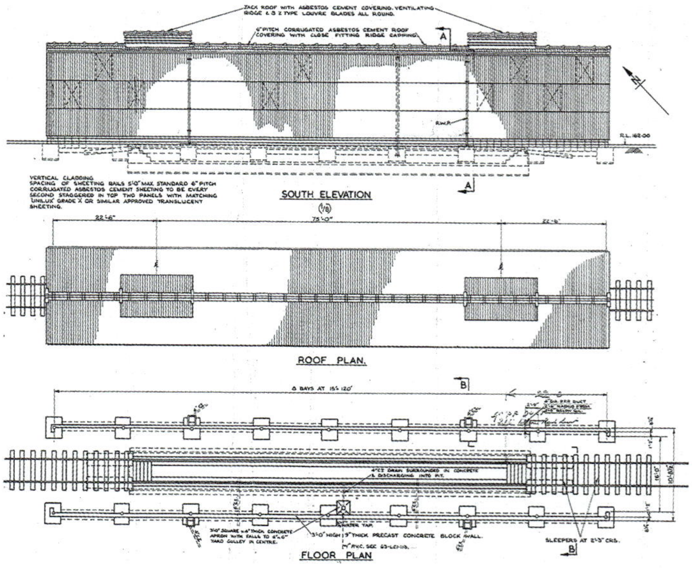

Above: Tinsley breakdown crane shed diagram.

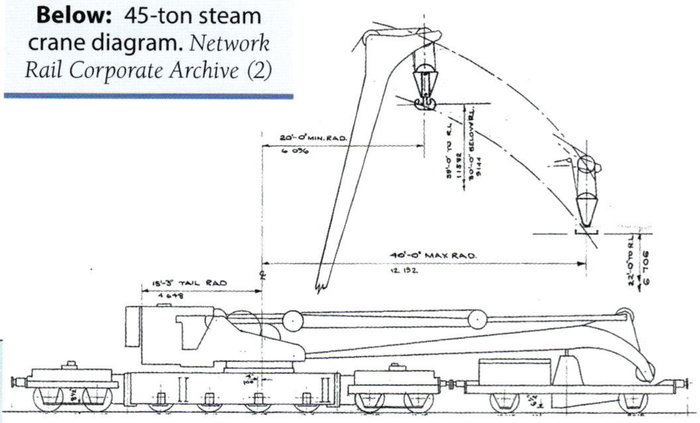

Below: 45-ton steam crane diagram. *Network Rail Corporate Archive (2)*

Left: Tinsley's 45-ton steam crane, ADRR95214, stands on the diesel depot in 1979, possibly awaiting repairs. Several steam cranes were converted to diesel-hydraulic operation, but Tinsley's would escape this programme and later enter preservation. *Roger Wainwright*

the south, and from Kiveton Park in the east to Edale in the west, but it could be called on to help in other areas as necessary.

To house the breakdown train, a dedicated single-road shed was provided close to the servicing shed.

In the mid-1980s Tinsley was allocated a steam crane numbered DE330102, this being part of an order for two cranes placed by the Ministry of Supply in 1942. These were direct replacements for two LNER cranes (951515 and 941600) which had been requisitioned by the War Department in 1941 and sent overseas. Built by Ransomes & Rapier of Ipswich in 1943 to order No. F4991-3, this crane was given LNER No. 941601 and allocated new to King's Cross depot.

Upon nationalisation in 1948 the crane was renumbered DE330102. It was transferred to Grimesthorpe in 1961, moving over to Tinsley in 1965. It would be renumbered again as ADRR95214, although the latter number never appeared on the crane itself. It ended its days with its number sign written – incorrectly – as "ADE 330102".

The breakdown crane unit comprised the crane itself, a jib runner to support the jib and two "Stokes bogies". These last are small weight-relieving trucks found at either end of the main crane, which supported some of the weight when travelling to and from site. When the crane was working, its full weight was restored to the crane unit or "carriage". Four heavy outrigger "props" were fully extended and screwed down against wooden packing to help steady the crane as it lifted. This wooden packing was carried in the tool van and sometimes on the jib runner; it gave the crane its maximum "block base", thus allowing it to work at its full rating.

Above: A job for the crane! Men are busy removing the overhead wires to allow the crane to access these derailed wagons in the reception sidings in 1981.

With a route availability of 3, DE330102 had a maximum haulage speed of 45 mph. Once on the site of a derailment the crane was able to propel itself, power coming from two cylinders on the body sides via a series of gears and dog-clutches. However, it could only move at a much reduced speed.

Withdrawn from BR service in 1986, it was acquired by the North Yorkshire Moors Railway, where it remains today. As for the other LNER replacement crane, ADRR95215, this is now preserved on the Bluebell Railway (source Breakdown Crane Association *www.bdca.org.uk).*

Below: Steam crane ADRR95214 was called out to re-rail an errant 76011 at Rotherwood sometime in 1979 and is seen here being set up for the lift. *Both Steve Ryszka*

Above: A night time photo of ADRR95214 in the reception sidings with plenty of steam to spare! The occasion was the re-railing of a wagon that had arrived in the yard overloaded with steel slab. The track beneath the wagon had spread, so before the wagon could be re-railed its load had to be transferred to other wagons. It was then lifted onto another line and the track subsequently repaired. *Steve Ryszka*

Below: The steam crane was superseded by diesel crane ADRC96704, which is seen re-railing 56081 at Barrow Hill on 12th September 1987. The locomotive derailed all wheels in the early hours of the morning and ended up listing heavily, necessitating attendance of the heavy gear! Re-railing took several hours to complete. *Andrew Johnson*

A New Diesel Crane

Tinsley's steam crane was superseded by a Cowans Sheldon 75-ton diesel crane. This had been built as a steam operated crane in 1962 and converted to diesel-hydraulic operation in 1978. Its original number was RS1094/75, later becoming ADRC96704. It was sent to Stratford immediately after conversion from steam to diesel hydraulic, remaining there until 1988 when it was transferred to Tinsley. Upon

Left: Besides derailment duties, the breakdown crane was also used for power unit exchanges at Tinsley. 47201 is undergoing an engine swap in July 1990, with 08879 being used to move the Class 47 into position. *Andrew Johnson*

Below: Behind Class 20 Nos. 20093 + 20148, the breakdown train is en route to another job as it trundles along the "up" goods Line at Hasland, near Chesterfield, in this shot from the 1980s. *Chris Booth collection*

withdrawal the crane entered preservation and is now with the Midland Railway, Butterley.

Staff Coaches

The final breakdown crane at Tinsley was accompanied by BTU staff coach ADB975470, which was subsequently preserved and is being returned to passenger use by the Somerset & Dorset Railway Heritage Trust at Midsomer Norton station, Somerset.

Also in use in the Tinsley breakdown train were riding coach ADE320912 (originally E9012E), an ex-LNER diagram 10C RFK (corridor first with restaurant) and tool van DB975249, originally M32998, which was an ex-LMS six-wheeled gangwayed brake van (Stove R), transferred to departmental stock in 1972. After withdrawal from BR service the tool van was preserved at the Midland Railway, Butterley, but the body was later broken up and the chassis re-used beneath the rebuilt body of Midland Railway coach No. 1260. As for ADE320912, this met its end at Thornaby.

Snow Ploughs

According to the British Railways Eastern Region, Sectional Appendix to the Working Timetable and Books of Rules and Regulations – Southern Area, 1st October 1972:

"BR standard independent snow ploughs and heavy snow ploughs are mounted on vacuum-braked tenders. Some medium snow ploughs are mounted on vacuum-braked tenders and others on Class 08 shunting locomotives. The latter are regarded as vehicles and are fitted with air brakes. Their maximum speed must not exceed 40 mph and when being conveyed a dual-braked locomotive must be provided".

The Sheffield area had snow ploughs allocated to the following locations:-

- Shirebrook: 1 Cowlairs Medium DE.330917
- Shirebrook: 1 Gorton Large DE.330966
- Tinsley: 1 BR Standard DB.965211
- Tinsley: 1 Darlington Medium (ADE980) DE.330980
- Wath: 1 Darlington Medium DE.330910
- Wath: 1 Darlington Medium DE.330918

Several withdrawn Class 08s were converted into snow ploughs at Doncaster depot in late 1973, the first ones being D3006, D3035 and D3078 which were renumbered ADB966507/08/06 respectively. A further two, D3037 and D3069 were converted in 1974 to become ADB966510/09. More were converted in 1977, 08111, 08117 and 08119 being renumbered to ADB966512/13/11 respectively, later becoming ADB968012/10/11 respectively. Their usual locations following conversion were Doncaster, Frodingham, Grantham, Lincoln, March, Thornaby, Tinsley and Wath.

Tinsley received the former 08117 which was initially numbered ADB966513, but this was changed to ADB968013 in error whilst at Doncaster depot; the correct number should have been ADB968010. Once the error was realised the final digit was painted over but the correct digit '0' was not applied until a couple of months later! Subsequently moved to Tinsley depot, it was first observed there on 28th December 1977; the final confirmed sighting at Tinsley was on 1st April 1978, the unit being scrapped at Doncaster works in 1979.

Below: ADB965309 "SNOW QUEEN" at Tinsley on Sunday 23rd May 1993. *Dave Hitchborne. Creative Commons Attribution-Share Alike 2.0 Generic Licence*

The Class 08 to snow plough conversion project was deemed a failure due to unsteady riding; all the former locos were out of use by the spring of 1979 (source RCTS).

SNOW joKING!

Tinsley was allocated two BR standard snow ploughs, numbered ADB 965308 and ADB 965309; during November 1992 they received an exam to ready them for the winter period. After the exam was completed, Phil Hodgkiss wielded his magic paintbrush and gave them a full repaint!

Both snow ploughs were transformed from plain black to Railfreight Distribution colours of dark grey roof, grey top half with lighter grey bottom half, the underframe being black. A small yellow panel was added at the back, with "wasp" stripes on the plough blade. Cast Tinsley "Yorkshire Rose" plaques were added, along with the painted names ADB 965308 "SNOW KING" and ADB 965309 "SNOW QUEEN".

Both ploughs had been moved to Buxton by 1994 but did not remain there long, being moved to RFS Doncaster where they are thought to have been used for spares in Railtrack's snow plough refurbishment programme.

Below: No. 980 was a former steam locomotive tender converted to a snow plough; it is seen here in the company of a pair of Class 08s (one of which looks to be withdrawn) at the rear of the shed in 1973. Later renumbered DE 330980, the tender is a GNR Ivatt Type B, which would have been coupled with a steam loco from the following classes – C1, C2, K2, K3, O1, O2 and miscellaneous smaller classes. *Alan Hazell*

Chapter 17

Tinsley in the 1960s and 1970s

At Tinsley, as at any large marshalling yard, it was necessary to pre-determine which roads would be used for each traffic flow, thus enabling the most efficient method of working. Details of the siding allocations and the designations of each siding at the time of opening, are shown in Table 1.

Note that the yard opened before the completion of Aldwarke Junction, which allowed traffic to enter the yard from both directions, so it was still necessary to group traffic for destinations on either the former GCR or former LMS lines. When Aldwarke was completed, the siding allocations were duly revised.

Siding Allocations

Some of the siding allocations listed in Table 1 are worthy of further explanation:

Siding No. 6 held traffic for Masborough sorting sidings and the Rotherham Westgate branch, this being anything bound for C.F. Booths scrapyard and any wagonload traffic

Table 1: Designation of Sidings

Siding	Main Yard	"Addo" Code			
	Siding Allocation	Main Sort	Second Sort	Fitted Heads	Special Codes
1	Toton unfitted	101	801		
2	Toton fitted	102	802		
3	Chapeltown South, stations to Barnsley	103	803		
4	Norton Junction fitted	104	804		
5	Norton Junction unfitted	105	805		
6	Masborough sorting sidings (including Westgate branch), Holmes, C.H. gas, I&I Walker SYCC	106			
7	Parkgate I.&S. Co. (Rawmarsh)	107			
8	Barrow Hill, Sheepbridge, stations to Stanton Gate	108			
9	Washwood Heath unfitted	109			
10	Washwood Heath fitted	110	810		
11	Treeton, Orgreaves, Woodhouse Mill	111			
12	Healey Mills fitted	112			
13	Healey Mills unfitted	113			
14	Chaddesden unfitted	114			
15	Stourton unfitted	115			
16	Feed Road – Traffic for:				
	Derby St. Marys			514	
	Chaddesden fitted			514	
	Stourton fitted.			515	
	Merchandise traffic for:				
	Ducie Street			516	20
	Nottingham			516	21
	Hunslet			516	22
17	Empty unlabelled Hifits	117			
18	Lincoln	118			
19	Hull fitted	119			
20	Hull mineral	120			
21	York mineral	121	821		
22	York fitted	122			
23	Feed Road – Traffic for:				
	Lincoln fitted			618	
	New England fitted			625	
	Frodingham fitted			626	
	Goole			619	
24	Parkway, Leicester, Brent,	124			
24 (cont)	St. Pancras, King's Cross, Carlisle, Sighthill				
25	New England unfitted	125	825		
26	Frodingham unfitted	126			
27	Rotherham Road (including Kilnhurst, Roundwood 11" Mill, Aldwarke New Mill)	127	827		
28	Empty unlabelled vanfits etc.	128			
29	Empty unlabelled steel carrying wagons	129			
30	Empty unlabelled mineral wagons	130			
31	Harvest Lane, Neepsend Gas, Oughty Bridge	131	831		
32	Grimesthorpe sundries	132			
33	Grimesthorpe full wagon loads	133			
34	Barnsley Junction	134	834		
35	South Wales	135			
36	Deepcar	136			
37	Wadsley Bridge	137			
38	Darnall C&W shops	138			
39	Mottram fitted	139	839		
40	Mottram unfitted	140			
41	Guide Bridge/ Birkenhead	141			
42	Brunswick	142			
43	Feed Road – fitted traffic for:				
	Colwick			744	
	Doncaster			747	
	Whitemoor			748	
44	Colwick unfitted	144			
45	Broughton Lane, Ecclesfield East, Smithywood	145	845		
46	Ickles (including Meadow Hall, Wincobank station yard)	146	846		
47	Doncaster unfitted	147	847		
48	Whitemoor unfitted	148			
49	Brake vans for main yard west	149			
50	Mechanical feed road				
51	Emergency feed road				
52	ESC scrap	152			
53	ESC other	153			

for Westgate station. It also held traffic for Holmes sidings, the Isaac and Israel Walker private siding to the north of Masborough station (manufacturers of foundry blacking), Carr House gas works and the South Yorkshire Chemical Company (SYCC).

Siding No. 7 was designated for traffic to the Parkgate Iron and Steel Company sidings around Parkgate and Rawmarsh station, adjacent to Aldwarke Junction.

Whilst some destinations, such as Toton and Mottram, had separate sidings for unfitted and fitted wagons, most destinations didn't generate enough traffic to warrant such a distinction.

To simplify the preparation of trains with fitted heads, sidings 16, 23 & 43 were nominated as fitted feed roads into which brake-fitted wagons would be routed for later sorting and marshalling. For example, siding No. 43 held braked wagons for Colwick, Doncaster and Whitemoor. The west end pilot loco would sort and marshal the fitted wagons into the west end of their requisite road i.e. No. 44 siding for Colwick, No. 47 for Doncaster and No. 48 for Whitemoor. Therefore, when the train was ready for departure, as many braked wagons as possible were assembled to give the train a "fitted head". At a time when there were still large numbers of unbraked wagons, the addition of a fitted head permitted an increase in the train's permissible speed and load limit.

Siding No. 46 held traffic for Ickles sidings (on the former GCR line between Rotherham Central station and Tinsley East) along with traffic for Wincobank station yard, which was one of the yards slated for closure. Wincobank's "down" sidings had a capacity of 836 wagons and its "up" sidings, 465 wagons; even after Tinsley opened, Wincobank's sidings were still used for overspill traffic from Meadow Hall and the Blackburn Valley line.

Traffic from Tinsley to Meadow Hall would include wagons for the East Midlands Gas Board (the former Sheffield and District Gas Company works) and the Yorkshire Engine Company.

Siding No. 50 was the mechanical feed road. Traffic routed here was moved automatically by the Dowty hydraulic units for further sorting in the secondary sorting sidings, Nos. 60 to 84. Should the Dowty system be out of order, traffic would be routed along siding No. 51 and a pilot locomotive would move the wagons to the required siding.

The secondary yard had its own set of siding designations, shown in Table 2.

Table 2: Secondary Yard Siding Allocations 1965

Sdg No.	Allocation	"Addo" Code
60	Tinsley Park Cripple Sidings	960
61	Carlton and Normanton, Manvers	961
62	Woodhouse and Orgreaves GC	962
63	ESC River Don Works	963
64	Rotherham Central	964
65	West Tinsley, Shepcote Lane	965
66	Empty unlabelled specially-constructed wagons	966
67	Grimesthorpe Gas Works	967
68	Stations to Earles sidings (Hope Valley)	968
69	Firth Browns, Upwell Street Wharf	969
70	ESC Grimesthorpe Works	970
71	Tinsley Yard service departments	971
72	Darnall, Handsworth	972
73	Nunnery Yard, Blast Lane	973
74	Buxton	974
75	Brake Vans for secondary yard west end	975
76	Neepsend Power Station	976
77	Dronfield, Queens Road	977
78	Heeley, Millhouses, Sheffield Midland	978
79	Aldwarke "Naylors"	979
80	Wath Yard	980
81	Attercliffe	981
82	Blackburn Meadows	982
83	Warsop, Langwith, Spink Hill, Creswell, Mansfield	983
84	Worksop	984

Telex System

February 1966 saw the installation of a Telex machine in the express freight sidings to allow staff to keep in touch with other Eastern Region yards and ensure traffic coming in and out, was kept moving speedily.

When a train arrived in the yard a number taker would dictate the details of the contents of each wagon and its destination etc., by means of a UHF radio linked directly to the Telex office. In the Telex office this was recorded directly onto a reel-to-reel tape recorder.

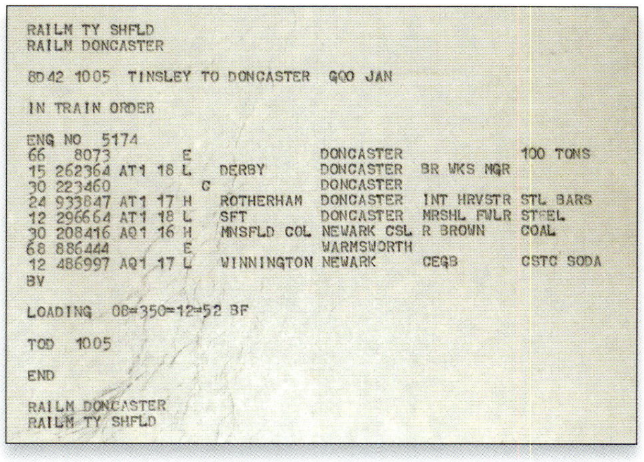
```
RAILM TY SHFLD
RAILM DONCASTER

8D42 1005  TINSLEY TO DONCASTER  GOO JAN

IN TRAIN ORDER

ENG NO  5174
66   8073        E              DONCASTER                100 TONS
15 262364 AT1 18 L   DERBY       DONCASTER  BR WKS MGR
30 223460          C             DONCASTER
24 933847 AT1 17 H   ROTHERHAM   DONCASTER  INT HRVSTR STL BARS
12 296664 AT1 18 L   SFT         DONCASTER  MRSHL FWLR STEEL
30 208416 AQ1 16 H   MNSFLD COL  NEWARK CSL R BROWN    COAL
68 886444        E               WARMSWORTH
12 486997 AQ1 17 L   WINNINGTON  NEWARK     CEGB       CSTC SODA
BV

LOADING  08=350=12=52 BF

TOD  1005

END

RAILM DONCASTER
RAILM TY SHFLD
```

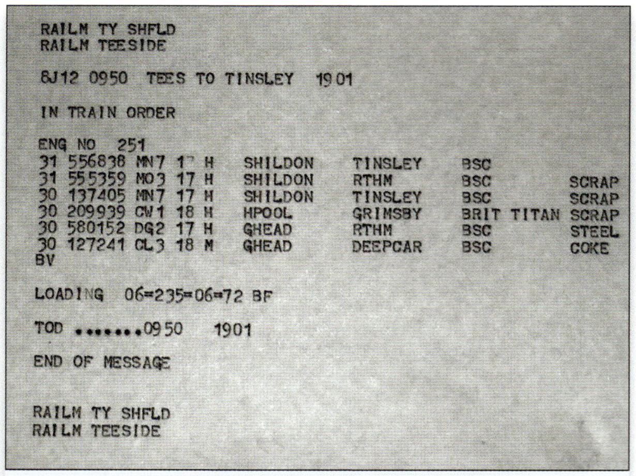
```
RAILM TY SHFLD
RAILM TEESIDE

8J12 0950  TEES TO TINSLEY  1901

IN TRAIN ORDER

ENG NO  251
31 556838 MN7 17 H   SHILDON    TINSLEY    BSC
31 555359 MO3 17 H   SHILDON    RTHM       BSC          SCRAP
30 137405 MN7 17 H   SHILDON    TINSLEY    BSC          SCRAP
30 209939 CW1 18 H   HPOOL      GRIMSBY    BRIT TITAN   SCRAP
30 580152 DG2 17 H   GHEAD      RTHM       BSC          STEEL
30 127241 CL3 18 M   GHEAD      DEEPCAR    BSC          COKE
BV

LOADING  06=235=06=72 BF

TOD .......0950   1901

END OF MESSAGE

RAILM TY SHFLD
RAILM TEESIDE
```

Left & above: Examples of Telex printouts. The example on the left shows train 8D42, 10.05 Tinsley to Doncaster, locomotive number 5174, then eight wagons and a brake van (BV). The example above shows train 8J12, 09.50 Tees to Tinsley, loco number 251, then six wagons and BV.
Alan McKenna collection

The Telex operator would listen back to the recording and type the details to a punched tape, using codes to denote the different types of wagon, e.g. a bogie bolster was number 48, and a 16-ton mineral wagon number 30. Further details included: wagon number, route code, label date, wagon weight, special traffic notes, sending station, destination station, consignee and commodity. A typical printout for Tinsley would thus read:-

30 208416 AQ1 16 H MNSFLD COL NEWARK CSL R BROWN COAL

MNSFLD COL = Mansfield Colliery,
NEWARK CSL = Newark Castle

Tinsley Depot on 13th May 1964.

Above right: A view across the depot at platform level. Two Class 20s, a Class 37 and a Class 47 are in residence. The furthest Class 20 is D8052, the Class 47 is D1542, and the other two locos are unidentified. *BR LS.589/Chris Booth collection*

Right: A view through the workshop area towards the eastern end of the shed with two fitters at work. A Brush Type 2 (Class 31) is being attended to at the eastern end of the shed. *BR LS.592/Chris Booth collection*

Below: Class 37 D6966 and Class 47 D1798 receive attention at the east end of the shed. D6966 has had its rocker box covers removed, possibly via the open door in the bodyside. Two fitters are working at floor level, one of them holding a hose of some description. *BR LS622/ Chris Booth collection*

The main control tower contained a slave Telex which received inward train consists for use by the main control tower supervisor. Details of the train would be sent to other yards, along with copies to the divisional manager and the sending station (known as "Rail TY Sheffield"), all before the train had even left the yard.

In 1968/69 other regions adopted this method of transmitting advance traffic information (ATI), which caused the Telex office at Tinsley to be enlarged. Sending and receiving an increasing number of consists on a single machine soon became impossible and a method of using ATI to improve the performance of connecting and through traffic was devised by the Regional HQ at York.

The layout of the main control tower precluded all the elements of ATI being in one place, so the individual offices required were scattered around the yard.

From 1st March 1971 the yard manager's function was changed to that of area manager, who subsequently had to deal with traffic movement over a much wider area.

At the same time it was decided to form a closer working relationship with the occupiers of private sidings in the area, in the hope of improving customer satisfaction. The theory was that an information service would help private firms anticipate traffic arriving, allowing them better preparation for unloading their wagons and thus improve wagon turnaround times.

And so, with the aim of bringing all the traffic functions into one building, an area traffic office (ATO) was established, opening on 29th January 1973. The office worked continuously from 06.00 on Monday to 14.00 on Saturday with a staff complement of:

- Traffic Supervisor 'C'
- Charting Supervisor 'A'
- C.O.2 Charting
- C.O.2 Information
- C.O.1 Telex Operator (2 on night turn of duty)

Tinsley Diesel Maintenance Depot

Once Tinsley diesel maintenance depot was fully operational, its complement of diesels could be seen far and wide. The following is a glimpse of some of those workings through the 1960s and 1970s.

Allocation in the 1960s

In the 1960s, Tinsley's allocation could be split into four distinct categories – pilot locomotives, local trip workings, main line work and out-based locos. In practice no such titles were assigned and the locos moved transiently through the different types of work as required by the ever-changing operating requirements. Table 3 shows Tinsley's allocation of diesels in October 1965.

Besides the more common locomotive classes, Tinsley had a brief association with the Class 17 "Claytons". Twelve locos were allocated from new; the first, D8604, arrived on 15th September 1964 with the final one, D8615, arriving on 10th March 1965. They were reallocated to Barrow Hill (41E) from 10th April 1965, but all were moved to Scotland in May 1966.

Below: 5V03, consisting of mixed steel, leaves Tinsley Yard at 17.56 for Port Talbot on 24th August 1965, hauled by Class 47 No. D1744. The train would have come from the express freight sidings, and is heading along the by-pass line towards Catcliffe and Treeton Junction, where it will take the "Old Road" southwards to Wales. Note the sole person watching the proceedings in his long black coat – no hi-visibility clothing back then! *BR TFD 127 Chris Booth collection*

Table 3: Tinsley Locomotive Allocation – October 1965

Class 08/10
D3086 D3127 D3129 D3131 D3251 D3252 D3253 D3254 D3288 D3289 D3293 D3326 D3330 D3336 D3439 D3440 D3443 D3444 D3452 D3475 D3574 D3575 D3576 D3659 D3660 D3661 D3662 D3664 D3685 D3702 D3703 D3707 D3727 D4028 D4029 D4035 D4036 D4037 D4038 D4039 D4040 D4041 D4042 D4043 D4044 D4045 D4046 D4047 D4048 D4049 D4050 D4051 D4052 D4054 D4055 D4056 D4058 D4059 D4062 D4063 D4064 D4065 D4068 D4070 D4071 D4072 D4073 D4074 D4089 D4090 D4091 D4092 D4093 D4094
Class 13
D4500 D4501 D4502
Class 20
D8058 D8059 D8060 D8061 D8062 D8063 D8064 D8065 D8066 D8067 D8068 D8069
Class 25
D7624 D7625 D7626 D7627 D7628 D7629 D7630 D7631 D7632 D7633 D7634
Class 31
D5528 D5533 D5536 D5538 D5539 D5540 D5541 D5542 D5543 D5546 D5549 D5550 D5552 D5555 D5557 D5558 D5561 D5565 D5567 D5568 D5579 D5583 D5584 D5585 D5586 D5665 D5670 D5677 D5680 D5681 D5682 D5683 D5684 D5685 D5686 D5690 D5691 D5692 D5804 D5805 D5806 D5840 D5841 D5842 D5843 D5844 D5845 D5846 D5847 D5848 D5849 D5850 D5855 D5856 D5858 D5859 D5860 D5861 D5862
Class 37
D6728 D6729 D6742 D6743 D6744 D6745 D6746 D6747 D6748 D6749 D6750 D6751 D6752 D6753 D6754 D6796 D6797 D6798 D6799 D6800 D6801 D6802 D6803 D6804 D6805 D6806 D6807 D6808 D6809 D6810 D6811 D6812 D6813 D6814 D6815 D6816 D6817 D6818 D6965 D6966 D6967 D6968
Class 47
D1525 D1527 D1531 D1552 D1553 D1574 D1576 D1768 D1769 D1770 D1771 D1772 D1773 D1774 D1775 D1776 D1777 D1778 D1779 D1780 D1781 D1782 D1783 D1784 D1785 D1786 D1787 D1788 D1789 D1790 D1791 D1792 D1793 D1794 D1795 D1796 D1797 D1798 D1799 D1800 D1801 D1802 D1803 D1804 D1805 D1806 D1862 D1863 D1864 D1865 D1866 D1867 D1869 D1870 D1871 D1872 D1873 D1874 D1875 D1876 D1877 D1878 D1879 D1881 D1882 D1888 D1890 D1891 D1892 D1895 D1896 D1897
Class 48
D1703

Local Trip Work

For local trip and pick-up traffic, locomotives of Types 1 to 3 were preferred, with Class 31s dominating. Their diagrams took them to steel plants and collieries in the immediate Tinsley and Sheffield area.

Trip workings would generally consist of empty trains out of the secondary yard to local sidings, returning as loaded

Below: An unidentified Class 25 trundles along the north arrival line and heads to the reception sidings with 9T46, comprising a mixed rake of empty unfitted wagons in this un-dated early 1970s period shot. Behind the train, a Class 31 approaches TY258 signal on the north-west arrival, to await its turn to enter the reception sidings. The yard is quite busy at this time, with another rake of hopper wagons in the process of being humped. *Chris Booth collection*

trains to the reception sidings, from where the traffic would be humped and sorted.

The second group of locos were higher-powered Types 3 and 4, which tended to work main line traffic out of Tinsley or, in many cases, in further flung areas.

Examples of workings further afield included English Electric Type 3s (Class 37) covering the Cleethorpes/Grimsby to King's Cross express passenger services and Brush Type 4s (Class 47) working as far as Birmingham on through steel trains from Frodingham to South Wales. There was also a daily diagram to Banbury.

Although most diagrams were of a single day's duration, some were spread over two days; locos provided for the High Dyke to Frodingham iron ore trains had a diagram which allowed for the loco to be changed once a week.

Trainload Working Begins

The nature of freight traffic was changing from wagonload to trainload well before Tinsley's opening.

June 1965 saw Brush Type 4 No. D1794 (47313) working MGR trials in the Sheffield and Doncaster divisions. Coal workings certainly weren't the mainstay of Tinsley's work, but such trials were a portent of the dwindling requirement for wagon marshalling, both at Tinsley and nationally.

Block coal workings around the Sheffield Division were some of the first such workings in the country, notably when Bevercotes colliery began sending MGR trains to service the newly-opened West Burton Power Station.

The Midland Main Line (MML) also saw heavy flows from the Manvers coking plant when production was at its height during the 1960s. These were normally handled by Eastern Region "Peaks" and Tinsley-based Brush Type 4s.

Passenger Turns

Tinsley-based locos were routinely placed in charge of passenger and parcels trains in the early years. The following selection are a few notable examples.

Tinsley was home to the five Class 48s (Class 47s fitted with V-form engines, of which more later) from new. D1703 arrived on 17th September 1965 with the other four locomotives arriving between 4th November 1965 and 21st July 1966. These locos were often noted on the flagship "Master Cutler" from Sheffield to St. Pancras, before they were all rebuilt with conventional Sulzer engines and reclassified as Class 47s.

Perhaps the furthest afield a Tinsley loco would have ventured on a passenger turn at that time, would have involved the Brush Type 4 that acted as passenger train pilot at Doncaster Station. This was often used replacing a defective or failed or locomotive, regularly working up to Scotland.

Darnall, and later Tinsley locomotives, also worked a large number of secondary services. D6799 (37099) was noted at the head of a Sheffield to Cleethorpes via Doncaster stopping service on 26th July 1966, whilst Brush Type 4 D1537 (47010) was noted with the Sheffield Pullman.

Main line express services over the MML from Sheffield to London were hauled by locos from the London Midland Region, in the main allocated to Toton.

Tinsley also provided locomotives for a range of regular parcels workings. For example, the 12.34 Sheffield–Leeds was noted with D6812 (37112) at the head of the train on 20th July 1967. The last Bournemouth–Newcastle via the former GCR main line was hauled by Tinsley-allocated D1572 (47018) on 3rd September 1966.

Out-based Locomotives

Tinsley-allocated locos could sometimes be out-based at other locations as required, thereby not making it back to Tinsley every week. Shirebrook, Barrow Hill, Wath and Worksop all predominantly used Tinsley locos (although it should be borne in mind that they were not sub-sheds in the manner of steam motive power depots and their sub-sheds). The odd-one-out in the Sheffield Division was Wath, which was responsible for its own allocation of main line locomotives, for which a brand new three-road maintenance depot was built. Table 4 lists the allocation for Wath in 1965.

Table 4: Wath Locomotive Allocation as at 30th October 1965

350 hp Shunters	D3060 D3061 D3062 D3063 D3064 D3329 D3333 D3726 D3430 D3431 D3432 D3433 D3434
Class 31	D5807 D5808 D5809 D5810 D5811 D5812 D5813 D5814 D5815 D5816 D5817 D5818 D5819 D5820 D5821 D5822 D5823 D5824 D5825 D5826 D5827 D5828 D5829 D5830 D5831 D5832 D5833 D5834 D5836 D5837 D5838 D5839
Class 37	D6959 D6969 D6961 D6962 D6963 D6964

The other stabling points would have seen at least a dozen Tinsley main line locos stabled on any given weekend. For example, Worksop usually saw locos stabled over a weekend in either the "down" side coal dock or the loco holding sidings at Worksop East. Table 5 shows those stabled on 21st April 1968.

Table 5: Locomotives Stabled at Worksop on 21st April 1968

Locomotives	Allocated to
D1564 D1566 D1568	31B Immingham
D1887 D1897 D5535 D5582 D5678 D6703	41A Tinsley
D5814	41C Wath
D3701, D4066	41J Shirebrook West

Langwith and Shirebrook

Langwith Junction depot, near Shirebrook, initially had its own allocation of diesel shunting locomotives, and later began using main line diesel locomotives. All were initially based in the old steam depot – a situation far from ideal but which was repeated nationwide during the transition from steam to diesel. To improve matters, the old Midland Railway goods shed at Shirebrook West station was converted into a small servicing shed with fuelling point. This wasn't without its own problems, being cramped and with little room for stabling. As a result, the ER built a brand new Shirebrook West diesel depot, which opened on 6th February 1966, the same date that Langwith Junction depot closed.

After opening, six Brush Type 4s and three Brush Type 2s were out-stationed from Tinsley at Shirebrook West, to work Class 7* coal trains over the former LD&EC route between Mansfield Concentration Sidings and Whitemoor or New England (near Peterborough).

HS4000 "Kestrel"

Perhaps the best-known locomotive to be out-based from Tinsley was the Hawker Siddeley/Brush prototype diesel-electric locomotive, HS4000 "Kestrel". This was loaned to BR for appraisal, by its manufacturer Brush Electrical Machines, from 14th May 1968. It was sent to Shirebrook

Above: Two unidentified Class 47s and a Class 08 receive attention inside Shirebrook West depot on 24th June 1965. Any Sunday would have seen at least a dozen locomotives stabled at Shirebrook. *Chris Booth collection*

West to be trialled between Mansfield Concentration Sidings and Whitemoor on heavy freight trains, usually of 1,450 to 1,600 tons. These were well within the capabilities of this powerful loco; indeed, due to operating restrictions in the area, it was extremely difficult to find a train of sufficient weight to really test its capabilities! The workings were:

7J31 SX 11.09 Mansfield Colliery Sidings–Whitemoor
7P31 SX 15.10 Whitemoor (Norwood Yard)–Mansfield Colliery Sidings
7J07 SX 19.40 Mansfield Colliery Sidings–Whitemoor
5P07 SX 23.45 Whitemoor (Down Yard) –Thoresby Colliery Junction.

Kestrel made much-publicised trials on 7th August 1968 between Mansfield Concentration Sidings and Lincoln. Comprising of 52 loaded MGR wagons with a gross trailing weight of 2,028 tons, this was announced at the time as the heaviest train ever to operate over BR. The wagons' speed limit was 45mph and the test included controlled starts on rising gradients, in particular the 1 in 50 Boughton bank and the 1½ miles of 1 in 120 Markham bank; from a standing start, 20mph was reached in 6½ minutes, the loco maintaining that speed until easier gradients were reached. Re-starting on the slippery rails really tested the loco's automatic wheel slip detection system.

HS4000 left Shirebrook in late August 1968 for tests over the Derby–Crewe–Nuneaton–Derby and Crewe–Nuneaton–Leicester routes but returned in late September to continue the Whitemoor runs. By the end of 1968 it had accumulated a total of 26,000 miles, 22,000 of which were revenue-earning.

Below: Class 31 D5804 stands with a Class 37 at Wath on a Sunday in 1969. Note the painted 41A allocation below the number. *Alan Smalley*

Left: Class 31 5805 stands with various other locomotives of Classes 20, 31 and 47 at Barrow Hill on a Sunday in 1969. A Class 37 is just visible inside the shed which, with its turntable, is still in existence today. *David J. Smith*

Below: The Hawker Siddeley prototype HS4000 stands on the "up" line adjacent to Shirebrook West depot in 1968. *Chris Booth collection*

After a spell on the ECML working passenger turns, whilst manned by Shirebrook crews and inspectors, Finsbury Park crews were trained on the locomotive and a more intensive schedule of trips between King's Cross and Newcastle began in October 1968.

In May/June 1969, Class 47 bogies were fitted to bring the overall weight down and to allow Kestrel's use on higher-speed passenger workings. Regular diagrammed working commenced on 25th October with the 09.00 King's Cross–Newcastle and return, and later the 07.55 King's Cross–Newcastle and return.

On 10th November 1969, HS4000 worked the 16.20 King's Cross–York stopping service, and was also noted at Stratford depot, laying over whilst handling a Hull–Stratford freightliner.

Maintenance for Kestrel during this period was handled by Tinsley. On 29th May 1970, Class 15 No. 8213 hauled Kestrel from Tinsley to the Vickers Armstrong works at Barrow, where it would remain until early September 1970. HS4000 subsequently returned to Shirebrook for use on freights until it again visited Vickers Armstrong in June 1971, for engine removal and overhaul.

After another brief period of operation at Shirebrook, HS4000 was requested to appear at an international exhibition "Railroad Rolling Stock-71" held in Scherbinka, near Moscow, After amassing a total of 136,646 miles of service in Britain, Kestrel was sold to Russia for £127,000 and departed these shores, never to return.

Depot Visits

Many enthusiasts would cycle around the area to visit depots at weekends, in the hope they would be afforded access. A visit to local depots undertaken by John Woolley on Wednesday 17th May 1967

Above: Class 47s D1787 + D1546 stand adjacent to the servicing shed at Tinsley in October 1967. D1546 was new to Darnall (41A) on 18th October 1963, later moving to Finsbury Park before returning north to Tinsley from week-ending 23rd April 1966. Moving again in 1968, it returned for spells at Tinsley in 1969 and 1988. Renumbered 47016, it was withdrawn on 12th December 1998 and was cut up at Wigan on 4th April 2000.

D1787 was new to Tinsley on 26th November 1964, moving to Immingham in June 1966. It was still an Immingham loco at the time of this photo. Renumbered 47306, it would later become part of the Tinsley RfD fleet and be named "The Sapper". Upon withdrawal on 31st May 2006, it would enter preservation. *The Transport Interchange* (instead of a weekend) rewarded him with locos in residence as shown in Table 6, the main line locos at Shirebrook being Tinsley-allocated. Examples of locos seen on shed at Tinsley can be found in Chapter 23.

Table 6: Locomotives on Depots – 17th May 1967

Wath	
Class 08	D3060 D3061 D3064 D3082 D3726 D4031 D4033
Class 25	D7625 D7626 D7632 D7633 D7635 D7636 D7638
Class 31	D5633 D5810 D5812 D5814 D5815 D5818 D5823 D5825 D5826 D5829 D5838
Class 37	D6744 D6799 D6964 D6968
Class 47	D1788 D1890
Class 76	E26004 E26006 E26023 E26024 E26029
Barrow Hill	
Unclassified	D2402 D2403
Class 08	D3252 D3440
Class 20	D8059 D8060 D8130
Class 31	D5536
Class 37	D6815 D6818
Shirebrook West	
Class 08	D4060 D4061 D4085
Class 20	D8017
Class 31	D5538 D5581 D5585 D5804 D5864
Class 47	D1703 D1704 D1798 D1872 D1888

Freightliner Trials

On 26th–28th April 1967, Tinsley-allocated D1867 (47217) worked trials in preparation for a new Sheffield Freightliner Terminal to King's Cross service, making two "up" and two "down" runs daily. The same loco repeated the trials from 2nd to 5th May.

The Class 48s

Late-1967 saw Tinsley's Class 48s, Nos. D1702–06 (47114–118) working exclusively from Shirebrook West depot, usually on coal traffic, although D1702/04 were frequent power for the 1N09 King's Cross–Leeds and 1E85 return. D1705 (47117) was noted on the turn for the first time on 21st December 1967.

Although the locos were well utilised, BR decided not to continue with the 12LVA24 engine experiment; the locos were re-engined with standard 12LDA28 engines and reclassified as Class 47 in line with their sisters. D1702 was converted at Crewe works between July 1968 and December 1969, using parts from D1908 which had been withdrawn on 12 August 1969 with severe accident damage.

All five locos had been converted by early 1971, the redundant 12LVA24 engines being returned to France where they were fitted into 68000 series locos.

The 1970s

By the end of 1969, official figures showed that the investment in Tinsley Yard, the diesel depot and the Grimesthorpe freight concentration depot, originally forecast to make savings of £1.43 million, had not quite reached its target, but had at least realised £1.322 million in net revenue. This was maybe a sign that the future of the yard was not looking rosy.

Mention has already been made of Class 15, 8213, a type not normally associated with Tinsley. The tenuous connection was that, during 1970, Stratford dispatched several locos to Tinsley for mechanical attention and for load bank testing.

The Sheffield area had seen locos of this type before, as ten of the class were ordered for evaluation under the British Railways Modernisation Plan "pilot scheme". The class was a collaboration between British Thomson Houston (electrical equipment), Paxman (power units) and Clayton (bogies and

Above: Ex-works Class 46, D164, is seen on the servicing shed in October 1967. This loco was new on 19th April 1964 and subsequently allocated to the Leicester Division in week-ending 6th March 1965. Moving to Holbeck (55A) at the end of December 1969, it would become 46027 in January 1974. Withdrawal came on 25th November 1984, and it met its end at Vic Berry's scrapyard in Leicester in December 1986. *The Transport Interchange*

Table 7: Treeton Junction Friday 1st May 1970

Train	Direction From	At	Line	Towards	Headcode
3-1-1	Beighton	0025	G	Tinsley Yard	4E15
1-4-3	Beighton	0035	G	Tinsley Yard	9T33
4-1	Beighton	0040	M	Canklow	7E63
1-4-3	Beighton	0045	G	Tinsley Yard	9T32
2-3	Tinsley Yard	0046	M	Canklow	0V60
3-2-2-1	Beighton	0103	G	Tinsley Yard	8Z62
1-4	Beighton	0125	M	Canklow	9E89
2-3	Beighton	0134	M	Canklow	0D03
4-1-4	Beighton	0139	G	Tinsley Yard	7E68
5	Beighton	0235	M	Canklow	6E50
2-3-2-1	Beighton	0242	G	Tinsley Yard	0J03
2-3-2-1	Beighton	0252	G	Tinsley Yard	0J03
3-2	Beighton	0256	M	Canklow	8Z70
2-3-2-1	Beighton	0313	G	Tinsley Yard	0J03
3-2-2-1	Beighton	0319	G	Tinsley Yard	8E40
3-2-2-1	Beighton	0331	G	Tinsley Yard	8E23
1-4-3	Beighton	0353	G	Tinsley Yard	9T94
4-1	Beighton	0353	M	Canklow	7E63
2-3-2-1	Beighton	0403	G	Tinsley Yard	0J03
3-2	Beighton	0431	M	Canklow	8E65
1-1-3	Beighton	0444	G/M	Canklow	
5	Beighton	0441	M	Canklow	6E31
3-2	Beighton	0456	M	Canklow	8F33
2-3-2-1	Beighton	0447	G	Tinsley Yard	0J03
2-3	Beighton	0503	M	Canklow	

G=Goods Line M=Main line

superstructure), with frame construction and final assembly taking place at the Yorkshire Engine Company works at Meadow Hall.

Numbered D8200–D8209, these locomotives entered service between November 1957 and November 1958, most of them having test runs between Meadow Hall and Chinley.

Following completion of the load bank work at Tinsley in 1970, reports are that all but one were given test runs on local freights, the other loco being condemned at Tinsley. One confirmed sighting was on 14th July 1970 when 8209 worked the 09.55 Tinsley Yard to Peak Forest, returning with the 13.57 Peak Forest to Tinsley Yard. After testing was completed, the locos returned to Stratford to work out the remainder their (short) lives.

Treeton Junction in 1970

Treeton Junction signal box controlled the eastern arrival and departure line connections from the "Old Road"; a train register book from the signal box dated Friday 1st May 1970 reveals the actual density of traffic. Table 7 shows the bookings between midnight and 05.00; note the routing codes in the train description e.g. a light engine from Beighton direction towards Canklow as 0D03 is a 2-3, whereas 0J03 (a light engine towards Tinsley Yard) is a 2-3-2-1. These routing codes informed the signalman at Treeton which way to send the approaching trains.

Table 8: Firms and Sidings Local to Tinsley in the 1970s

Commodity	Proprietor	Location
Scrapyards	Smith's	Ecclesfield
	Roe Brothers	Ecclesfield
	BSC Meadowhall	Meadow Hall
	T W Ward	Brightside
	Parkin and Johnson	Brightside
	Coopers Metals	Attercliffe
	Cohen	Tinsley
	Booth	Rotherham
	Arnott Young	Aldwarke
	T W Ward	Beighton
Steel Distribution	Walker Wright	Darnall
Steelworks	East Hecla	Wincobank
	BSC River Don	Brightside
	Firth Brown	Grimesthorpe
	Hadfield's	Attercliffe
	Davy Loewy	Darnall
	BSC Templeborough	Meadow Hall
	BSC Ickles	Rotherham
	BSC Thrybergh	Aldwarke
Coke	NSF Smithywood Coking Plant	Ecclesfield
	NSF Orgreaves	Orgreaves
	NSF Brookhouse	Brookhouse
Oil	BP Ecclesfield West	Chapeltown
	BSC Aldwarke	Aldwarke
Collieries	NCB Orgreave	Orgreave
	NCB Treeton	Treeton
	NCB Brookhouse	Brookhouse
	NCB Silverwood	Silverwood
Chemical	Orgreave	Orgreave
	BOC	Broughton Lane
	Brookhouse	Brookhouse
Other	Coal Depot	Nunnery
	Attercliffe Goods	Attercliffe
	Parkway Market Goods	Darnall

BOC – British Oxygen Company

BP – British Petroleum

NSF – National Smokeless Fuels

In the early the 1970s there was still substantial traffic from small local rail-served yards and premises, as summarised in Table 8.

The yard was still very busy, and in October 1972 a simplifier was issued to yard staff to help them. This showed details such as train destination, loadings, crewing information and so on. This is reproduced in Tables 9 and 10.

In 1974 figures showed that BR had carried 175 million tons of freight using an average of 2,700 locomotives per day

Above: Tinsley as viewed from Wood Lane bridge on 31st December 1987, after lifting of the hump and arrival lines. The yard is much reduced with all sidings to the left of the main control tower now cut off. *Chris Booth*

Above: A view from the main tower on 6th August 1974 shows a still-active yard.
Nigel Tout Creative Commons Attribution-ShareAlike 4.0 International License

on freight work. An average day saw 4,500 planned freight trains and a further 3,400 "trip" workings which, all told, gave an estimated 150,000 daily wagon movements.

Although rationalisation was beginning to take hold, there were still 49 marshalling yards, 500 freight terminals and 2,200 private sidings countrywide – but things were about to change at Tinsley.

Reviews and Changes are Afoot

By July 1976 it was being reported in the Sheffield "Star" newspaper that Tinsley Yard was only operating at two-thirds of its capacity, and questions were being asked about its future, with three other large yards in the area (Wath, Barrow Hill and Healey Mills) all competing for a reduced level of traffic

Mr George Johnson, the area manager, was quoted as saying that "before the recession the yard was handling up to 3,500 wagons over a twenty-four hour period, but this had dropped to an average of 2,000". There were fears that BR was to close the yard as part of an economy drive, but BR and the National Union of Railwaymen (NUR) were keen to allay these fears even though the NUR Transport Review magazine had said that closure was on the cards.

An Eastern Region spokesman was quoted as saying that "a review was being made of the functions of the yard, but these took place regularly. If it was decided that alterations were desirable, they would be discussed under normal procedures with staff representatives before any decision was made".

Unfortunately, because Tinsley had suffered such a significant drop in traffic against a high level of capital investment, it was considered costly to run.

There were claims that the two main problems with the yard were the cost of bringing it up to standard and the continued use of the Dowty retarder system. It was said that oil leakages from the Dowty equipment were causing subsidence and earth movement in the soft grey shale on which the yard was built, with BR putting the cost of repair work at £600,000. Mr Colin Schofield, the civil engineer in charge of track maintenance in Tinsley Yard, was quoted as saying "when British Railways built the yard they used ash ballast, and this effectively ensured that the shale remained stable. However, as the ash no longer became available, BR had to move on to the use of stone ballast and this did not hold the shale base, a situation that could not be envisaged when the yard was built". Roy Calvert, the divisional manager, said that "even if the track was re-laid where necessary they would still have to use the Dowty retarding system with its attendant oil losses".

According to the Star newspaper the review had been going on for some months, with no date set for its completion, but the Chairman of BR, Mr Peter Parker, said that "they had to review marshalling yard strategy in order to maintain the railways". He went on to say "in Sheffield

Table 9: Tinsley Marshalling Yard Departure Simplifier   Commencing 02/10/1972 - Sheet 1

Headcode	Depart Time	To	Days Run	Out Via	Loco Class	Basic Tons	Loco Men	Guard	Siding
7L78	00:10	Washwood Heath	MX	SE	46		304	304	9&10
9T31	00:20	Deepcar	SO	SW	76	837	RWD93	RWD93	40
3H08	00:20	Hull	MX	NW	2x20	1050	BG122	BG122	73, 74
9T49	00:24	SFT	MX	BTS	08	720	418	418	32,33,28
8L38	00:30	Warrington	MX	SW	76	837	GB6	GB6	45
7M08	00:35	Bescot	MX	SE	2x25	950	311	311	3&4
8M73	00:45	Dewsnap	MSX	SW	76	839	RWD25	RWD25	34
9T31	01:00	Deepcar	MX	SW	76	837	RWD93	RWD93	40
7N26	01:00	Tees	MX	NW			YK176	YK176	17,18
9T44	01:21	Ickles	MX	NW	37	800	403	403	29,30
3N74	01:30	Toton	MX	SE	25		317	317	1&2
8M56	01:40	Ashburys	MX	SW	76	837	RWD22	RWD22	42
7M97	02:05	Leicester	MX	SE	45	950	LE89	LE89	82,83
6M48	02:10	Teesport	MX	NW					
8M39	02:30	Derby St Marys	MX	BTS	2x20	1500	BTN30	BTN30	76
7D04	02:35	Grimsby	MX	SW	47	950	WSP9	WSOP9	62
8M09	02:45	Washwood Heath	MX	SE	25	900	DBY10	DBY10	9&10
7M29	02:50	Nottingham	MX	BTS	25	600	NOTTS585 SO 320 SX	602 SO 320 SX	84
9T94	03:20	Barrow Hill	MX	SE	37	1030	BH275	BH275	8
8D99	03:25	Scunthorpe	MX	NW	2x31	950	FROD29	FROD29	20
8D42	03:40	Doncaster	MX	NW	2x31	1050	523	523	23
9T31	03:40	Deepcar	SO	SW	76	837	RWD93	RWD93	40
8E00	03:50	Hull	MX	NW	37	950	BG121	BG121	73&74
9T69	04:11	Wath	MX	NW	37	800	WTH159	WTH159	39
7D05	03:55	Lincoln	SX	SW	37MX 31MO	650 550	323 MX 324 MO	323 MX 324 MO	66
8M26	04:00	Godley Junction	MX	SW	76LE	750	GB11	GB11	43,44,45
8D76	04:10	Worksop	MX	SW	31	550	WSP10	WSP10	67
9T49	04:15	SFT	MX	BTS	08	720	418	418	32,33
9T31	04:20	Deepcar	MSX	SW	76	837	RWD93	RWD93	40
8L11	04:45	Healey Mills	MX	NW	37	900	HM332	HM332	36,37
8P51	05:10	Whitemoor	EWD	SW	37MO 31MX	650 550	356 MO RET4 MX	603	64,65
9T30	05:35	Wadsley Bridge	SX	NW	76	837	RWD90	RWD90	41
7D04	05:40	Grimsby	MO	SW	31	550	332	332	62
9T50	05:45	Neepsend	SX	SW	47	1140	419	419	46
9T49	05:50	SFT	SX	BTS	08	720	416	416	32,33,28
8J97	06:00	Mansfield	SX	SW	31	550	329	329	81
9T41	06:00	Treeton	MSX	SE	2x08	1040	392	392	11
9T38	06:05	West Tinsley	SX	BTS	08	720	383	383	68,69,70,72
9T41	06:07	Broughton Lane	SO	NW	31	810	392	392	7
8M89	06:10	Bescot	MO	SE	45	950	303	303	3&4
8M26	06:20	Beswick	MO	SW					
6T39	06:20	Skiers Spring	SX	BTS	37AB	700	386	386	EF
9T35	06:20	Broughton Lane	SX	SW	2x08	1060	374	374	7
9T50	06:30	Neepsend	MO	SW	47	1140	419	419	46
9T40	06:35	Tinsley West	EWD	NW	37	E&B	389	389	BK.BANK
9T30	06:35	Wadsley Bridge	MO	SW	76	837	RWD90	RWD90	47
9M48	06:35	Earles	EWD	BTS	37AB	630	413	413	78
9T34	06:40	Treeton	MO	SE	2x31	1040	370	370	BK.BANK
8L12	06:55	Healey Mills	MO	NW	37	900	346	346	36,37
9T41	06:55	Woodhouse Mill	MO	SE	2x08	1040	392	392	11
8M76	07:00	Godley Junction	SX	SW	76	837	RWD47 MO GB403 MSX	RWD47 MO GB403 MSX	43,42,44
9T42	07:00	Darnall & Woodhouse	SX	SW	31	650	395	395	71
8M18	07:05	Carlisle	EWD	NW	31MO 45MX	800 900	357	357	14,15
9T49	07:10	SFT	SX	BTS	08	720	416	416	28,32,33
9T44	07:15	Aldwarke New Site	EWD	NW	37	800	401	401	26
6E37	07:20	Pontefract	MX	NW	AB				
6M03	07:30	Nottingham	MX	SE	45	850	305	305	EF
9T38	07:40	Brightside	SX	BTS	08	720	383	383	S.YD

Table 9: Tinsley Marshalling Yard Departure Simplifier Commencing 02/10/1972 - Sheet 2

Headcode	Depart Time	To	Days Run	Out Via	Loco Class	Basic Tons	Loco Men	Guard	Siding
8M48	07:45	Dewsnap	MO	SW	76	837	RWD42	RWD42	34
8M58	08:00	Ashburys	MO	SW	76	837	GB408	GB408	42
9T39	08:20	Masborough	SX	NE	2x08		377	377	12
9T35	08:45	Nunnery	SX	SW	2x08		374	374	79
7V63	09:05	Bescot	SO	SE	47	800	300	300	16
9T49	09:30	SFT	SX	BTS	08	720	416	416	28,32,33
8M48	09:30	Dewsnap/Ashburys	MSX	SW	76	837	GB1	GB1	34&35
8M24	09:50	Toton	SX	SE	2x20	1520	104	104	EF
8M10	10:00	Warrington	SX	SW	76	837	GB6 MO RWD37 MSX	GB6 MO RWD37 MSX	45
8D43	10:05	Doncaster	EWD	NW	2x31	1050	524	524	23
9M59	10:10	Peak Forest	SX	BTS	2x20	650	315	315	75
9T94	10:10	Barrow Hill	EWD	SE	37	1030	BH276	BH276	8
9T40	10:15	Meadow Hall	EWD	NW	37	690	389	389	38
8N83	10:20	Tees	MX	NW	2x20		533	533	17,18
7V63	11:15	Severn Tunnel Junction	SX	SE	47	800	300	300	16
8N85	11:35	Tyne	SX	NW	37	950	342	342	5&6
9T44	11:39	Rotherham Road	SX	NW	37	800	401	401	27
6D14	11:50	Healey Mills	SX	NW	31	850	348	348	36&37
9T42	13:06	Dunston & Barlow	SX	SW	31	670	396	396	25
9T36	13:10	Brightside	SX	BTS	08	720	384	384	68,69,70,72
8M70	13:10	Bescot	SX	SE	45	950	306	306	3&4
0T44	13:26	Aldwarke	SX	NW	37	DBV	402	402	BK.BANK
0T40	13:30	Tinsley West	SX	NW	37	DBV	390	390	BK.BANK
8D56	13:50	Scunthorpe	SO	NW	2x20		FROD110	FROD110	19,20
9T30	13:40	Deepcar	SX	SW	76	837	RWD91	RWD91	40
0T48	13:50	Burton MO Toton MSX	SX	SE	37	DBV	414	414	BK.BANK
0T41	14:00	Treeton	SX	SE	2x08	DBV	393	393	BK.BANK
8H02	14:05	Hull	SX	NW	37AB	950	340	340	73,74,77
9T38	14:20	Brightside	SX	BTS	08	720	384	384	68,69,70,72
9T69	14:34	Wath	SX	NW	37	800	WTH161	WTH161	39
9T35	14:41	Broughton Lane	SX	SW	2x08AB	1060	375	375	7
8L33	14:50	Godley Junction	SX	SW	76	837	GB25	GB25	42,43,44
9T49	15:15	SFT	SX	BTS	08	720	417	417	28,32,33
8M10	15:20	Ashburys	SX	SW	76	837	GB12	GB12	42
0T36	15:45	Masborough S/S	SX	NE	2x08	DBV	378	378	BK.BANK
8V85	15:47	Severn Tunnel Junction	SX	SE	45	1050	DBY223	DBY223	EF
6M99	15:55	Widnes	WTHFO	SW	76AB		GB203	GB203	EF
8M52	16:10	Edge Hill	SX	SW	76	837	RWD43 MSX	RWD43 MSX	43&44
9T44	16:20	Rotherham Road	SX	NW	37	800	402	402	26&27
9T30	17:00	Deepcar	SX	SW	76	837	RWD91	RWD91	40
9T40	17:30	Meadow Hall	SX	NW	37	690	390	390	38
8F67	17:30	Tees	SX	NW	31	700	343	343	17&18
8M40	18:00	Toton	SX	SE	2x20	750	316	316	1&2
8K80	18:50	Warrington	SX	SW	76	837	RWD23	RWD23	45
9T94	19:25	Barrow Hill	SX	SE	37	1030	BH278	BH278	8
6M59	19:50	Widnes	MO	SW	76				EF
8D74	20:00	Worksop	SX	SW	2x20	950	WSP8	WSP8	67
7S78	20:10	Glasgow	SX	NW	31	650	HK694	HK694	14&15
8M03	20:25	Dewsnap	SX	NW	76	837	RWD21	RWD21	34
8L16	20:26	Healey Mills	SX	SW	31	900	349	349	36&37
8T44	21:08	Ickles	SX	NW	37	DBV	403	403	BK.BANK
8D38	21:10	Scunthorpe	SX	NW	2x31	950	FROD27	FR27	19
8P37	21:15	Whitemoor	SX	SW	37	650	LIN37	LIN37	64&65
7M52	21:20	Washwood Heath	SX	BTS	25	600	302	302	9&10
6L35	21:55	Ellesmere Port	SX	SW	2x76AB		RWD	RWD	
8D41	21:55	Doncaster	SX	NW	37	950	335	335	23,75
9M60	22:00	Peak Forest	SX	BTS	40	720	BUX1	BUX1	75
8D45	22:40	Immingham	SX	NW	31	650	DCN149	DCN149	24
7C80	22:45	Temple Mills	SX	SW	37/31	950	328 SX	328 SX	63
0T49	22:50	SFT	SX	BTS	08	DBV	418	418	
6V72	23:15	Bristol	SX	SE	46	650		601	13
6E61	23:30	Tyne	SX	NW	37	700	34	34	5
8H17	23:35	Healey Mills	SX	NW	31	250	353	353	36&37
7E10	23:59	Edge Hill	SX	SW	76	600	GB5	GB5	43&44

Loco men explanation: BH=Barrow Hill, BUX=Buxton, DBY=Derby, DCN=Doncaster, FROD= Frodingham, GB=Guide Bridge, HM=Healey Mills, LIN=Lincoln, NOTTS=Nottingham, RWD=Rotherwood, WSP=Worksop, WTH=Wath. All remaining crews were Tinsley-based.

Above: On Friday 22nd April 1977, 31259 passes Lincoln Central with brand new 56025/026 in tow, en-route from Harwich to Tinsley. *Paul Cooke*

the surest future for railway workers was to have an efficient, successful and wanted railway system" and that "when the yard opened the container revolution had hardly started, but we now not only have to accept change, but to welcome it – we've actually got to have an appetite for it"

In October 1976 the review process was still ongoing. Mr Johnson was saying that "BR was moving into the 'whole train complex' – a goods train comprised of one company's products – an example being the pit-to-power station MGR trains, although they would still regard mixed traffic as an essential part of their business". He also commented on the arrival at Tinsley of seven new high-powered diesel locomotives (Class 56) in line with this policy of having heavier trains.

In November 1976, Mr Roy Calvert, BR Sheffield divisional manager, was quoted as saying "that he was meeting unions in the next three weeks to give them a progress report, and he claimed that marshalling yards throughout the region were under review and they were expecting that it would be completed by the autumn (although this was in a November newspaper), when a decision on Tinsley would be made". He said that "millions of pounds would be required to update the yard and cover ever-increasing operating costs".

Another Nail in the Coffin

In 1977 the Government began to invest in the former Firth Vickers stainless steel rolling mills at Shepcote Lane, the investment giving rise to Stainless Melting And Continuous Casting (SMACC), which would allow casting of steel to take place at Shepcote Lane rather than importing it from other works. The construction of SMACC also meant steel traffic would no longer need to use the yard as wagons could run directly into the plant, thereby hitting another nail into Tinsley Yard's coffin.

Class 56s Arrive

The Class 56 story began in 1973 when troubles in the Middle East saw a rise in oil prices, triggering a Government policy shift towards the greater use of coal. Such was the urgency for a new design of locomotive in such a short timescale, BR itself could not hope to have the design finalised in time.

Subsequently, the BRB put the design out to tender. Interest was forthcoming from American and European firms, but in the event Brush won out and an order for 30 locomotives was placed in September 1974, with an order for a further 30 locos going to BREL Doncaster.

Brush had built the prototype HS4000 so had experience with the type, but they were also in collaboration with the Romanian railway industry to develop a high-powered diesel-electric loco for possible use in Western Europe and Eastern Bloc countries. Much design work had already been undertaken so Brush was able to offer a realistic delivery date for the new locomotives, which were effectively based on the Class 47 bodyshell, but with a more powerful Ruston Paxman 16RK3CT 3520 hp engine, de-rated to 3250 hp, with state-of-the-art electrical equipment.

Testing the Engine

To test the Ruston engine and control equipment, accident-damaged 47046 was rebuilt in 1975 and equipped with a 16-cylinder Ruston 16RK3CT engine rated at 3,250 hp. Renumbered 47601, it was supposed to be the test-bed for the class but was not released from Crewe works until 4th January 1976, by which time 56001 had already arrived in the country, meaning that very little experience was gained from the exercise.

Based at Tinsley, 47601 was used on heavy coal trains so its performances could be monitored. It would later be rebuilt again as a test-bed for the Class 58 programme, becoming 47901.

Left: Prior to the arrival of the Class 56, the Class 47 was the mainstay on MGR duties in the area. Accompanied by Barry Armstrong, Roger Wainwright hangs on to a Class 47 in the Servicing Shed. *Roger Wainwright*

Below: A Class 20 stands on the fuelling point outside the main shed with Steve Holley, Richard Butcher, and Roger Wainwright on the buffer beam. *Roger Wainwright*

Class 56 Arrives at Tinsley

56001 was completed in April 1976 and tested on trains around Bucharest and Craiova, after which it was hauled (along with sister 56002) from Romania to Belgium, arriving in Zebrugge on 3rd August 1976. Both locomotives were handed over to BR the following day, before being loaded onto the train ferry for Harwich.

After being examined for fitness to run, they were hauled from Harwich to Tinsley TMD by 31318 on 7th August 1976.

At Tinsley, the locos were thoroughly examined with a fine tooth comb, which revealed a whole host of faults arising from poor workmanship, along with major bogie problems. Extensive rectification work was required before the locos could be put into traffic, and whilst awaiting attention they were stored at Barrow Hill, Shirebrook and Darnall.

On 24th September 1976, 56001 worked a test train of 26 loaded MGR hoppers from Tinsley to Appleby via Treeton Junction and Settle, thus becoming the first class member to work in this country. Unfortunately, it developed axle box problems which required an axle change. The loco was sent to Doncaster works for rectification. 56001 worked between Doncaster and Peterborough with a test train on 22nd/23rd February 1977 and was taken into stock at Tinsley on 28th February 1977, being sent to Derby for further testing in June, before again returning to Tinsley. The loco would finally enter service in November 1978.

56006 and 56007 arrived at Harwich on 10th September 1976 and were towed to Tinsley for storage. 56006 worked a commissioning trial on 6th December with 6Z25, 10.10 Tinsley to Blea Moor, consisting of a train of loaded hoppers, and was sent to Shirebrook for artisan training purposes. After modifications it was accepted into 75% BR ownership on 25th February 1977, and became the first of the class to haul a revenue earning service on the national network on 27th February 1977.

The locos were also sent out on test runs with a train consisting of Pullman car E330 and 12 BBA wagons (air-braked bogie bolsters) loaded with rails. This would run as 4X31, 10.00 Broughton Lane to Peterborough West Yard.

56025/026 were noted at Lincoln on 22nd April behind 31259, en-route from Harwich to Tinsley, stopping overnight at the former Lincoln steam depot sidings.

The class continued to be rigorously checked and sent for modifications for months to come, with the final loco (56030) arriving at Harwich on 26th August 1977, although it wasn't moved to Tinsley until 14th September 1977. It would be taken into stock on 21st October 1977.

Once commissioned into traffic at Tinsley, the 56s began taking over from Classes 20, 37 and 47 on coal trains in the Nottinghamshire and Derbyshire coalfields, out-based at Shirebrook, Barrow Hill and Worksop. The 56s' additional horsepower permitted increased loadings, and before long they were typically hauling between 36 and 42 HAA wagons.

Drivers Between Duties

Above left: Top shed cabin. Bill Barlow (left) and Dick Osbourne (otherwise known as fizzing Dick!). *Roger Wainwright*

Top right: From left to right: Les Craig, Les Middleton (ex-Rotherwood driver) George Hirst (train crew supervisor), Paul Kenworthy (guard). *Roger Wainwright*

Above right: David Haigh, Alan Beardshaw and Les Craig undergo a Saturday afternoon bike-washing session on Tinsley top shed. These were shed drivers waiting for locos to come on shed or for shunting on top shed. *Roger Wainwright*

Below: Sleeping beauties in Tinsley train crew mess room! Left to right: Unknown, guard Paul Kenworthy, guard Roger Marsh and an unknown ex-Wath guard. *Roger Wainright*

Left: The opening of Tinsley TMD in 1964 saw drivers transferred from Darnall, Grimesthorpe and Millhouses to the new depot. This caused issues for staff who did not have their own transport and lived close by their former depots. To alleviate the problem, BR set up a bus service from the outlying areas to Tinsley in 1964. Initially restricted to railway and construction staff only, this restriction was removed in 1968. There were several routes and different operators over the years, route 287 operating from July 1973 on a 24-hour basis, departing from Sheffield Midland station. Here, a Sheffield Corporation double-decker bus waits for passengers in the TMD car park. Note the classic Hillman Imp in the car park. *Peter White*

Right: Another operator was National Travel, which apart from its own coaches used East Midlands Motor Services and Mansfield District vehicles on hire, both also part of the National Bus Group. Route 899 ran from the front of Sheffield Midland station to Wood Lane and was available to any staff. Very often there would only be one person on it! Here, in 1981, Mansfield District fleet No. 548, Bristol RE GVD 548K waits at Midland station for passengers, which on this day by the looks of it, were very few! *Steve Ryszka*

Left: East Midlands fleet No. 18, UAU 18S, waits for passengers on Wood Lane in 1981. This was a 49-seat bus with a Duple body, new in August 1977, which later moved to Rainworth Travel. It was sold on to Toohey in Nenagh, Ireland in December 1991 and re registered as 77 TN 53. Sold on again to Lesclacha, Newmarket-on-Fergus, Ireland in May 1993, it had been scrapped by April 1999. *Steve Ryszka*

Chapter 18

Total Operations Processing System (TOPS)

In 1968 British Rail became interested in the Total Operations Processing System (TOPS) used in the USA, and in the ongoing effort to increase efficiency decided to purchase the system – with Government approval.

TOPS was a computer-based data information system that recorded the movement and location of all locomotives and rolling stock on the railway network. The idea was that every train would have its details committed to the computer system, and could be tracked wherever it went. Each train would have a consist, showing the train headcode, the loco number and all the wagon numbers in the order they were placed in the train. The consist would also show train length and weight, details of the commodities being carried and details of any dangerous goods being transported. It would also show any wagons with faults that required rectification. TOPS could give details of all the wagons in a marshalling yard along with all the locos on a diesel depot.

When TOPS became fully operational it gave real time information as to where every item of rolling stock was located. This immediately reduced the number of "lost" wagons on the network and highlighted the overabundance of available wagons, precipitating the mass withdrawal of wagons, especially older unfitted ones.

In the early 1970s TOPS was being readied for nationwide coverage and on 27th October 1975 journalists were invited to Tinsley Yard to take a look around the new TOPS office which, on the day of commissioning, was manned by Lawrence Hollan, Robert Whitham and Eirleen Thompson.

The TOPS process began with a shunter compiling a list of wagon numbers making up a train, which he would then fax or telephone to the TOPS clerk in the office, who in turn would input the information into the computer. A train was not allowed out of the yard until the TOPS list was completed and a copy obtained by the guard of the train.

Illustrated is a typical enquiry sheet as printed for each train, in this case for 8E73, Dewsnap to Tinsley, hauled by 76035. The train consists of 18 loaded and 14 empty wagons of types MCV, MCO, HTO and HTV, with CAO brakevan B953603 at the rear. The total train weight was 507 tons and the train 682 feet long, or 732 feet including the locomotive.

A single Class 76 loco was permitted to haul a maximum load of 750 tonnes over Woodhead. Just 43 tonnes brake force indicates there were brakes on the loco only, so the loco would have to use its regenerative braking system on the descending gradients. Two Class 76s in multiple could handle 1,500 tonnes comfortably when working fully-braked MGR trains destined for Fiddlers Ferry Power Station.

To enable the train to run as Class 8, with a maximum speed of 35 mph, the wagon "Instanter" couplings would be set in the short position, ensuring the wagons were close-coupled. The train would be able to load more wagons if it were to run totally unfitted as Class 9, but the couplings would have to be set in the long position and the train would have a maximum speed of 25 mph.

The following is an explanation of the codes used, reading across the top line of wagons:

- B - Wagon Initial
- 593064 - Wagon Number
- H - Loading Status (Heavy/Medium/Light/Empty)
- MCV - Wagon Type
- 20 - Wagon Weight in Tonnes
- 35 - Maximum Speed
- 1 - Route Availability
- 23 - Wagon Length
- 11 - Tonnes Brakeforce, (0 = unfitted)
- 250 - Tag (a routing instruction) 250 is the area code for Tinsley,
- 25511 - Destination (BSC River Don Works)
- ISSUNG - Commodity Code (Iron and Steel Scrap ungraded) ISS171/72/76 Iron and Steel Scrap of various grades
- 33046 - Originating Point
- G - Indicates that it is in need of a minor repair
- BSC – Customer – British Steel Corporation

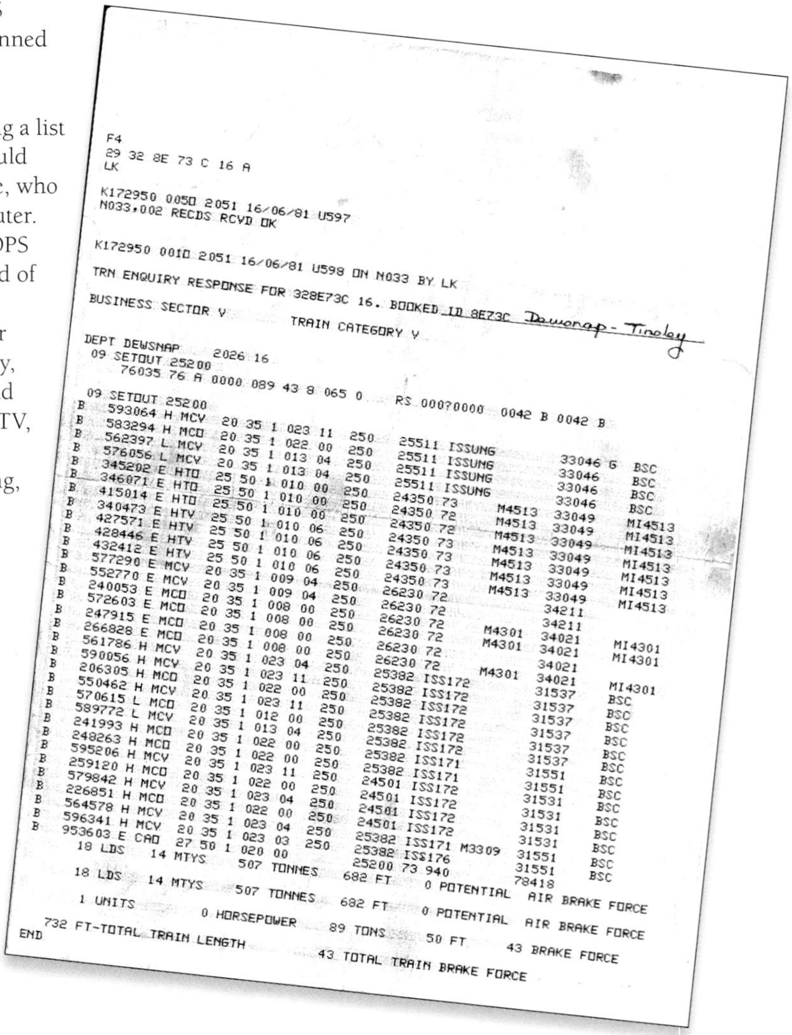
F4
29 32 8E 73 C 16 A
LK

K172950 0050 2051 16/06/81 U597
N033,002 RECDS RCVD OK

K172950 0010 2051 16/06/81 U598 ON N033 BY LK
TRN ENQUIRY RESPONSE FOR 328E73C 16. BOOKED ID 8E73C Dewsnap - Tinsley
BUSINESS SECTOR V TRAIN CATEGORY V
DEPT DEWSNAP 2026 16
09 SETOUT 25200
76035 76 A 0000 089 43 8 065 0

09 SETOUT 25200 RS 00070000 0042 B 0042 B
B 593064 H MCV 20 35 1 023 11 250 25511 ISSUNG 33046 G BSC
B 583294 H MCO 20 35 1 022 00 250 25511 ISSUNG 33046 BSC
B 562397 L MCV 20 35 1 013 04 250 25511 ISSUNG 33046 BSC
B 576056 L MCV 20 35 1 013 04 250 25511 ISSUNG 33046 BSC
B 345202 E HTO 25 50 1 010 00 250 24350 73 M4513 33049 MI4513
B 346071 E HTO 25 50 1 010 00 250 24350 72 M4513 33049 MI4513
B 415014 E HTO 25 50 1 010 00 250 24350 72 M4513 33049 MI4513
B 340473 E HTV 25 50 1 010 06 250 24350 73 M4513 33049 MI4513
B 427571 E HTV 25 50 1 010 06 250 24350 73 M4513 33049 MI4513
B 428446 E HTV 25 50 1 010 06 250 24350 73 M4513 33049 MI4513
B 432412 E HTV 25 50 1 010 06 250 24350 73 M4513 33049 MI4513
B 577290 E MCV 20 35 1 009 04 250 26230 72 34211
B 552770 E MCV 20 35 1 009 04 250 26230 72 34211
B 240053 E MCO 20 35 1 008 00 250 26230 72 M4301 34021 MI4301
B 572603 E MCO 20 35 1 008 00 250 26230 72 M4301 34021 MI4301
B 247915 E MCO 20 35 1 008 00 250 26230 72 34021
B 266828 E MCO 20 35 1 008 00 250 26230 72 M4301 34021 MI4301
B 561786 H MCV 20 35 1 008 00 250 25382 ISS172 31537 BSC
B 590056 H MCV 20 35 1 023 04 250 25382 ISS172 31537 BSC
B 206305 H MCO 20 35 1 023 11 250 25382 ISS172 31537 BSC
B 550462 H MCV 20 35 1 022 00 250 25382 ISS172 31537 BSC
B 570615 L MCO 20 35 1 023 11 250 25382 ISS172 31537 BSC
B 589772 L MCV 20 35 1 012 00 250 25382 ISS172 31537 BSC
B 241993 H MCO 20 35 1 013 04 250 25382 ISS171 31551 BSC
B 248263 H MCO 20 35 1 022 00 250 25382 ISS171 31551 BSC
B 595206 H MCV 20 35 1 022 00 250 24501 ISS172 31531 BSC
B 259120 H MCO 20 35 1 023 11 250 24501 ISS172 31531 BSC
B 579842 H MCV 20 35 1 022 00 250 24501 ISS172 31531 BSC
B 226851 H MCO 20 35 1 023 04 250 24501 ISS172 31531 BSC
B 564578 H MCV 20 35 1 022 00 250 25382 ISS171 M3309 31551 BSC
B 596341 H MCV 20 35 1 023 04 250 25382 ISS176 31551 BSC
B 953603 E CAO 20 35 1 023 03 250 25200 73 940 78418
18 LDS 14 MTYS 27 50 1 020 00 507 TONNES 682 FT
18 LDS 14 MTYS 507 TONNES 682 FT 0 POTENTIAL AIR BRAKE FORCE
1 UNITS 0 HORSEPOWER 89 TONS 50 FT 0 POTENTIAL AIR BRAKE FORCE
732 FT-TOTAL TRAIN LENGTH 43 BRAKE FORCE
END 43 TOTAL TRAIN BRAKE FORCE

The country was divided up into TOPS Responsibility Areas (TRA), the system giving every siding, station, junction, yard and depot a unique five digit STANOX number (or station number). Numbering commenced in the far north of Scotland at Thurso, with a STANOX of 01001, and finished at Dollands Moor with 89748; numbers 00001–00950 were used for continental locations. The Sheffield Division was split into the TRAs of Tinsley, Wath, Barrow Hill and Shirebrook; the accompanying diagram shows Tinsley TRA, along with the STANOX numbers for each siding in the area. All STANOX numbers within a TRA began with the same first two digits for example those within the Worksop TRA started 26 while those in the Tinsley TRA started 25.

Table 1 shows the STANOX numbers for the wagons in the featured train list.

Table 1: STANOX Numbers In 8E73 Train Enquiry

24350	Wath Yard
24501	Aldwarke New Site
25200	Tinsley
25511	BSC River Don Works
25382	Deepcar
26230	Worksop "Down" Reception

The really clever part of TOPS was the use of "tags" to provide shunting staff with information about how a wagon was to get from its originating siding to its destination.

An example might be a wagon released from Quidhampton (near Salisbury) STANOX 86124, destined to Corpach paper mill near Fort William, STANOX 06011. When the wagon is released, TOPS applies a 3-digit tag number which tells the staff where to send it. An air-braked 60mph (i.e. Speedlink capable) vehicle will get an ABN (air-braked network) tag, probably 862. The tag number is made up of the first two digits of the next yard the wagon will be heading for – in this case Eastleigh, STANOX 86090, then "2" denoting a Speedlink service.

As soon as the wagon leaves Quidhampton it will pick up a 652 tag, which tells the shunter at Eastleigh to marshal the wagon with Speedlink services destined for Bescot, STANOX 65714. Similarly, Bescot TOPS office would receive a consist showing an 072 tag on this wagon, indicating it is destined for Mossend, STANOX 07361. On arrival at Mossend the wagon would probably get a 9xx tag meaning it is a local movement within their TRA.

In terms of marshalling trains there were strict guidelines in place, such as those applying to 6S73 from Bescot, which would be marshalled with 072 tags next to the loco, followed by 092s for Carlisle (STANNOX 09151), then 352s for Warrington Arpley (STANOX 35545).

Tagging, control of the order of wagons and the means to request specific numbers of particular types of empty wagons made for highly sophisticated utilisation, routing of loads and redistribution of empties that few outside the TOPS function really appreciated.

TOPS as an efficiency tool helped to improve wagon fleet utilisation. The figures speak for themselves; whilst tonnage carried by the network had only dropped by 4% from 178 million in 1974 to 170 million in 1977, the wagon fleet had been reduced by 30%, from 241,000 to just 167,000.

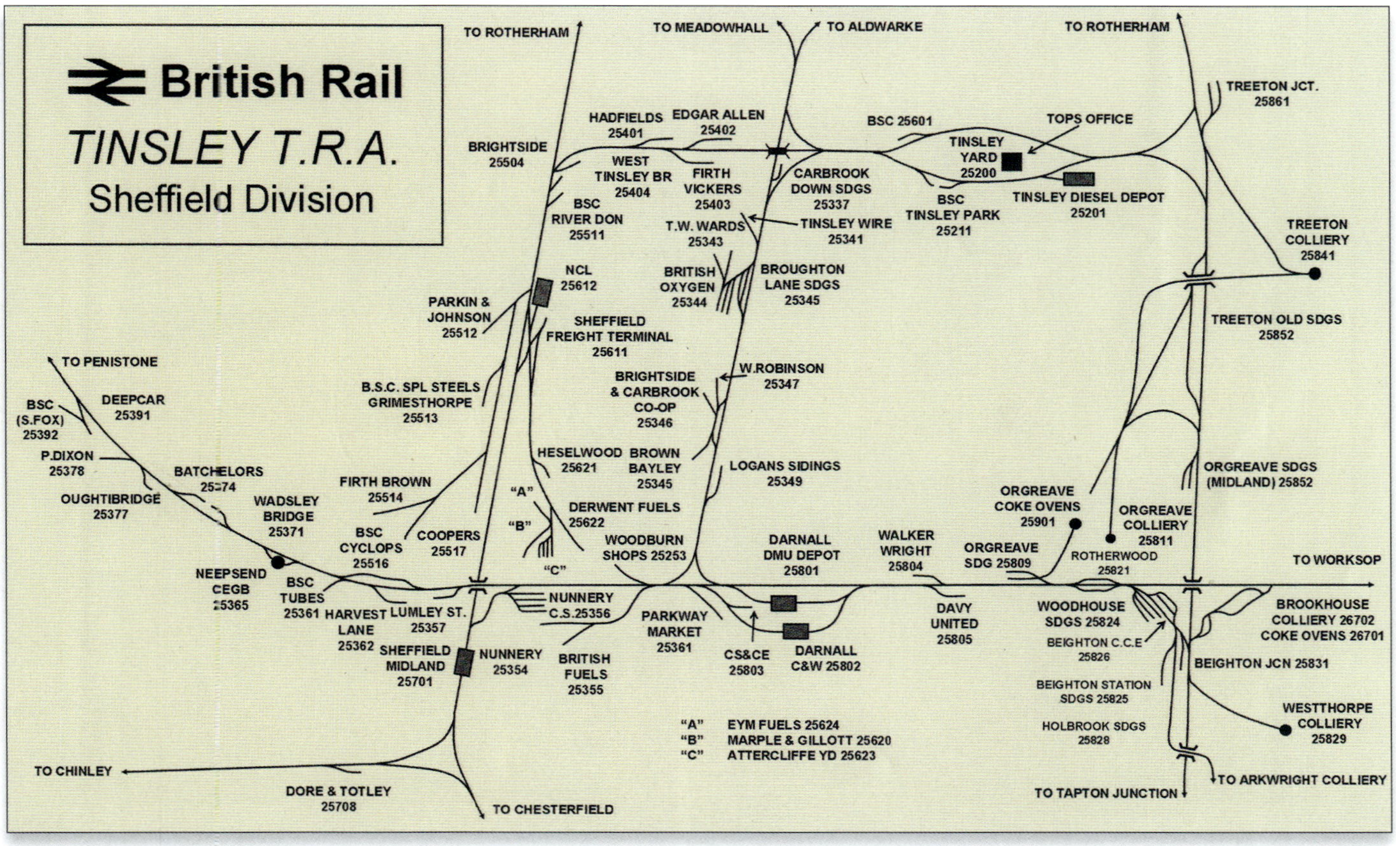

Above: Plan of the Tinsley TOPS Responsibility Area showing STANOX numbers for each location. *Chris Booth collection*

Above: An unidentified Class 25 trundles along the main yard east departure line on a snowy day in 1983. The secondary yard east departure line comes in from its dive-under to the left. On the right a train is on the east departure line heading towards Catcliffe. *Steve Ryszka*

Above: 40034 and now preserved 40118 in the reception sidings at Tinsley on 28 August 1983. The gradient towards the hump can be seen. *Steve Ryszka*

Chapter 19

Tinsley in the 1980s

After the huge investment of the 1960s and steady decline of the 1970s, the 1980s were a time for taking stock of infrastructure and asset requirements in the Sheffield area.

Scrap Metal

By the end of the 1970s the problem of an ageing scrap-carrying wagon fleet had come to a head and, although a prototype air-braked wagon had been built in 1976, it wouldn't be until 1982 that a production batch of 20 would be built by the Standard Wagon Co for owners, Railease.

Scrap traffic was initially gathered from over a dozen different collection points, but by May 1983 the wagonload network was diminishing; by May of the following year, what scrap flows remained were transferred to Speedlink.

It was estimated that the cost of replacing 2,000 unbraked and vacuum-braked scrap-carrying wagons would have been £32 million in the early 1980s, equivalent to £16,000 per new wagon. BR would only agree to buy wagons if a 10-year contract for transporting scrap metals was signed by the Scrap Federation.

In 1984, United Engineering Steels' plants at Aldwarke and Stocksbridge (Deepcar) won a £2.52 million grant towards the cost of 160, two-axle scrap-carrying wagons. These new wagons received the TOPS code of POA and were pooled with a further 21 wagons already in service. A further 163 51-tonne wagons for scrap traffic were due in service by October 1984, enabling the elimination of older life-expired vacuum-braked wagons, other than those used in the immediate Sheffield/Rotherham area. From 1st November 1984, all scrap conveyed in Speedlink services was in air-braked wagons, whilst the few trainload flows in operation tended to use vacuum-braked stock. A notable example was the flow from Halewood to Aldwarke, which was still using vacuum-braked MXV wagons as late as 1987.

In 1989, Trainload Metals set up its own network of less-than-trainload traffic which covered the majority of scrap metal flows. The network was unsustainable however and by 1993 all "common-user" metals services had been discontinued. Since then, only trainload services have run.

The Railease wagon fleet was disbanded in 1990 and all 181 wagons passed into BR ownership under TOPS code SSA. The small fleet of ex-HEA domestic coal hoppers also used for carrying scrap were recoded HSA, these being used on a variety of flows. Until 1991 the operating base for these workings was Tinsley Yard, with trip workings to Aldwarke and Deepcar, but this then altered to concentrate on Aldwarke with workings to Halewood and Workington ceasing.

The Last Years of Electric Trains and Closure of the Woodhead Route

The final years of the MSW route saw relatively few electric-hauled services originating or terminating at Tinsley.

One example was 6Z44, Northwich to Tinsley which, on 3rd July 1981, was headed by 76007 and 76012. The train consisted of three empty wagons originating from Oakleigh sidings at Northwich and destined for Wath Yard (24350), plus 23 loaded MCV wagons originating from Dee Marsh BSC, all destined for either Deepcar Stocksbridge works (25382) or Tinsley Park works (25211). Typically, this train would have worked over Woodhead, through the closed Sheffield Victoria station to Woodburn Junction, where it would have been routed towards Tinsley. After taking the electrified south western arrival line around the south side of Tinsley Yard, and under Wood Lane bridge, it would have entered one of Tinsley's electrified reception sidings, numbered six to eleven. The Class 76s would have been released from the train and run into the train engine/pilot loco spur behind Tinsley Yard signal TY213, from where they would have gone via the engine line towards Tinsley Park signal box. From here they could have returned light engine, possibly to Rotherwood, or gone to pick up another electric freight from the main yard's electrified sidings 34 to 46, or from the express freight sidings which were also electrified. The wagons themselves would be hump shunted, the Deepcar traffic being routed into siding No. 36 for a later timetabled service to take them forward. Wath Yard also had its own siding, No. 80 in the secondary yard, so the three empty wagons would have been routed via siding No. 50, the continuous mechanical feed road, and the hump summit of the secondary sorting sidings. They would be tripped to Wath as part of Tinsley's trip diagrams for local traffic.

Table 1 lists a few other electric-hauled services in June and July 1981.

```
K172950 0010 2148 03/07/81 U327 ON N080 BY KC

TRN ENQUIRY RESPONSE FOR 376Z441 03. BOOKED ID 6Z441  Northwich - Tinsley

BUSINESS SECTOR V          TRAIN CATEGORY V

DEPT GODLEY       2112 03
 09 SETOUT 25200
     76007 76 X 0000 089 43 8 065 0    RS 00070000  0071 A 0147 C
     76012 76 X 0000 089 43 8 065 0    RS 00070000  0048 B 0048 B

 09 SETOUT 25200
B   565970 M MCV  20 35 1 017 04  250   25382 ISS171          39003    BSC
B   558573 M MCV  20 35 1 017 11  250   25211 ISS171          39003    BSC
B   589576 M MCV  20 35 1 017 11  250   25382 ISS171          39003    BSC
B   574519 M MCV  20 35 1 017 04  250   25382 ISS171          39003    BSC
B   555710 M MCV  20 35 1 017 11  250   25211 ISS171          39003    BSC
B   551649 M MCV  20 35 1 017 11  250   25382 ISS171          39003    BSC
B   577974 H MCV  20 35 1 023 04  250   25382 ISS171          39003    BSC
B   590239 M MCV  20 35 1 017 11  250   25211 ISS171          39003    BSC
B   564[illegible]4 M MCV  20 35 1 017 04  250   25211 ISS171          39003    BSC
B   596[illegible]32 M MCV  20 35 1 019 03  250   25211 ISS171          39003    BSC
B   589402 M MCV  20 35 1 017 11  250   25211 ISS171          39003    BSC
B   593196 H MCV  20 35 1 023 11  250   25211 ISS171          39003    BSC
B   568144 H MCV  20 35 1 023 04  250   25382 ISS171          39003    BSC
B   573381 M MCV  20 35 1 017 04  250   25382 ISS171          39003    BSC
B   550351 M MCV  20 35 1 017 11  250   25382 ISS171          39003    BSC
B   556783 M MCV  20 35 1 017 11  250   25382 ISS171          39003    BSC
B   574956 M MCV  20 35 1 017 04  250   25382 ISS171          39003    BSC
B   266331 M MCV  20 35 1 017 11  250   25382 ISS171          39003 G  BSC
B   552568 H MCV  20 35 1 023 11  250   25382 ISS171          39003    BSC
B   552054 H MCV  20 35 1 023 11  250   25382 ISS171          39003    BSC
B   552358 H MCV  20 35 1 023 11  250   25382 ISS171          39003    BSC
B   561013 M MCV  20 35 1 017 04  250   25382 ISS171          39003    BSC
B   550251 M MCV  20 35 1 017 11  250   25382 ISS171          39003    BSC
B   428893 E HTV  25 50 1 010 06  250   24350 72       M4513  37122    MI4513
B   421925 E HTV  25 50 1 010 06  250   24350 72       M4513  37122    MI4513
B   419216 E HTV  25 50 1 010 06  250   24350 73       M4513  37122    MI4513
      23 LDS    3 MTYS     459 TONNES   535 FT  207 POTENTIAL  VAC BRAKE FORCE

      23 LDS    3 MTYS     459 TONNES   535 FT  207 POTENTIAL  VAC BRAKE FORCE

      2 UNITS        0 HORSEPOWER   178 TONS    100 FT     86 BRAKE FORCE

  635 FT-TOTAL TRAIN LENGTH        293 TOTAL TRAIN BRAKE FORCE
```

Above: A typical train enquiry, this example shows 6Z44, Northwich to Tinsley. *Kim Collinson collection*

Table 1: Other Examples of Electric-Hauled Trains in June/July 1981

09/06/81	76009+76011	8M11	Tinsley to Ashburys
11/06/81	76051	8M38	Tinsley to Warrington
12/06/81	76014	8M48	Tinsley to Ashburys
16/06/81	76035	8E73	Dewsnap to Tinsley
17/06/81	76054	8E73	Dewsnap to Tinsley
19/06/81	76040	8M11	Tinsley to Ashburys
19/06/81	76051	8E73	Dewsnap to Tinsley
22/06/81	76054	8E73	Dewsnap to Tinsley
25/06/81	76051	8E73	Dewsnap to Tinsley
25/06/81	76006+76024	8M11	Tinsley to Ashburys
26/06/81	76026+76021	8E11	Dewsnap to Tinsley
26/06/81	76007+76012	8M38	Tinsley to Warrington
29/06/81	76040	8M11	Tinsley to Ashburys
01/07/81	76014	8E07	Dewsnap to Tinsley
01/07/81	76025+76027	8E46	Warrington to Tinsley
03/07/81	76028	8E73	Dewsnap to Tinsley
03/07/81	76007+76012	6Z44	Northwich to Tinsley
09/07/81	76028	8E07	Dewsnap to Tinsley

Another example of an electric-hauled train, on 19th June 1981, was 8E73 Dewsnap to Tinsley hauled by 76051. This consisted of 17 loaded and 13 empty wagons and included: a single OHV steel open wagon from Horwich Works (30301) to Darnall (25802), 16 MCVs and MCOs from Pendleton (31551) which were all destined for either Stocksbridge steelworks (25382) or Aldwarke New Site (24501), and 11 wagons from Ardwick (32007) for Seymour Junction (27121), near Barrow Hill.

After closure of the Woodhead Route, the working timetables were recast and many services around the Tinsley area ceased. All of the Ashburys, Dee Marsh, Dewsnap, Garston, Glazebrook, Guide Bridge and Warrington to Tinsley services were discontinued. Mansfield Concentration Sidings to Garston Docks services were diverted via the Hope Valley and thus no longer served Tinsley or Rotherwood yards; all of the Rotherwood trains were curtailed.

Table 2 shows the net change in traincrew requirements after the closure of Woodhead. Healey Mills was one of the main beneficiaries, as this depot took over MGR traffic that had formerly travelled via the Woodhead Route. Tinsley depot also gained from Woodhead's closure, as some services switched to diesel haulage over the Hope Valley route; previously all electric services out of Tinsley had been worked by Guide Bridge or Rotherwood-based train crews.

Table 3 shows the effect the closure of Woodhead had on the number of train crew employed. Present

Table 2: Net Effect Upon Traincrew Requirement Arising from Woodhead Closure

Trunk Hauls	Drivers	Assistants	Guards
Sheffield	-	-	-
Shirebrook	-2	-	-2
Tinsley	+5	+1	+7
Masborough	-1	-	-1
Rotherwood	-9	-4	-9
Wath	-7	-1	-7
Healey Mills	+6	-	+6

Trip Working	Drivers	Assistants	Guards
Sheffield	-	-1	-
Shirebrook	-	-	-
Tinsley	+2	+2	+6
Masborough	-	-	-
Rotherwood	-3	-3	-7
Wath	-4	-3	-1
Healey Mills	-	-	-

establishment total is the number of men a depot should have had on its books prior to the closure of Woodhead, whilst the proposed number was what they should each have had post-closure, thus giving the difference in a depot's establishment. In reality, the actual number of men at a depot was often out of kilter with the required number and the actual difference post-Woodhead closure would often be much greater, due to depots being over-staffed initially.

There were changes to the opening times of signal boxes at Deepcar and Sheffield Victoria No. 3, which saw positions removed and subsequent savings in wages, along with the closure of Rotherwood Sidings signal box. The closure of Rotherwood Yard also saw the loss of the following positions:

1 x Liaison Clerk Class C
3 x Supervisor Grade C
6 x Chargemen
3 x Leading Railman
1 x Junior Railman (Messenger)

along with relief staff based at Woodhouse.

Right: 45044 passes through Tinsley with a rake of depressed centre Presflo PCA wagons on 14 June 1983. *Steve Ryszka*

Table 3: Proposed Closure of MSW – Effect on Train Crew

	Establishment				Actual	
Drivers	**Present**	**Proposed**	**Difference**	**£ Variation**	**Present**	**Actual Difference**
Tinsley	172	183	+11	+64,2070	183	-
Rotherwood	28	-	-28	-163,436	28	-28
Wath	77	70	-7	-40,859	99	-29
Other depots*	667	672	+5	+29,185	688	-16
Total	944	925	-19	-110,903	998	-73
	Establishment				Actual	
Drivers Assistants	**Present**	**Proposed**	**Difference**	**£ Variation**	**Present**	**Actual Difference**
Tinsley	75	78	+3	+14,778	78	-
Rotherwood	17	-	-17	-83,742	11	-11
Wath	35	34	-1	-4,926	36	-2
Other depots*	315	316	+1	-4,926	297	+19
Total	442	428	-14	-68,964	422	+6
	Establishment				Actual	
Guards	**Present**	**Proposed**	**Difference**	**£ Variation**	**Present**	**Actual Difference**
Tinsley	91	106	+15	+68,250	90	+16
Rotherwood	28	-	-28	-127,400	23	-23
Wath	56	49	-7	-31,850	58	-9
Other depots*	397	402	+5	-22,742	383	+19
Total	572	557	-15	-68,258	554	+3

* Other depots includes: Shirebrook, Sheffield, Doncaster, and Healey Mills

Sheffield Victoria No. 3 was closed altogether from 5th November 1983 and the former Deepcar lines to Penistone were renamed Stocksbridge "up" and Stocksbridge "down". Sheffield Victoria No. 4 box remained open to control movements into Nunnery carriage sidings, but use of the sidings diminished, and with the loss of its Class 08 pilot in September 1983, staff within the sidings were confined to just one shift. The sidings officially closed on 5th February 1984 but continued to see use as a store for condemned coaching stock.

Below: The reduction in traffic through Tinsley Yard allowed the secondary yard sidings to be used for storage of condemned rolling stock, many vehicles containing blue asbestos. In this view taken on 21st November 1983, lines of withdrawn DMUs, including examples of Gloucester RC&W Class 100, keep company with some Mark 1 sleeping cars and a Mark 2 coach. The stock would be stripped of spares and completely boarded up at Darnall to prevent access, then tripped into the yard to await later movement to Snailwell. By the end of 1983 over 400 coaches had been stored in the yard, together with 30 non-blue asbestos coaches that had been offered for sale, but not found a buyer. These eventually went to Horwich works for disposal. *Les Nixon*

Left: Having deposited a rake of "seacow" wagons in the yard, 25195 leads an unidentified Class 25/0 at Tinsley Park in April 1983. This loco entered service on 6 May 1965 as D7545 and would be withdrawn on 14 June 1985, being cut up at Vic Berry's, Leicester on 31st December 1986. *Steve Ryzska*

Sectorisation

From 4th January 1982 a re-organisation process known as "Sectorisation" was instigated which it was said "would turn BR into a more efficient, customer-focused organisation through shaping the services it offered in response to customer demands, whilst at the same time moves would be made to correlate costs and revenues more closely to give each sector and their managers greater accountability".

Whilst chronic under-usage from the start and continued loss of traffic from BR as a whole may be at the heart of Tinsley's decline, BR's policy of Sectorisation accelerated its ultimate demise, as Tinsley's locomotive fleet went through a significant transformation.

Depot allocations began to reflect specific sectors. Toton, for example, became principally Railfreight-orientated, whilst Tinsley leant towards Freightliner. As a result, most of Tinsley's Class 37s for steel and general traffic were transferred away, leaving just a handful to work Freightliner diagrams until Class 47s had been fitted with multiple-working equipment.

As Railfreight split further into sub-sectors, the newly introduced Class 58s were all allocated to Toton and by late-1986 began dominating coal traffic at the expense of Tinsley-allocated Class 56s. Tinsley's Class 56s subsequently became part of Trainload Coal and were thus re-allocated to Toton.

As more Class 47s were transferred to Tinsley so more Class 45s became surplus to requirements. 45105/132/133 were all withdrawn on 11th May 1987 and were to join 45111/112/114/118/119/122/125/126/136/138/139/143/146/148 stored in the yard to await their fate. In late May 1987 the first of six convoys of retired "Peaks" moved from Tinsley to March depot for storage.

Settle & Carlisle Line Sale

During the 1970s, the Settle & Carlisle Line had suffered from a lack of investment, with many viaducts and tunnels falling into a state of disrepair. By the early 1980s, the line was carrying only a handful of trains per day, and British Rail decided the cost of renewing the viaducts and tunnels would be prohibitively expensive; line closure notices were issued in 1984. A huge campaign was fought to save the line but, as late as August 1988, the British Railways Board posted advertisements stating they had 'appointed Lazard Brothers to advise on the sale of the line'. This purported sale was to include at least five Class 45 locomotives, which were to be refurbished and re-classified as Class 45/2.

Tinsley's involvement extended to keeping these locomotives maintained in running order; at least two examples, 45115 and 45128, were started up with official approval in late 1988. The government finally refused consent to close the line in 1989, and British Rail started to repair the deteriorating tunnels and viaducts.

End of the "Peaks"

The paintbrush was wielded on 45106 at Tinsley and it became the depot's celebrity loco! It was repainted into a green livery and saw use on Sheffield to St Pancras InterCity services, along with many enthusiasts' specials and open day visits. Unfortunately, it sustained fire damage whilst working the 07.12 Derby–St Pancras service, and was withdrawn on 20th February 1989. Tinsley depot staff were keen for a "Peak" to continue in service, so 45128 was reinstated the same day, but after undergoing an unsuccessful load bank test the locomotive was withdrawn again on 22nd April; that was the end of "Peaks" in everyday service.

Condemned Stock

The reduction in traffic allowed some sidings to be used for storing withdrawn rolling stock, awaiting the call to either the Mayer Newman scrapyard at Snailwell or Vic Berry's yard in Leicester. The large amount of stock being stored led to the introduction of a special 9P40, Thursdays-only, 19:45 Tinsley to Snailwell service. This ran via Darnall and Worksop to Retford, then via the ECML to Peterborough; it was the only Class 9 service allowed to run on the ECML, running at 60 mph. If, however, the train had DMU vehicles in the consist, it would run as 9X40 and be routed via Lincoln and Sleaford to Peterborough. The "X" designation indicated out-of-gauge vehicles in the consist, as the DMU vehicles rode higher after being stripped of engines and gearboxes. Table 4 lists the DMUs and coaching stock stored in the yard in June 1983.

Table 4: Withdrawn DMUs and Coaching Stock in Tinsley Yard on 3rd June 1983

DMUs	50340/372/385/526/543/585/788/792/815, 51117, 55989, 56118/157/159/163, 59147/158/161/162/163/164/179/180/181/182/198
Coaching Stock	1620, 4298, 4321, 4323, 4518, 4569, 4592, 9376, 15955, 15967, 25344, 26213, 35411

Table 5: Speedlink Trunk Services through Tinsley 13th May 1985 to 28th September 1985

Headcode	Train	Arrive	Depart	Days
6E38	21.33 Bescot to Doncaster Belmont	00.17	02.00	SX
6E64	19.15 Severn Tunnel Jcn to Haverton Hill	00.28	01.51	SX
6E76	23.40 Toton to Tinsley	01.03		SX
7J29	00.46 Doncaster to Tinsley	01.33		MX
6M87	02.31 Tinsley to Toton		02.31	MX
7M88	01.27 York Dringhouses to Toton	02.56	03.50	MX
6J62	20.26 Parkeston Quay to Tinsley	03.28		SX
6L81	04.10 Tinsley to Hull Saltend		04.10	MX
6E75	21.24 Willesden to Hunslet	04.36	05.16	SX
7E63	02.10 Warrington Walton Old Jcn to Tinsley	05.33		MX
6J60	06.30 Tinsley to Barrow Hill		06.30	SX
6J61	13.11 Barrow Hill to Tinsley	13.34		SX
6V06	16.48 Healey Mills to Severn Tunnel Jct	17.54	18.41	SX
7E49	16.54 Toton to York Dringhouses	18.13	19.14	SX
7D63	17.49 Healey Mills to Worksop	19.05	20.40	SX
6M91	20.55 Tinsley to Whitemoor		20.55	SX
7D19	21.44 Tinsley to Doncaster		21.44	SX
6M70	21.56 Tinsley to Warrington Walton Old Jcn		21.56	SX
6M85	20.18 Tees to Bescot	23.18	00.18	SX

Miners' Strike

On 12th March 1984, Arthur Scargill, president of the National Union of Mineworkers (NUM), declared that the various strikes then ongoing in the individual coalfields would become a national strike; he called for action from NUM members in all coalfields. That month, Tinsley had seen the weekly number of wagons entering the yard totalling just over 5,000, of which 3,000 were hump-shunted. As the miners' strike wore on, the attendant disruption and loss of Railfreight traffic meant that by November 1984 only 1,500 wagons a week were being humped, out of the 3,000 wagons passing through the yard. Annual figures showed an average of 356 wagons were dealt with on a daily basis by the express freight yard, and just 429 wagons hump-shunted each day. Only 785 wagons a day were passing through the yard on average – a far cry from the yard's capabilities.

Whilst the miners' strike meant an immediate loss of coal traffic, the longer-term trend saw the loss of smaller consignments of steel, scrap and coal, especially for domestic use. MGR coal trains were the new norm, along with block loads of steel routed directly from Scunthorpe to the BSC plants at Templeborough and Aldwarke; oil was also increasingly moving in block loads which by-passed the yard. There simply wasn't the requirement to split trains of 16-ton coal wagons, or wagon loads of steel for their respective destinations any more.

Speedlink

In an effort to offer a better service to customers, BR began operating a fast freight service between Bristol and Glasgow in 1972 using air-braked wagons which could run at 75mph instead of the usual 45mph. The success of this service saw further air-braked wagons and services introduced in the early 1970s, and by 1977 there were 29 such services running per day. September 1977 saw the formal launch of the "Speedlink" network.

The Speedlink network was more restrictive than a traditional wagonload service which used marshalling yards. Fixed timetables between a limited number of destinations resulted in faster services with greater reliability of delivery times. By 1979 BR had reduced its wagon fleet to 137,000 and closed a third of its marshalling yards. Freight depots were reduced by nearly one fifth.

By 1984 Speedlink was operating 150 trunk services per day and carrying eight million tonnes per year. By this time traffic was based on just twelve main marshalling yards: Mossend, Carlisle, Tyne Yard, Dringhouses, Warrington, Healey Mills, Doncaster, Toton, Whitemoor, Bescot, Willesden and Severn Tunnel Junction. Note that Tinsley was not amongst them. Each yard received at least one train per day from each of the other yards. A series of smaller yards also provided services for non-Speedlink trains such as crew reliefs and the like; Tinsley was amongst those yards.

In 1981 there were over two hundred of these smaller yards, but numbers fell to less than a hundred over the ensuing years. The Speedlink plan called for each train to spend no more than an hour shunting at each depot. Wagons were collected by trip workings in the afternoons and evenings, hauled through the trunk network at night, and delivered by trip workings before lunch the following morning.

Speedlink also saw changes to the operation of Tinsley Yard. Long-wheelbase Speedlink VDA wagons were not suited to hump shunting, thus rendering Tinsley's hump redundant. There were also differences in how trains were made up: Prior to Speedlink, wagons for the same destination would be routed into a specific siding; when that siding was full, other wagons for that destination would be routed into a second siding. The next onward train would convey the wagons from the first siding, leaving those in the second siding to be formed into another train, to depart possibly as much as 24 hours later. If a customer had goods on the second train, delivery could be a day later. To avoid such delays, BR developed a reservation system.

By the mid-1980s Tinsley had 19 daily Speedlink services using the yard, coming from far and wide. Table 5 lists those calling at Tinsley after hump shunting had ceased.

Sheffield Freight Terminal Destroyed

On 14th December 1984, Sheffield freight terminal was severely damaged by a major fire which started in Hoults Removals warehouse. The cause of the fire was traced to a faulty electrical heater igniting newspaper. The freight terminal was by then part of the National Carriers empire and the building where the blaze began was attached to the main depot. After the fire the site was cleared and relegated to being a steel-loading facility.

Below: Tinsley East Junction signal box in 1980. Viewed towards Ickles, the bridge over the River Don is just beyond the box. The signals are Tinsley East's home signal with Ickles No. 25 distant signal beneath it. *Roger Wainwright*

Left: Tinsley East Junction signal box after the fire which destroyed the roof. As can be seen it now sports a flat roof. Also note that as Ickles signal box has now closed, the distant signal has been removed. Class 101 DMU set No. C380 passes the box with a diverted train for Leeds in May 1989. *Ivan Stewart*

Tinsley East Junction Signal Box Fire

When signal boxes had coal fires, roof fires were a common occurrence. The chimney pipe would often become red hot during cold days and, as the chimney went up through the box roof, it did not take much to ignite the old paint-covered wooden ceiling. Tinsley East Junction signal box was severely damaged by such a fire in April 1985 and temporarily taken out of use. All points operated by the box were secured in the normal position and the area operated by hand-signalling. The box was later repaired and given a new flat roof; after re-opening it remained in service until 1992 when the area was re-signalled and control passed to Woodburn Junction.

The Demise of Speedlink and Creation of Railfreight Distribution

The concept of dedicating locomotives to specific business sectors developed further as Railfreight was split into three Trainload Freight businesses. In November 1986 the metals sub-sector transferred a number of Classes 20, 37 and 47 to Thornaby for exclusive use on metals trains. This model was then adopted by the other Trainload Freight business sectors but was deemed inappropriate for the Speedlink network. However, as each business sector garnered its own allocated fleets, this left Speedlink competing for resources; it became imperative for Speedlink to have its own tailored and dedicated fleet of locomotives.

Left: A pair of Class 37s bring a rake of mixed-livery HEA wagons past Shepcote Lane Junction into the yard during 1984. This was probably a Speedlink working conveying domestic coal. *Roger Wainwright*

In 1987/88 Speedlink's air-braked freight network could see trains of between 1,200 and 1,400 tonnes working to a maximum speed of 60 mph; a dedicated fleet, working to these requirements, could be finely tuned through concentrated maintenance regimes. Speedlink's chosen dedicated depot had to be well-situated and of high quality – Tinsley was the perfect choice.

Speedlink's locomotive requirement stretched almost the entire length of the UK. Whilst there were some localised pockets, such Class 26s used in Scotland's and Class 33s on the Southern Region, the bulk of its allocation was centred upon Tinsley. From May 1988 it consisted of 25 x Class 31, 34 x Class 37 and 110 x Class 47. Speedlink was now responsible for resourcing its own services and any additional power, if available, would have to be bought in from other operators.

Fuelling and light servicing of Speedlink locomotives could essentially be done at any suitable depot, whilst diagrams were arranged to ensure Tinsley's fleet returned to the depot at the required heavy service intervals. Tinsley's Class 47s had to cover 85 diagrams; at least 88 locos were needed for everyday service, giving an 80% availability requirement.

After many months of planning and sanction from the Secretary of State for Transport, the go-ahead was given to merge Freightliner, a wholly owned subsidiary of the BRB, with Speedlink. The merger was to take affect from 10th October 1988 and the combined business would be known as Railfreight Distribution (RfD). The business would have 27 Freightliner terminals, 200 dedicated locomotives, 3,800 containers, 100 staff ex-Speedlink and 1,400 from Freightliner. The name Speedlink was dropped, but Freightliner branding would continue on that part of Railfreight Distribution's network.

Tinsley Driver Links

Like most depots Tinsley's drivers, drivers' sssistants and guards were organised on a "link" structure. A link was a group of rostered turns of duty which were based around a small group of destinations or specific requirements, such as shunting and trip working. The links would start with local shunting jobs, often with the most unsociable hours, through to the highest links which had the better jobs, often with greater distances that attracted additional mileage payments for the drivers. As each subsequent link tended to mean a higher payment, progression through the links was on a seniority basis and thus, in general, the higher the link the longer the service of the trainmen in them. In 1982 Tinsley depot had a set of four drivers' assistants links alongside those for drivers and guards.

Drivers' assistant link 1 included the most unsociable hours for starting work – at 02.00 on a Monday morning as a spare drivers' assistant. This meant they could be commandeered by the traincrew supervisor for any turns, dependent on their route and traction knowledge. There were 24 different weeks of shift patterns in link 1, the working week varying from one week of 32 hours with rest days on Friday and Saturday, through to a 42½ hour week with the rest day on Saturday. Other weeks contained a single rest day which varied throughout the week.

As the drivers' assistant rose through the links, so the hours of work became more sociable. Links 2 and 3 also had 24 weeks of turns, whilst the rest days of link 3 included a higher proportion of rest days on Saturdays compared to links 1 and 2.

Link 4 was the preferred link with 20 turns covered, but by only 10 men who rotated around the 20-week roster. This link was best for time off, in that almost all mid-week rest days were on the Monday following a rest day on a Saturday, thus giving six three-day weekends and, on two occasions, a four-day weekend as an additional rest day was taken on a Friday.

As for drivers, they had 12 links in which links 1 to 4 also had 24 weeks of turns with varying hours per week. As the drivers rose through their links, most of the shifts were of eight hours and many did not require a secondman. By link 10, every Saturday was a rest day and shifts started at the more sociable hours of either 06.00 or 13.20.

Drivers in link 11 continued to enjoy all their Saturdays off, but began their working day at 07.45. When the driver reached the final top link he had an 8-week rota, every Saturday off, one with an additional Friday every eight weeks and every eight-hour shift started at 12.15.

Drivers' Duties

Anyone climbing the ranks to become a driver would first have to be a secondman (renamed drivers' assistant in later years) and learn the appropriate duties. After passing a rules, traction and train handling examination they would be promoted to passed secondman (renamed relief driver in later years) and would have to be passed as competent on the controls, braking and fault finding of various different types of traction The person would also have to have a detailed knowledge of each type of locomotive.

Roger Wainwright was promoted to relief driver on 2nd June 1980 as per the letter illustrated. Whilst at Tinsley, relief

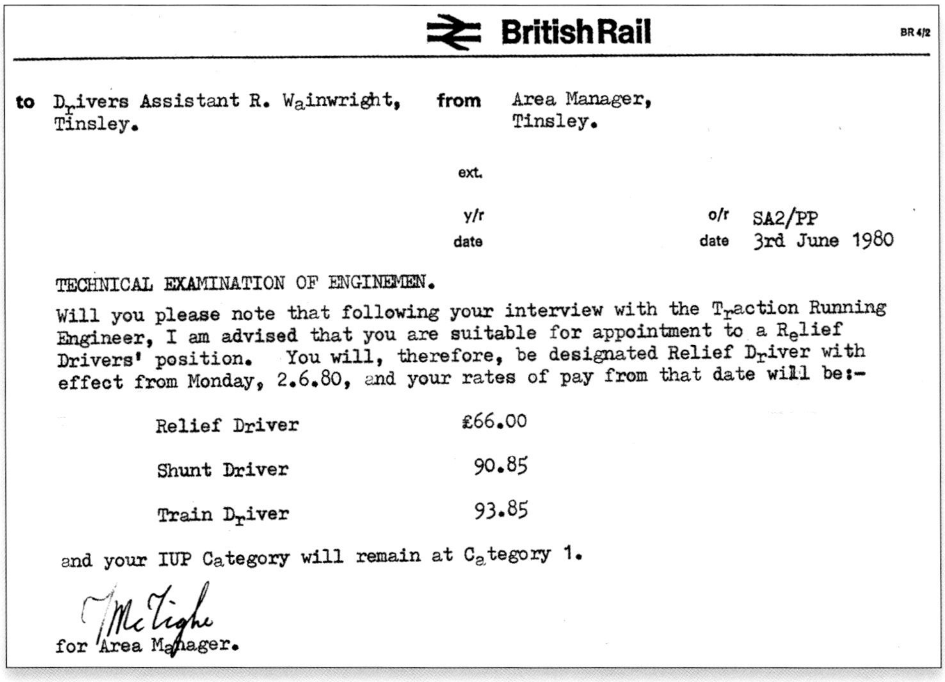

British Rail BR 4/2

to Drivers Assistant R. Wainwright, Tinsley.

from Area Manager, Tinsley.

ext.

y/r date

o/r SA2/PP
date 3rd June 1980

TECHNICAL EXAMINATION OF ENGINEMEN.

Will you please note that following your interview with the Traction Running Engineer, I am advised that you are suitable for appointment to a Relief Drivers' position. You will, therefore, be designated Relief Driver with effect from Monday, 2.6.80, and your rates of pay from that date will be:-

Relief Driver	£66.00
Shunt Driver	90.85
Train Driver	93.85

and your IUP Category will remain at Category 1.

T McTighe
for Area Manager.

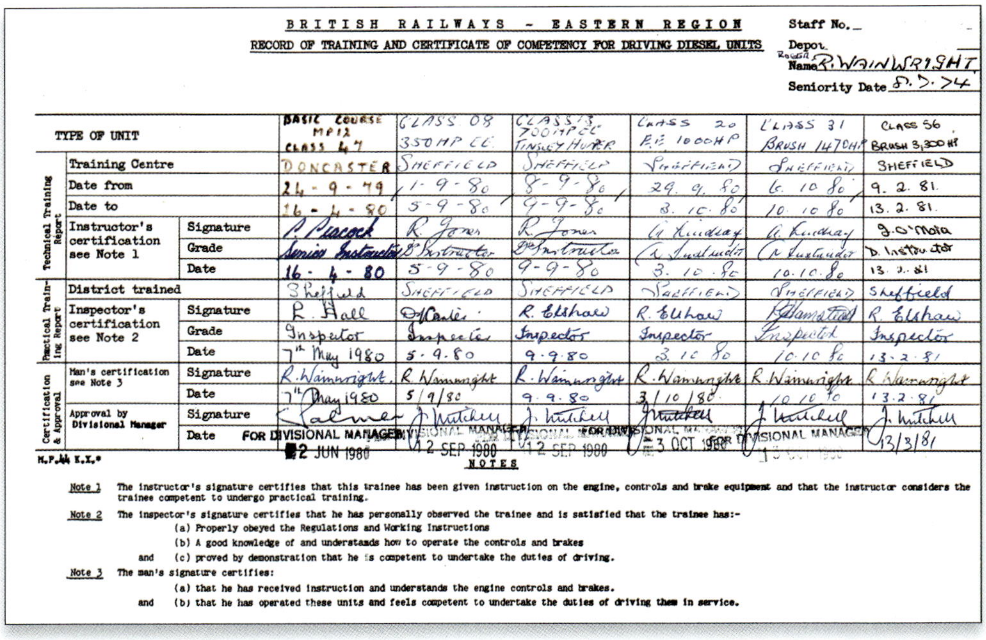

BRITISH RAILWAYS - EASTERN REGION

RECORD OF TRAINING AND CERTIFICATE OF COMPETENCY FOR DRIVING DIESEL UNITS

Staff No.

Depot

Name R. WAINWRIGHT

Seniority Date 8.7.74

TYPE OF UNIT			BASIC COURSE MP12 CLASS 47	CLASS 08 350HP EE	CLASS 13 700HP EE TINSLEY HUMPER	CLASS 20 EE 1000HP	CLASS 31 BRUSH 1470HP	CLASS 56 BRUSH 3,300HP
Technical Training Report	Training Centre		DONCASTER	SHEFFIELD	SHEFFIELD	SHEFFIELD	SHEFFIELD	SHEFFIELD
	Date from		24-9-79	1-9-80	8-9-80	29.9.80	6.10.80	9.2.81
	Date to		16-4-80	5-9-80	9-9-80	3.10.80	10.10.80	13.2.81
	Instructor's certification see Note 1	Signature	P. Peacock	R. Jones	R. Jones	A. Lindsay	A. Lindsay	J. O'Mara
		Grade	Senior Instructor	D. Instructor	D. Instructor	A Instructor	A Instructor	D. Instructor
		Date	16-4-80	5-9-80	9-9-80	3.10.80	10.10.80	13.2.81
Practical Training Report	District trained		Sheffield	SHEFFIELD	SHEFFIELD	SHEFFIELD	SHEFFIELD	Sheffield
	Inspector's certification see Note 2	Signature	R. Hall	[illegible]	R. Elshaw	R. Elshaw	[illegible]	R. Elshaw
		Grade	Inspector	Inspector	Inspector	Inspector	Inspector	Inspector
		Date	7th May 1980	5.9.80	9.9.80	3.10.80	10.10.80	13.2.81
Certification & Approval	Man's certification see Note 3	Signature	R. Wainwright	R. Wainwright	R. Wainwright	R. Wainwright	R. Wainwright	R. Wainwright
		Date	7th May 1980	5/9/80	9.9.80	3/10/80	10.10.80	13.2.81
	Approval by Divisional Manager	Signature	Palmer	J. Mitchell	J. Mitchell	J. Mitchell	J. Mitchell	J. Mitchell
		Date	FOR DIVISIONAL MANAGER 12 JUN 1980	FOR DIVISIONAL MANAGER 12 SEP 1980	FOR DIVISIONAL MANAGER 12 SEP 1980	FOR DIVISIONAL MANAGER 3 OCT 1980	FOR DIVISIONAL MANAGER [illegible]	13/3/81

M.P.44 E.X.

NOTES

Note 1 The instructor's signature certifies that this trainee has been given instruction on the engine, controls and brake equipment and that the instructor considers the trainee competent to undergo practical training.

Note 2 The inspector's signature certifies that he has personally observed the trainee and is satisfied that the trainee has:-
(a) Properly obeyed the Regulations and Working Instructions
(b) A good knowledge of and understands how to operate the controls and brakes
and (c) proved by demonstration that he is competent to undertake the duties of driving.

Note 3 The man's signature certifies:
(a) that he has received instruction and understands the engine controls and brakes.
and (b) that he has operated these units and feels competent to undertake the duties of driving them in service.

Left: Part of Roger Wainwright's record of training and certificate of competency card, which ran to three sides. *Roger Wainwright*

Moving on to 1985, on 26th February, acting as relief driver on 47278, he took 6G16, 07.45 Tinsley–Wath–York, then on 25th and 26th June he was with 37030 and 37031 working 6T54 from Dodworth to Doncaster. On the same duties on 18th July he was in charge of 37041 and 37047. An entry made on 25th June 1986 showed he was acting as traction conductor on 08500 as it was taken from Tinsley to Barnborough Colliery on loan to the NCB.

driver Wainwright's traction knowledge covered the classes shown in Table 6. Note that he was passed to drive Classes 40 and 46, despite these classes never being allocated to Tinsley.

Table 6: Roger Wainwright's Traction Knowledge Record

Traction	Approved
Class 47	2 June 1980
Class 08	12 September 1980
Class 13	12 September 1980
Class 20	3 October 1980
Class 31	13 October 1980
Class 56	13 March 1981
Class 37	26 January 1982
Class 40	29 September 1982
Class 46	20 December 1982
Class 45	2 July 1984

With the closure of Wath depot, Tinsley men began to work former Wath diagrams; Relief driver Wainwright's diaries show that, during a week in February 1983, he worked 6T54, the 15.30 departure from Dodworth, with 56112 on the 7th and 8th, 56002 on the 9th, 56112 on the 10th and 56007 on the 11th.

Other 1983 excerpts show that he worked the following turns:

- Sunday 27th February: 10.00 ballast trip from Tinsley to Wath, hauled by 20011
- Wednesday 15th June: 9G02, 13.00 ex-Penistone with 20057 + 20212
- Monday 20th to Friday 24th June: Wath area conducting and driving 0T54, with 37031 + 37078
- Friday 24th June: Acting as relief driver, he brought what was possibly the last train from Darfield Colliery at 10.20, driving 37031 + 37032 on 6T54
- Monday 27th June: Worked 8T57 then 6J66, 13.50 Dovecliffe Colliery to Earles sidings with 47244
- Monday & Tuesday 21st/22nd November: At Dunford Bridge with 8G17, 11.30 Dunford ballast
- Thursday 1st March 1984: Last train at Westhorpe colliery. Acting as secondman, with Driver F Northcliffe, the train ran as 6T01, 07.47 Westhorpe to Tinsley hauled by 37122

Trip Workings

In January 1984, the trip notice booklet issued to staff listed 15 trip diagrams and 12 pilot diagrams which were numbered as shown in Table 7.

Table 7: Tinsley Trip Diagrams from 23rd January 1984

Trip No	Minimum Loco Required	Trip No	Minimum Loco Required
T30	2x20	No 1 Pilot	13
T32	08 (AB)	No 3 Pilot	08 (VB)
T33	08	No 4 Pilot	08 (VB)
T34	2x20	No 13 Pilot	08 (AB)
T35	2x20	No 15 Pilot	08
T36	2x20	No 18 Pilot	08 (DB)
T39	08	No 19 Pilot	08
T41	08	No 20 Pilot	08
T42	08 (AB)	No 26 Pilot	08 (DB)
T54	37	No 27 Pilot	08 (DB)
T55	47	No 28 Pilot	08 (DB)
T56	56	No 60 (Wath)	08 (AB)
T57	31		
T58	2x20		
T59	31		

AB=Air Braked, VB=Vacuum Braked, DB=Dual Braked

These workings served destinations such as the local steel industries at Shepcote Lane, Deepcar (Stocksbridge), Aldwarke, Roundwood, Templeborough, Masborough and Ickles, conveying scrap metal and finished products, along with trips to Sheffield Freight Terminal. The diagrams show the minimum power requirements for each working. This meant that, for instance, two Class 20s would be required for a trip between Tinsley and Deepcar, although a single more powerful loco could be allocated to the job if necessary. Table 8 describes the trip diagrams from May 1988.

T 58

NO.1 BALLAST WORKINGS

MIN POWER REQ. (2 x20)

DRIVER	ASSIST	GUARD
TIN.866 06 05 - 14 05SX	06 05 - 13 55SX	06 35 - 14 35SX
TIN.867 11 30 - 19 30SX	11 30 - 19 20SX	11 30 - 19 20SX
TIN.868 NOT USED		

		ACTIVITY	ARR	DEP	WTT	FREQ
TIN.866	PL	Tinsley Depot		07 00	LD	SX
	TTC	Works No.1 Ballast until				
	RELD	by TIN.867				
TIN.867	REL	TIN.866				SX
	TTC	Works No.1 Ballast until				
	DISP	Tinsley Depot	19 00			

T 59

No.2 BALLAST WORKINGS

MIN POWER REQ. (31)

DRIVER	ASSIST	GUARD
TIN.869 07 30 - 15 30SX	07 30 - 15 20SX	07 50 - 15 20SX

		ACTIVITY	ARR	DEP	WTT	DAYS RUN
TIN.869	PL	Tinsley Depot		08 15	LD	SX
	TTC	Works No.2 Ballast until				
	DISP	Tinsley Depot	15 00			

No.18 Pilot (Darnall)

Min.Power Requ: (08) DB

TIN.935 NOT USED
TIN.936 12 10 - 20 10 SX
TIN.937-938 NOT USED

		Activity	Arr.	Dep	WTT	Frequ.
TIN.936	PASS	Three Magpies/ Attercliffe Rd.		12 55	24 Bus	SX
		Staniforth Rd.	13 14	13 22	71 Bus	
		Garth Rd.	13 25			
	PL	Darnall		14 05		MO
	PC	Darnall		13 45		MSX
	FS	until				SX
	DISP	Darnall Loco Spur	18 30			
	PASS	Garth Rd.		18 59	71 Bus	
		Staniforth Rd/ Attercliffe Rd.	19 02	19 26	24 Bus	
		Three Magpies	19 45			
	PNB	between 3rd & 5th hours as convenient.				

T35

MIN POWER (2 x 20)

DRIVER	ASSIST	GUARD
TIN.812 20 45 - 05 26SX	—	20 45 - 05 16SX
813 07 23 - 15 11SX	07 23 - 15 01SX	07 53 - 15 01SX
814 15 03 - 22 03SX	15 03 - 22 03SX	15 33 - 22 33SX

		ACTIVITY	ARR	DEP	WTT	FREQ
TIN.813	PL	Tinsley Depot		08 18	LD	SX
		Tinsley MY	08 27	09 08	9*	
		Masborough Stn Yd	10 01	10 34	9	
		Tinsley RL	11 02	11 18	LD	
		Tinsley MY	11 31	12 18	9	
		Aldwarke NS	12 53	13 25	9	
		Tinsley RL	14 02	14 18	LD	
	DISP	Tinsley Depot	14 41			
TIN.814	PL	Tinsley Depot		15 58	LD	
		Tinsley MY	16 07	16 48	9	
		Darnall C&W	17 08	17 53	9	
		Tinsley RL	18 13	18 29	LD	
		Tinsley MY	18 42	19 23	9	
		Aldwarke NS	19 58	20 30	9	
		Tinsley RL	21 07			
	RELD	by TIN.812 at 21 10				
TIN.812	REL	TIN.814 at 21 10				
		Tinsley RL		21 15	LD	
		Tinsley MY	21 28	22 09	9	
		Aldwarke NS	22 44	23 16	9	
		Tinsley RL	23 53	00 09	LD	
	PNB	Tinsley MY	00 21	01 41	9	
		Wath Yd	02 42	03 27	9	
		Tinsley RL	04 24	04 40	LD	
	DISP	Tinsley Depot	04 56			

* RUN VIA BRIGHTSIDE AND RUN ROUND.
BRAKE VAN TO BE PROVIDED AT EACH END

T.55

MIN POWER REQ. (47) AB

DRIVER	ASSIST	GUARD
TIN.858 06 45 - 14 45SX	06 45 - 14 35SX	07 15 - 15 15SX
TIN.859 13 45 - 21 45SX	13 45 - 21 35SX	14 15 - 22 15SX
TIN.860 23 45 - 07 45SX	23 45 - 07 35SX	00 15 - 08 15MX

Work M.G.R. trips from Treeton, Brookhouse and Westthorpe Collieries to West Burton, Cottam, Carlton Thorpe Marsh or Rotherwood and return with empties as required, or conventional coal trains from Treeton Colliery to local power stations - T.T.C.

Left and above: A selection of extracts from the 24th January 1983 trip notice book for Tinsley.
Kim Collinson collection

Left: An unidentified Class 37 crosses from the "down" to the "up" line as it departs from Deepcar with 9T01 trip to Tinsley Yard, consisting of 16-ton wagons. The photo is dated 1978 meaning the GCR Deepcar signal box was still open at this time, as the Woodhead Route was still very much in operation. The trackwork was rationalised in 1994 with all the sidings and connections removed, and the former main line, by now single track, connected directly to the Stocksbridge railway. The box actually survived until 1994, when it succumbed to an arson attack. Deepcar station is hidden behind the signal box, but one of its platforms can be seen on the right of the image. Note the large pile of ashes from the signal box fire surrounding the buffer stop!
Roger Wainwright

Left: The 4-lever ground frame to the north of Deepcar box is prominent in this photo, which shows a pair of Class 20s in the "down" siding on 0T01 trip from Tinsley to Deepcar. The locos will traverse the points and head towards the yard behind the camera to collect wagons. The Stocksbridge railway line heads off to the left of the locos. Although undated, the lack of overhead wires dates it to post-1981. Some wagons can be discerned in the shunt neck beyond the signal, which is unlit. Although not quite visible, there are sleepers chained across the "up" and "down" main line tracks about 100 yards beyond the signal, a sure sign that the route towards Penistone is out-of-use and awaiting track recovery. *Roger Wainwright*

Table 8: Tinsley Trip Diagrams for May 1988

Trip	Days	Train	Arrive
T01	MO	06.47 Tinsley to Deepcar	07.30
2x Class 20 minimum power required		08.13 Deepcar to Tinsley (LD)	08.43
	MSX	06.17 Tinsley to Deepcar	07.00
		07.43 Deepcar to Tinsley	08.13
	SX	14.36 Tinsley to Deepcar	18.20
		19.49 Deepcar to Tinsley	20.32
T02	MO	06.56 Tinsley to Aldwarke	07.31
1x Class 20 minimum power required		07.51 Aldwarke to Tinsley (LD)	08.14
		08.54 Tinsley to Deepcar	09.37
		11.06 Deepcar to Tinsley	11.49
	MSX	07.11 Tinsley to Aldwarke	07.46
		08.06 Aldwarke to Tinsley (LD)	08.28
	SX	14.48 Tinsley to Aldwarke	15.23
		15.55 Aldwarke to Tinsley	16.32
T03	MO	07.20 Tinsley to Aldwarke	07.55
1x Class 20 minimum power required		08.15 Aldwarke to Tinsley (LD)	08.37
	MSX	06.49 Tinsley to Aldwarke	07.24
		07.44 Aldwarke to Tinsley (LD)	08.08
T04	MO	06.35 Tinsley to Sheffield FT	06.52
1x Class 08 minimum power required. Service Sheffield Freight Terminal (FT) and Attercliffe Sidings		12.09 Sheffield FT to Tinsley	12.31
	MSX	06.16 Tinsley to Sheffield FT	06.33
		11.49 Sheffield FT to Tinsley	12.11
	MO	13.36 Tinsley to Sheffield FT	13.53
		15.30 Sheffield FT to Tinsley	15.52
		16.02 Tinsley to Sheffield FT	16.13
		19.09 Sheffield FT to Tinsley	19.31
	MSX	13.16 Tinsley to Sheffield FT	13.33
		15.30 Sheffield FT to Tinsley	15.52
		16.02 Tinsley to Sheffield FT	16.13
		19.19 Sheffield FT to Tinsley	19.41

LD = Light Diesel

The End of Hump Shunting

With the changes to traffic it was perhaps inevitable that the yard would reduce in size and hump shunting would cease. Initial plans would see the closure of the main yard and secondary yard control towers over the weekend of 27th/28th October 1984, but the work was postponed and they eventually closed on 15th December 1984; all their equipment was decommissioned as stage 1 of the yard rationalisation plans. This work also saw changes to the express freight sidings, and in the secondary yard stop blocks were erected at the eastern end of sidings 60 to 84 inclusive; the track to the east of the stop blocks was secured out of use. In the main yard, sidings 1 to 5 and sidings 50 to 53 were taken out of use and stop blocks erected at the eastern end of sidings 6, 7, 8 and 22 to 49. The track to the east of the stop blocks was secured out of use and later removed.

A new section of track was installed to connect sidings 16 to 21 to the main yard east departure line, meaning sidings 9 to 21 were connected to this line. All the point work was reduced to hand operation in the yard and all signalling equipment to the west of Tinsley Yard TY274 outlet signal was decommissioned.

Right: The trip notice for trip T54 between Tinsley and Wath, following closure of the Woodhead Route.

Below: Two west end yard pilots go about their work on 20th May 1985. On the left is 08543, with 08857 in the distance shunting a couple of oil tanks. *Michael Rhodes*

```
                          T 54

MIN POWER REQ. (2 x 20) AB

DRIVER                       ASSIST                GUARD
TIN.855 04 42 - 12 42MO  04 42 - 12 32MO      05 12 - 12 42MO
    855 04 15 - 11 42MSX 04 15 - 11 32MSX     04 45 - 11 45MSX
    856 11 20 - 18 20MO  11 20 - 18 20MO      11 20 - 18 20MO
    856 10 20 - 17 20MSX 10 20 - 17 20MSX     10 20 - 17 20MSX
    857 16 30 - 23 30MO  16 30 - 23 30MO      16 30 - 23 30MO
    857 15 30 - 22 30MSX 15 30 - 22 30MSX     15 30 - 22 30MSX

Work M.G.R. trains (Loads/Empties) as programmed on the
weekly circular from Dodworth, Darfield, Dovecliffe and
Manvers to Wath Yard.

             ACTIVITY          ARR    DEP    WTT    FREQ
TIN.855 PL   Tinsley Depot            05 37  LD     MO
             Prove the line to Silverwood Colly
             Wath Yard         07 05
        PL   Tinlsey Depot            05 10  LD     MSX
             Wath Yard         05 45
             Work as programmed until
        RELD Wath Yard by TIN.856 at 12 00MO, 11 00MSX
        PASS Wath Yard                12 02  TAXI   MO
             Tinsley           12 32
        PASS Wath Yard                11 02  SC'E'  MSX
             Tinsley           11 32

TIN.856 PASS Tinsley                  11 30  TAXI   MO
             Wath Yard         12 00
        PASS Tinsley                  10 30  SC'E'  MSX
             Wath Yard         11 00
        REL  TIN.855 at 12 00MO, 11 00MSX
             Work as programmed until
        RELD Wath Yard by TIN.857 at 17 10MO, 16 10MSX
        PASS Wath Yard                17 12  TAXI   MO
             Tinsley           17 42
        PASS Wath Yard                16 12  SC'E'  MSX
             Tinsley           16 42

TIN.857 PASS Tinsley                  16 40  TAXI   MO
             Wath Yard         17 10
        PASS Tinsley                  15 40  SC'E'  MSX
             Wath Yard
        REL  TIN.856 at 17 10 MO, 16 10 MSX
             Work as programmed until
             Wath Yard                21 00  LD     MO
        DISP Tinsley Depot     21 43
             Wath Yard                20 00  LD     MSX
        DISP Tinsley Depot     20 43
```

East End Sidings and Connecting Lines

Further track rationalisation saw the following lines and sidings taken out of use and all associated signalling decommissioned:

- North arrival line
- South-west arrival line between the BSC departure line and Tinsley Yard signal box
- Hump engine return line
- Reception sidings 1 to 11 inclusive, together with all associated spurs
- Reversing siding
- Engine by-pass line

Many routes and lines had their designations changed; the yard became merely flat-worked and was no longer hump shunted.

Left: The guard leans nonchalantly on his veranda as 20003 + 20066 depart light engines and brake van towards Shepcote Lane Junction with another trip to the Sheffield Freight Terminal on 23rd September 1980. *Michael Rhodes*

Below: Looking east towards the main yard on 26th September 1985. 20040 + 20043 depart with trip T36 to Sheffield Freight Terminal whilst 20086 + 20150 wait at the head of trip T35 to Aldwarke. The latter is made up entirely of POA wagons with scrap. *Michael Rhodes*

Table 9: Main Yard Siding Allocation 1985

No	Allocation	No	Allocation
1	Spare POAs	27	No allocation
2	Spare 16-ton minerals	28	Spare STVs
3	Minerals	29	No allocation
4	Scunthorpe	30	No allocation
5	SPA + Aldwarke	31	Smithywood
6	Deepcar + spare BBA & BDA	32	Treeton
7	Scunthorpe empty BDAs	33	Barrow Hill
8	Sheffield Freight Terminal	34	Marple & Gillott
9	Engine line	35	Cripple wagons
10	Speedlink south	36	No allocation
11	Speedlink south	37	Tinsley miscellaneous
12	Reception siding	38	Darnall
13	Reception siding	39	Marple & Gillott
14	Deepcar	40	No allocation
15	Aldwarke New Site	41	No allocation
16	Speedlink north	42	No allocation
17	Speedlink north	43	Cripples for grading
18	Empty POAs	44	Cripples
19	Reception siding	45	Cripples
20	Reception siding	46	Grading siding
21	Transfer	47	Grading siding
22	Beighton	48	Grading siding
23	Darnall	49	Grading siding
24	No allocation	50	Siding lifted
25	Spare BDAs	51	Siding lifted
26	No allocation	52	Siding lifted
		53	Siding lifted

The express freight sidings were now used for storage of block loads, and as arrival sidings for roads 1 to 19 in the main yard. Sidings 12, 13, 19 and 20 in the main yard were designated reception sidings, traffic being able to arrive in any of those. Sidings 34–49 were retained to stable condemned wagons en route to local scrapyards, whilst the secondary yard was fully turned over to the storage of condemned coaching stock, awaiting the call to the scrapyard.

More Rationalisation

In March 1985, stage two of the yard rationalisation plans took shape and the allocation of sidings in the yard was changed to that shown in Table 9.

A search of the TOPS computer in May 1985 revealed there were 6,000 wagons on hand in the Tinsley area, of which over half were in use carrying freight traffic. Although this seems a large number, only 750 of those wagons were shunted in the yard each day, the remainder used mostly in block loads which did not use the yard.

Table 10 shows the origin and destination of the trains still using the yard during the summer timetable of 1985.

The traffic was primarily scrap and steel products, with some steel workings destined for the remains of the Sheffield Freight Terminal. There was also fruit for the Parkway Market, primarily bananas.

The wagons in use for steel traffic were divided into pools, each pool being allocated to specific traffic. During a typical period in 1985 there were perhaps 60 BDA wagons allocated to Tinsley–Cardiff workings, with maybe 50 of those employed on that traffic each week, the remaining 10 being allocated to trips to local works such as Templeborough or Aldwarke. The previously mentioned pool of 181 POA wagons for scrap traffic was supplemented by 2,000 MXV 4-wheel 16-ton wagons. The latter were not suitable for Speedlink and, although there was a proposal to convert 21-ton MDV wagons for scrap traffic, very few were actually converted.

Although there could be as many as 2,000 wagons in Tinsley Yard at this time, around 1,800 of these were cripples awaiting attention; a full-time grading inspector

Above: Tinsley Yard from Wood Lane bridge in April 1985. The new main yard departure line has been laid and is in the process of being tamped. The hump line is still connected, but out-of-use and it will not be long before all connections to the sidings to the left of the closed control tower will be removed. Note the oil contamination where the hump locos would have stood; there is also contamination where the Dowty retarders would have been, in the centre of the image. Even though all the Dowty equipment is out of use, the test trolley is still in its neck to the left of the control tower.
Dave Stones

Table 10: Summer Timetable 1985

Arrivals & Departures at Tinsley

Head-code	Train	Arrive	Days Run
9J71	23.10 Doncaster–Tinsley	00.07	SX-Y
6E76	23.40 Toton–Tinsley	01.03	SX
7J29	00.46 Doncaster–Tinsley	01.33	MX
9J70	00.08 Dringhouses–Tinsley	01.49	MX-Y
7M68	02.07 Tinsley–Stanlow		MSX
6M87	02.31 Tinsley–Toton		MX
6J45	00.43 Immingham–Tinsley	02.46	TFO
6J62	20.26 Parkeston Quay–Tinsley	03.28	SX
6L81	04.10 Tinsley–Hull		MX
6E63	02.12 Warrington–Tinsley	05.33	MX
6J60	06.30 Tinsley–Barrow Hill		SX
8M81	07.52 Tinsley–Derby		MSX
6L46	08.32 Tinsley–Goole		SX
8055	12.30 Tinsley–Betteshanger		MTThO
6J61	13.11 Barrow Hill–Tinsley	13.34	SX
9D44	14.56 Tinsley–Scunthorpe		SX-Y
6J25	15.50 Scunthorpe–Tinsley	17.18	SX-Y
9J28	15.40 Healey Mills–Tinsley	17.12	SX-Y
6E68	14.16 Wolverhampton–Tinsley	17.39	SX
9L28	18.34 Tinsley–Healey Mills		SX
9H41	19.05 Tinsley–Snailwell		ThO
6D70	19.40 Tinsley–Scunthorpe		SX-Y
9D71	20.30 Tinsley–Doncaster		SX-Y
7D32	20.45 Tinsley–Immingham		MFO
6H91	20.55 Tinsley–Whitemoor		SX
7D19	21.44 Tinsley–Doncaster		SX
6M70	21.56 Tinsley–Warrington		SX
6E49	20.45 Stanlow–Tinsley	23.57	FSX

was employed to evaluate the future of each wagon. The remaining 200 wagons were used on Speedlink traffic, the traffic spending as little time as possible in the yard, thus maximising revenue potential.

Change of Use

Around that time there was also a change of use for the servicing shed in the main yard. Loco refuelling duties were transferred to the main shed and the building took on the servicing of DMUs, previously carried out by Darnall. DMUs would arrive for servicing and washing overnight, before resuming their duties from Sheffield Midland.

Stage three of the rationalisation plans came in March 1986, bringing about the removal of three additional routes and further track reductions. Speedlink services were still arriving in the yard; Table 11 shows those using the yard in 1986.

In 1987/88 Tinsley acted in a supporting role to Doncaster Belmont Yard, as the latter was particularly tight for space and operationally inconvenient for trains heading to Sheffield. Thus, Tinsley continued to deal with trains for the Sheffield area, primarily scrap. There was also was a degree of portion-swapping on trains using the north-east to south-west axis.

Tinsley's Celebrity Locomotives

Tinsley's first foray into customisation came with the application of small white painted snails on 45142 and 45128. The first two locomotives to receive unofficial names

Left: The closed reception sidings are in the process of being lifted as a Class 37-hauled tank train makes its way towards Catcliffe and a Class 08 trundles past light engine in this image taken on 6th March 1986. *Stuart Reeves*

Below: Hand-drawn diagram of the rationalised yard after hump shunting had ceased. This was from a book of track layouts issued to Alan Hazell, a former controller at York, who annotated the diagram to reflect the new layout. *Chris Booth collection*

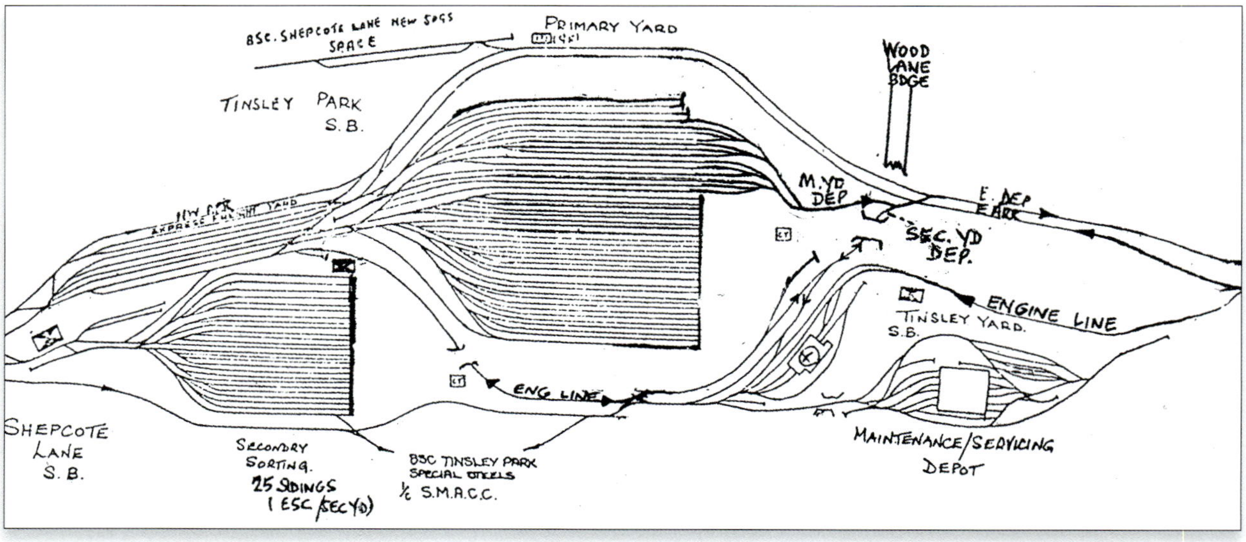

Below: 20016 + 20172 have passed Shepcote Lane and are bringing a local trip into the secondary yard on 28th October 1987. The train is comprised of eight SPA wagons and although it appears some have steel coils aboard, they are actually on the train to the right! *Dave Stones*

Above: As Dougie Buck, the driver of 08492 watches the cameraman, 76051 takes a rake of 16-ton wagons out of the express freight sidings as 8T30 to Deepcar on 23rd September 1980. *Michael Rhodes*

Table 11: Speedlink Trunk Services at Tinsley May 1986

Head-code	Train	Arrive/ Depart	Days Run
Route 2			
7E49	16.54 Toton to York Dringhouses	18.13/19.14	M-F
7E49	15.14 Bescot to York	18.51/19.54	M-F
6M85	19.34 Tees Yard to Bescot	22.12/00.40	M-F
6E38	21.33 Bescot to Doncaster	00.17/01.02	M-F
6E64	19.15 Severn Tunnel Junction to Haverton Hill	00.28/01.51	M-F
7M88	01.27 York Dringhouses to Derby St Marys	02.56/03.50	T-S
6L81	04.10 Tinsley to Hull Saltend	04.10	T-S
Route 6			
4R40	09.15 Tinsley to Whitemoor	09.15	M-F
6E96	11.45 Parkeston to Tinsley		M-F
6J62	20.26 Parkeston to Tinsley	03.37	M-F
6H91	20.55 Tinsley to Whitemoor	20.55	M-F
Route 7			
6M72	21.56 Tinsley to Warrington	21.56	M-F
7E63	02.23 Walton Old Junction to Worksop	05.12/06.12	T-S
Route 8			
6E75	Willesden to Leeds Hunslet	04.36/05.16	M-F

were the Class 20 pairing of 20030 and 20064, repainted green in early March 1987 and named "River Rother" and "River Sheaf" respectively. Railfreight red stripe-liveried 20118 followed as "River Don" in April 1987 and the three locomotives famously powered the "Three to the Sea" railtour from Sheffield to Brighton on 2nd May 1987.

Further names followed; 28 Class 08s, along with 09008 and 09013, all gained names – mainly associated with former depots in the local area, such as Wath ETD, Canklow, Millhouses and Grimesthorpe. The naming of Class 45s in 1987 was prolific, with examples such as "Phaeton", "Pegasus", "Eclipse" and "Taliesin", all fondly remembered. In total 22 Class 45s and one errant Class 46 received unofficial names. Of course, the Class 46 was the joker of the pack as it was actually 45104 renumbered as 46009 and named "Old Dalby". The real 46009 had actually met its fate in July 1984 in a staged nuclear flask test train crash at 100mph on the Old Dalby test track. The stand-out Class 45, however, was 45106 which received a repaint into an approximation of BR Green.

Just two Class 31s received names, as compared to 28 Class 37s, including four Class 37/4s which received volcano names, such as 37415 "Mt. Etna" and 37426 "Mt. Vesuvius". The greatest number of names were carried by 91 Class 47s – the first was applied in 1989 and the last on 24th February 1998 when 47523 was named "RAILFREIGHT".

The enigma of the Class was 47362, which carried the name "RIP Tinsley TMD 1965–1998" for just two hours on 23rd December 1997. The pride of both class and depot were 47375 "Tinsley Traction Depot - Quality Approved" and 47186 "Tinsley TMD Silver Jubilee", both officially named on 15th February 1990 and 9th March 1990 respectively.

Above: Newly painted for railtour duties, 20064 "River Sheaf" and 20030 "River Rother" are resplendent at Tinsley in March 1987. *Roger Wainwright*

Right: This caused some consternation amongst enthusiasts. 46009 "Old Dalby" – or was it? It was actually 1988-withdrawn 45104, but Phil Hodgkiss and his compatriot had renumbered it whilst it was stored at Tinsley. As the real 46009 had been destroyed in a test with a nuclear flask at Old Dalby, many enthusiasts were totally bemused! *John Webster*

Above: After the fire that ended its career, 45106 was stored at Woodhouse Junction for a period, along with several Class 20s and Class 47s. The loco is seen here on 25th February 1992 awaiting the call to the scrapyard. It was sold to C.F. Booth, Rotherham, along with several Class 47s. They arrived towards the end of March 1992, and by the end of April, 45106 was no more. *Chris Booth*

Above: Headed by 45052, a long line of withdrawn Class 45s are stored in the yard in the late 1980s. To the right, a Class 47 possibly awaits the same fate?

Originally D75, 45052 received two different unofficial names at Tinsley – "Satan" and "Nimrod" – before being withdrawn in June 1988. By August 1991, Tinsley Yard was still home to several withdrawn Class 45s, which were 45017/033/046/120/134/140/145/150, 97409/410/412. *Roger Wainwright*

Above: 45124 was sporting the unofficial name "Unicorn" when seen at Tinsley on 31st December 1987.

Originally D28, the loco was new to Derby (17A) on 4th May 1961. Withdrawn from service on 22nd January 1988, it was stored for just over four years before being cut up at MC Metals on 14th February 1992. *Chris Booth*

Below: Railfreight grey-liveried Class 31 No. 31147 has just passed beneath the Sheffield Parkway bridge and passes the lifted reception sidings with a Speedlink train on 20th July 1988. Note the yellow shunters' cabin still in situ, and the buffer stops on the embankment close to the Parkway. This is the eastern exit from the depot and the scene of several runaways over the years.

Originally D5565 and new to March (31B) on 19th November 1959, the loco was a Tinsley engine from week ending 9th October 1965. It enjoyed spells at Immingham, Tinsley, Immingham again and York, before returning to Tinsley from week ending 2nd May 1976. It received the name"Floreat Salopia"on 30th May 1993 and was denamed on 31st May 1997. The loco was cut up at Barrow Hill in April 2001. *John Webster*

Above: A different view of the partially-lifted reception sidings, as viewed from the A630 Sheffield Parkway overbridge, looking west towards Wood Lane bridge. The engine line is to the extreme left, and there is still at least one brake van on No. 10 road. Note the shunters' cabin still in situ, and an "Addo" kiosk lying on its side beyond a pile of sleepers. Note also that some of the lamp covers in the lighting towers are open – perhaps the bulbs have been removed for spares? *Michael Rhodes*

Right: An unidentified Class 45 is seen running alongside Tinsley Yard with a train of continuously-welded rail in the late 1980s. This photo shows disconnected siding Nos. 1 to 5 fitted with sleeper stop blocks and Nos. 6 to 8 fitted with buffer stops.

The sight of four Class 45s awaiting disposal on No. 5 road would be a sign of things to come. Over to the far left of the yard, condemned DMUs and coaching stock are stored awaiting the call to the scrapyard. A Class 08 is shunting a rake of steel wagons on No. 12 road whilst a Class 47 is attached to a train on No. 14 road. *John Webster*

Left: Looking in the opposite direction to that above. The line through Tinsley was often used by diverted passenger trains and specials. 45144 is seen working a football excursion (footex) on 12th April 1987. Behind the loco, the dismantling of the yard has seen roads 1 to 5 fitted with sleeper stop blocks, whilst roads 6 to 8 have conventional buffer stops. Roads 9 to 21 are still connected, but roads 22 to 27 are also fitted with sleeper stops. *Ted Evans*

Right: Another Footex on the same date sees an unidentified Class 47 which, so the photographer says, could be one of three…47411, 47454 or 47475! In the background, the former reception sidings have been lifted but the engine line is still in situ to allow access to the east end of the depot.

The trains were run in connection with the FA Challenge Cup semi-final which was held at Hillsborough in Sheffield, the first FA Cup semi-final to be held there since 1981. The train would be heading via Shepcote Lane and Broughton Lane, then via Woodburn Junction to Wadsley Bridge. The game, between Leeds United and Coventry City, resulted in a 3-2 win for Coventry. *Ted Evans*

Above: As viewed from the reception sidings beneath the A630 Sheffield Parkway overbridge looking towards Treeton, 45039 brings a long rake of 16-ton mineral wagons and four tanker wagons into the yard as 8J12, Tees Yard to Tinsley on 23rd September 1980. From the Treeton direction, the train could access any of the 11 reception sidings; this train is routed into one of sidings 8 to 11. This was as far as electrification of the yard extended – the wires terminated at the stanchions either side of the loco. *Michael Rhodes*

Above: There are not many photos of the yard like this one, viewed from the Sheffield Parkway overbridge. On 9th July 1986, 47355 has just crossed Catcliffe viaduct with 6V06, Healey Mills to Severn Tunnel Junction, as the rear of the train passes the site of Catcliffe station. The centre track has been lifted as the reception sidings are now out-of-use. Treeton Colliery headstocks can be seen to the extreme right. *Michael Rhodes*

Brightside Station Junction.
Left: In January 1985, 20105 and an unidentified sister round the curve from the Sheffield District Line at Brightside Station Junction, and head towards the "up" and "down" goods line, as viewed from Brightside Junction signal box. In June 2006, Sheffield PSB's No. 4097A points at Brightside Station Junction in the "up" and "down" goods line were removed and plain line substituted, thereby cutting off all access to Tinsley Yard from this direction; the route had not been used since resignalling of the area in 1992. After track lifting in 2007, the trackbed remained in situ along with all bridges. The first bridge to be removed was over Meadowhall Road, followed by the removal of short sections of embankment. In 2014 the construction of a new store required the removal of further sections of embankment. *Dave Stones*

Left: After a flurry of snow, an ex-works Class 31 rounds the same curve en route light loco from Tinsley Yard. The ground signal to the bottom left is B166, which protected the crossover road and controlled access to No. 1 and No. 2 reception sidings. The loco is passing over the interlaced sand drag, designed to stop runaways from the Sheffield District Line. The sidings to the right are Jessops Sidings, which served William Jessop & Sons' Brightside steelworks, out of view to the right. As with most of Sheffield's steelworks, there used to be a large internal railway system here, all of which has now gone, although the steelworks remains as part of Sheffield Forgemasters. *Dave Stones*

Above: Not the best of shots but included to illustrate the view from Brightside Junction signal box towards Tinsley, looking slightly more to the left from the image above. B166 signal is at the bottom, the sand drag can be clearly seen, and in the distance is B122 signal on the "up" Tinsley. The road on the right is Colliery Road, with the red brick Bridge Inn public house at the end on Brightside Lane. This public house was demolished in 2007. Whilst the bridge over Brightside Lane remains, the trackbed is now an urban forest. *Dave Stones*

Right: 40158 is seen at Tinsley Park in April 1983, hauling 76035/007/011/024/022 from Guide Bridge, with an ultimate destination of C.F. Booth's scrapyard at Rotherham. *Steve Ryszka*

Nocturnal Tinsley: Left and below: The job went on no matter what time of day or night! Here are a couple of photos of the hump and reception sidings, both taken in the early hours of 24 February 1981. *Both Steve Ryszka*

Chapter 20

The 1990s – Depot Closure and Further Rationalisation

The decade started well for Tinsley depot when, in its silver jubilee year of 1990, it became the third BR depot to attain the prestigious British Standard 5750 Part Two certificate, reflecting the success of the quality control system introduced in 1989. Specific categories of the award included production, servicing, purchasing, inspection and documentation and how they were all set out and met. The certificate was presented to area fleet manager, David Spillets, at the depot on 15th February 1990 by David Blake, BR's director of mechanical engineering. Phil Thickett, Tinsley depot engineer, described the commitment of his staff at all levels of the scheme as "staggering". In recognition of the award, plates bearing the name "Quality Assured" were attached beneath the nameplate of 47375 "Tinsley Traction Depot".

As the principal depot for Railfreight Distribution (RfD), these initiatives helped to ensure that Tinsley's 170 RfD and 16 Railfreight Construction locomotives were maintained to the highest standards. Tinsley also saw an increasing number of Classes 47/3 & 47/4 arriving for attention from Thornaby and Crewe depots; responsibility and, it may be said, affection for the class grew.

Change of Use for Servicing Shed

Wagons coming into Tinsley Yard would often require repairs; any wagons in need of attention would be designated as cripples and placed into a designated siding – one of 35 or 43–45. When several had accumulated, a trip from the yard to Darnall wagon repair depot would be undertaken. Tuesday 15th January 1991 saw the final run of cripple wagons from Tinsley Yard to Darnall, departing the yard at 09.15, hauled by 47332. After the wagons had been repaired and the depot cleared, the almost derelict Darnall depot was closed completely. To compensate for the loss of Darnall, the servicing shed in Tinsley Yard, having previously been turned over to the servicing of DMUs, now became a carriage & wagon repair depot. New local instructions were issued as shown in the following excerpt:

Local Instructions for Servicing Shed

1.1 The normal status of the service shed road is *"closed to rail movements"*.

1.2 In addition to the provisions of Rule Book Section B, Part III the following additional conditions apply and must be observed before any work on vehicles commences:

1.3 Protection for the two shed roads is provided by a moveable Stop Board in the four foot of both roads on each side of the shed. The normal position of the Stop Boards is upright. A red light will be provided in darkness or poor visibility. In addition, hand points will be clipped away from these roads.

2. Shunting will take place at the East End of the shed.

3.1 Before requesting a shunt the person-in-charge at the Service Shed must ensure that:
(a) All work on vehicles has stopped.
(b) The point clip at the East End has been removed and the Stop Boards have been lowered.

3.2 He should then request and advise the person in charge at Tinsley Yard that the shed is ready to be shunted.

3.3 Once this has been done no further work can take place on vehicles until the shunt is completed or has been cancelled by agreement with the person in charge at Tinsley Yard and the additional protection arrangements detailed above have been reinstated

The Creation of Railfreight Distribution and the End of Speedlink

By 1990 the Speedlink network was haemorrhaging money, with oft-quoted figures showing losses of £30 million per annum, on a £45 million turnover. In 1980 there were 12 trunk yards and 20 secondary yards, connected to over 200 sidings in the Speedlink network; by 1990 that network had shrunk to just eight main yards and only 100 freight sidings in regular use. Speedlink was now responsible for a mere 0.12% of all UK freight and only 2% of all rail-borne traffic.

With Government requiring the rail freight sector of BR to make greater profits, RfD Managing Director, Mr Ian Brown, consulted all Speedlink customers about the future of the network. Essentially it was deemed that Speedlink was unviable, and on 12th December 1990 RfD announced a date of 8th July 1991 for closure of the Speedlink network. A major problem for Speedlink was that individual Trainload Freight (TLF) companies were carrying much of the smaller traffic themselves. For example, in 1990/91 the metals sector moved 19 million tonnes of steel, but only 15% of this was in trainload quantities, thus depriving Speedlink of many of the more profitable flows. When Speedlink ceased in 1991, roughly 70% of its three million tonnes of traffic was retained by BR, equating to 125 trains per day.

At an exhibition in Birmingham in September 1992, RfD unveiled its plans to run international freight trains via the newly-opened Channel Tunnel from 6th May 1994. A contract was drawn up between RfD and Eurotunnel to guarantee a yearly payment until 2005, to the tune of a total net figure of £242 million. Unfortunately, the projected traffic levels were wildly optimistic; RfD recorded a financial loss of £64.5 million for the financial year ending March 1996.

Further Rundown of Sheffield Area Freight

Treeton Colliery and its coke and chemical plant closed on 7th December 1990; latterly, much of the colliery's production had been household coal. This was exactly the type of traffic that required shunting and sorting in Tinsley Yard, hence the loss of this colliery further reduced traffic in the Tinsley area.

Another loss was the Brightside Lane freight terminal (the remains of Sheffield Freight Terminal), which closed towards the end of 1990 with the loss of 20 jobs. At the time it was handling a very modest 70,000 tonnes of freight per year – mostly steel, and there were calls for the site to become a Channel Tunnel freight forwarding point. However, BR felt that more than 500,000 tonnes of freight would be needed to make the project viable and, as this would not be possible, ten acres of land at Brightside were sold off in 1992 to be used for a Royal Mail sorting centre. In small recompense, on 28th March 1992 an improved terminal for semi-finished steel products and other heavy goods was officially opened on the site of the former Freightliner depot near Masborough.

Things declined even further when efforts were made to run more trains directly between their originating points and final destinations, thus missing cut Tinsley Yard again. January 1991 saw several new and amended freight services introduced, with scrap traffic from the West Midlands running as block trains directly from Washwood Heath to the United Engineering Steels plants at Aldwarke, Deepcar, Roundwood, Templeborough and Thrybergh. Similarly, traffic from the North East came from Tees Yard; North West traffic came from Warrington and a block trainload of scrap from Crossley Evans at Shipley. At the same time a new twice-weekly trip from Aldwarke to Deepcar was introduced, to allow direct connections with the new trainload services. United Engineering Steel's own locomotives and staff shunted their yards and nearly all the traffic was carried in new SSA (ex-POA) wagons. From the beginning of 1992, traffic from Stocksbridge steelworks and Deepcar was ostensibly re-routed direct to Aldwarke New Site, although on 23rd March 1992, 56116 worked into the yard with the 08.25 Deepcar to Tinsley. By this time Tinsley Yard was seeing very little traffic other than that associated with the Stainless Melting And Continuous Casting (SMACC) and Stainless Plate And Coil Expansion (SPACE) plants.

Tinsley's Changing Locomotive Fleet

Former Southern Region Class 09s, 09008 and 09013, were reallocated to Tinsley for local trip working in August 1989, 09013 later being unofficially named "Shepcote". This loco was observed in use on the 9T38 trip from Rotherham Steel Terminal to Tinsley Yard on 2nd March 1990 hauling 3 wagons, a job ideal for the higher-speed shunter. Both locomotives stayed at Tinsley until September 1993, when they were re-allocated to Cardiff Canton. Other Class 09s noted as being allocated to Tinsley included 09011, 09021 and 09022 from March 1994 to August 1994, whilst 09021 had a second stint at Tinsley from November 1997 until the depot's closure in March 1998.

In August 1990, 18 of Tinsley's Class 47 fleet were stored serviceable, whilst 47096 and 47146 were reinstated to traffic. A number of the stored locos were Class 47/4s, which the Parcels and Provincial sectors were keen to acquire. In the run up to the end of Speedlink, RfD was looking to condemn 50 Class 47s, but in the event, many were either stored or transferred to other sectors. Other changes arising from Speedlink closures included the disbandment of all Scottish RfD pools except for Eastfield's fleet of 11 shared Class 37/4s, the other Class 37/0s being shared between Tinsley and Thornaby. Table 1 illustrates RfD's locomotive pools in March 1991.

In early May 1992 a new pool of high-engine-hours Class 47s was created, made up of 24 locomotives that RfD had deemed surplus to requirements. These were stored serviceable pending an upturn in traffic. Thirteen of the locomotives were technically still active to assist with restricted workings such as ballast duties. At that time Tinsley held 47249, 47070, 47317, 47345 and 47371 in restricted use.

Table 1 – RfD's Locomotive Pools March 1991 – as Reviewed by RfD for Speedlink Cessation

Pool	No.	Details
FDAT – RfD Main Fleet	110	Classes 47/0 and 47/3 with "all parallel" wiring and preferred machines for longer term use.
FDBT – RfD Reserve Fleet	37	High-engine-hours and overdue overhaul, "all parallel" 47/0 and 47/3s – to move to FDAT pool, replaced by 23 Class 47/4s
FDCT	12	"Series parallel" Classes 47/0 & 47/4s to be stored at Tinsley for spares for FDAT machines and then condemned.

Above: A general view of Tinsley taken from the embankment at the east end of the shed on 19th April 1993. RfD Class 47s dominate the scene. *Chris Booth*

Left: Metals sector 37248 on depot on 19th April 1991. This loco was delivered to Cardiff Canton on 30th October 1964 as D6948, moving to Healey Mills (55C) in August 1967. January 1972 saw it transfer back to Cardiff Canton before a move to Tinsley came in 1992 and a return to Cardiff in 1994. Transferred to Toton in 1995 as part of the Mainline Freight allocation and withdrawn in 2003, it was purchased by the Growler Group in conjunction with West Coast Railway Company, to whom the locomotive was loaned and entered WCRC service in 2006. The loan ended in 2009 and the loco entered preservation. Now based on the Gloucestershire Warwickshire Railway at Toddington, it has been restored to its original condition. *Chris Booth*

Left: Large logo blue 37128, unofficially named "Jupiter", is seen at Tinsley on the same date. New as D6828 to Cardiff Canton, this loco was transferred to Tinsley from 8th March 1975. It was not as lucky as its sister, being withdrawn on 1st February 1997 and broken up by T.J. Thompson and Son Ltd, Stockton by 1st November 2001. *Chris Booth*

The End of Tinsley's Class 37s

On 1st January 1989, Tinsley still had a respectably-sized fleet of Class 37s, which were mainly allocated to the following two pools:

- 14 in FAMT – (Freight Aggregates Tinsley) which were mainly outbased at Buxton for stone traffic from the Peak Forest area.
- 34 in FGWB – (Freight Speedlink Tinsley Class 37) which supplemented Class 47s on the former Speedlink and RfD workings.

September 1990 saw 27 Class 37s, formerly used predominantly on trainload workings, leave Tinsley for other depots.

Tinsley continued to lose its Class 37s throughout the early 1990s as re-allocations and withdrawals reduced the fleet still further. A few RfD Class 37s continued to be used in pairs until they were replaced by Class 47s, newly fitted with multiple-working equipment. A notable working booked for a pair of Tinsley Class 37s was one of RfD's longest block workings – the Spillers Petfood train, which ran weekdays along the ECML from Wisbech in Cambridgeshire to Deanside Transit near Hillington, west of Glasgow. Pairs of Tinsley Class 37s were also used on Freightliner services such as Southampton to Trafford Park. There were some allocations into Tinsley, notably a tranche of Class 37/4s in 1991 and 1992, but March and April 1994 saw another 14 locos transferred away. The final six Class 37s allocated to Tinsley are as listed in Table 2:

Table 2 - The Final Six Class 37s Allocated to Tinsley and Their Pool Codes From – To

Loco No.	Final Pool at Tinsley	Pool Reallocated to After Tinsley	Date
37154	MDTT – RfD Tinsley unrefurbished Class 37 locos (Extended Range)	DAYX – RfD (Stored Locos)	23/04/1994
37298	DART – RfD Tinsley (Restricted)	DAYX – RfD (Stored Locos)	24/07/1994
37131	DAXT – RfD Tinsley (Reserve Fleet)	FDYX – Trainload Freight North -Stored Locos	25/09/1994
37178	DAXT – RfD Tinsley (Reserve Fleet)	LNDK – Trainload Freight West - CF Class 37/0 (Departmental)	25/09/1994
37218	DART – RfD Tinsley (Restricted)	FDYX – Loadhaul - Stored Locos	10/09/1995
37238	DART – RfD Tinsley (Restricted)	FDYX – Loadhaul - Stored Locos	10/09/1995

During May 1991 RfD, undertook trials with a Class 59/1 and a Class 60 on 75mph 2,000-ton Freightliner trains, with a view to replacing its aging Class 47 fleet. Nothing came of the trials.

In the final months of Speedlink. RfD had been using seven Class 20 locomotives, two of which – 20137 and 20165 – had been noted a long way from home, working from Thornton Yard to Crombie RNAD in Fife. On 7th December 1992, Tinsley's long association with Class 20s came to an end when its final three – 20118, 20137 and 20165 – departed for Thornaby.

General Freight Workings

One small flash of silver lining in the cloud of Sectorisation came when Nottingham and Derby traincrew depots were assigned exclusively to Regional Railways and InterCity respectively. Trainload Freight pulled out of the depots and, whilst the bulk of their work went to Toton and Saltley, Tinsley also picked up some additional work.

Tinsley's RfD Locomotives on Passenger Turns

Besides RfD work, Tinsley's locos were still employed on a variety of passenger work. Of particular note was the Saturdays-only (SO) 08.47 Birmingham to Yarmouth service, which was booked for an RfD Class 47. 1992 also saw RfD Class 47/0 locomotives covering a SO Crewe–Holyhead diagram, whilst another couple covered SO Sheffield/ Birmingham New Street–Great Yarmouth diagrams. Tinsley RfD Class 37s also had summer-dated diagrams with two 37/4s covering Saturdays-excepted (SX) Blackpool–Liverpool turns, whilst a third had a SX Southport–Manchester Victoria diagram, with corresponding SO turns.

Network SouthEast (NSE) was also making use of Tinsley Class 47s, such as on 27th August 1991 when 47150 handled the 18.37 Paddington to Oxford and 21.40 return. Such locomotive-hauled services were soon to end as Regional Railways and NSE shed themselves of loco-hauled stock in pursuit of wholly DMU operation. However, the rescuing of failures was still a fairly common occurrence, such as when 47095 powered a pair of failed Class 108 DMUs on a Carlisle to Newcastle service on 20th June 1991. 7th July saw 47156 on a Class 1 duty, rescuing the 11.00 Carlisle to Euston from Warrington Bank Quay after overhead line problems. Other notable Tinsley-allocated workings included 47237 reaching Brighton with the 11.26 from Glasgow on 20th August and 47321 on a Dover to Liverpool working on 17th August 1991, the same loco reaching Holyhead on the 17.03 from Euston, forward from Crewe. Other locations visited included Blackpool, Llandudno, Exeter and Nottingham. Finally, Tinsley even lost a couple of its Class 47s to Regional Railways in late August 1991 as they were transferred to Inverness to help out on the Inverness to Aberdeen and Kyle of Lochalsh lines. Table 3 shows RfD's Tinsley-based active locomotive pools as at 8th July 1991.

Table 3 - RfD's Tinsley Active Locomotive Pools as at 8th July 1991

Pool	No.	Class	Details
MDRT	8	Re-geared Class 37	Manchester commuter and Teeside
MDST	13	Single Tank Class 37	Anglia Freightliner
MDTT	23	Twin Tank	General use
MDAT	104	Class 47	General use
MDCT	15	Class 47	Experimental automotive fleet

Below: In Railfreight Distribution livery, 37053 departs from Tinsley Yard via the main yard departure line with 6V14, 08.05 Hull Saltend to Baglan Bay acetic acid tanks (commonly known as the vinegar tanks due to the smell!). This was one of the surviving bulk workings after the demise of the Speedlink network. It ran on MWFO and arrived in Tinsley Yard at 10.37. The loco then ran round and departed at 11.06, via Treeton South and the "Old Road". The train returned as 6E33 TThSO, 20.14 Baglan Bay to Hull Saltend, arriving Tinsley at 02.48 and departing at 04.25 via the Shepcote Lane end of the yard. 6V14 is seen here heading towards Treeton on 23rd April 1992.

This loco was new to Darnall (41A) as D6753 on 14th September 1962, moving to Tinsley (41A) on 2nd May 1964. It then spent time at Ipswich, Stratford, March, Gateshead and Thornaby, before returning to Tinsley on 16th July 1989. Moving back to Thornaby in March 1994, it was renumbered 37344 on 10th November 1994, and was transferred to Immingham in November 1995. Stored in June 1996, it would be withdrawn on 1st July 1996 but would not be cut up until 2005 at Barrow Hill. *Chris Booth*

Supertram Works and Further Yard Rationalisation

The South Yorkshire Supertram system was authorised by a Parliamentary Act of 1985 and, on 19th March 1992, a short ceremony was held to mark the start of track-laying for the £240 million project. The system was opened in stages during 1994 and 1995, with a maintenance depot built on the 15 acre site of the former Nunnery carriage sidings, adjacent to the former GCR line from Sheffield Victoria to Worksop.

In May 1991 the area manager for Yorkshire freight issued a précis for consultation which read:

Freight Infrastructure in the Sheffield/Tinsley Area Including Proposals for Sheffield Supertram

1. Introduction

1.1 Proposals to build a light rapid transit system in Sheffield are being developed by South Yorkshire Supertram Ltd, a subsidiary company formed by the South Yorkshire PTE. The Supertram plans envisage using a route in the Lower Don Valley based on BR's existing line between Woodburn Junction and Broughton Lane.

1.2 The review of freight activity in the Sheffield area which has been carried out in the last year indicated a capability to encompass the Supertram proposals whilst at the same time enabling the continuance of BR's operations, utilising reduced infrastructure.

1.3 Separate proposals have already been formulated to reduce the size of Tinsley Yard in line with current and foreseen business requirements and the retention of 21 sidings in the present main yard forms the base on which track and signalling changes for the routes into and out of Tinsley Yard are to be developed.

2. Proposals

The main elements of the changes being proposed are as follows:

2.1 Singling of the line between Woodburn Junction and Tinsley East Junction to enable a parallel construction of the Supertram line.

2.2 Bi-directional signalling of this route, controlled from Woodburn Junction signal box.

2.3 Singling of the north east and south west curves between Broughton Lane Junction/Tinsley South Junction and Shepcote Lane Junction, both curves to be signalled for bi-directional working.

2.4 Remodelling of the throat areas at the west end of Tinsley (main) and (based on retention of 21 sidings) provision of a single bi-directional approach line from Shepcote Lane Junction, this area to be controlled from Tinsley Yard signal box.

2.5 Abandonment of the connections to the remainder of the main yard, the secondary yard, express freight sidings and engine line at Tinsley Park, but connections to the BSC SMACC site are retained, as is sufficient of the engine line at Tinsley Yard for access to/from the traction maintenance depot.

2.6 Remodelling of Woodburn Junction area to permit use by Supertram of the former Nunnery carriage sidings site and removal of other redundant connections.

3. Effect of Proposals on Traffic Staff

3.1 The effect of these proposals is to close the existing signal boxes, as follows:

Tinsley Park saving 3 x Class D Signalmen
Tinsley East saving 3 x Class A Signalmen
Shepcote Lane saving 3 x Class C Signalmen

Associated relief signalmen savings will also be achieved.

3.2 Signal boxes at Woodburn Junction and Tinsley Yard will take on additional areas of control as specified in (2) above. It will be necessary to reassess the grading of these signal boxes.

4. Progression and Staging of the Works

4.1 It is envisaged that physical work will start in September 1991 leading to a target date for commissioning the BR works in June 1992.

4.2 The exact staging of the works and the ability to meet the target date are still being considered by technical departments in conjunction with SYSL's contractors.

4.3 It is the intention to provide updated information on the staging of the scheme through the joint consultation procedure and at the same time consider the effects on the staffing at the locations involved.

Tables 4 and 5 show details of the signal boxes affected by the Supertram works..

Table 4: Opening and Closing Times of Signal Boxes in 1991 Consultation

Signal Box	Staff	Present	Proposed
Tinsley Yard	3 x Sig 'E'	Continuous	No change
Tinsley Park	3 x Sig 'D'	2200 Sun–0600 Mon 0600–1400 Mon to Sat 1400–2200 Mon to Fri RD Sat (NC) 2200–0600 Mon to Fri RD Sat (NC)	Box closed
Shepcote Lane	3 x Sig 'C' 1 x Train Register Lad	Continuous 0730–1530 FSX 0730–1430 FO RD Sat (NC)	Box closed Train Register Post withdrawn
Tinsley East	3 x Sig 'A'	2200 Sun to 0600 Mon 0600–1400 Mon to Sat 1400–2200 Mon to Fri 2200–0600 Mon to Fri RD Sat (NC)	Box closed

Table 5: Summary of Staffing and Financial Effects

Signalbox	A	B	C	D	E
Tinsley Yard – Sig 'E'	3	3	3	-	-
Tinsley Park – Sig 'D'	3	3	0	-3	-27,987
Shepcote Lane – Sig 'C'	3	1	0	-3	-25,008
Shepcote Lane Train Register Lad	1	2	0	-1	- 4,992
Tinsley East – Sig 'A'	3	2	0	-3	-21,406
RDR Signalman Class 3	2	2	1	-1	-
GPR Signalman Class 3	3	3	2	-1	- 9,329
Total	**18**	**16**	**6**	**-12**	**-88,722**

A = Present Establishment. B = Actual Establishment.
C = Proposed Establishment. D = Difference.
E = Financial Effect (£).

Yard Changes

February 1992 saw express freight sidings Nos. 1 and 2 taken out of use; March saw major permanent way alterations to the yard and approach lines in connection with the impending Supertram works. Shepcote Lane junctions saw a reduction in trackwork with points replaced by plain line, and on Sunday 9th August 1992 there were further track reductions at Tinsley Park. The arrival and departure lines between Brightside Junction and Shepcote Lane Junction were temporarily taken out of use on 20th September 1992 and Tinsley Park signal box was closed at the same time, with control of the area passing to Tinsley Yard signal box.

On 1st November 1992 the "down" line from Tinsley South Junction, through Broughton Lane Junction to Woodburn Junction was taken out of use, together with all associated signalling, to allow the Supertram lines to be built. On 8th November, Shepcote Lane signal box was decommissioned together with all its associated signaling. Then, on Thursday 12th November, the "down" line between Tinsley East Junction and Tinsley South Junction, and the north west arrival line between Tinsley South Junction and the Avesta SPACE sidings ground frame were taken out of use. These changes were completed by Sunday 6th December 1992; the line from Woodburn Junction to Tinsley East Junction was now single line. A new single to double junction was provided at Tinsley East Junction, with double track from there towards Rotherham. The "up" line between Tinsley South Junction and Tinsley East Junction was taken out of use and its signalling disconnected, as were the following lines and signaling:

- The north west departure line between Shepcote Lane and Tinsley South Junction.
- The south west arrival line between Shepcote Lane and Tinsley Yard with approximately 140 yards retained at the depot end for access to the depot.
- The east arrival line between Tinsley Yard signal TY.249 and Tinsley Park.
- Shepcote Lane Junction – all signalling disconnected and removed except for the routes from Tinsley South Junction to Woodburn Junction and from Tinsley South Junction to Tinsley Yard.

The double-track south west arrival and departure line formations between Broughton Lane Junction and Shepcote Lane Junction were re-laid as a single arrival/departure line, as were the north west arrival and departure lines between Tinsley South Junction and Shepcote Lane Junction. Shepcote Lane Junction itself was remodelled into a double line towards Brightside Junction and a single line towards Tinsley Yard. Tinsley East Junction signal box was decommissioned and the original Woodburn Junction signal box replaced by a new portable building-style signal box, which contained a GEC-Alsthom NX panel. This was erected adjacent to the old box and controlled new signalling in the aforementioned areas, along with the Worksop and Deepcar lines, as the old box had before it.

Completion of the resignalling saw Woodburn Junction become the fringe box to Sheffield PSB on the lines from Rotherham Central towards Tinsley, as well as from Worksop to Sheffield. Track circuit block regulations were worked throughout, except between Woodburn Junction and Woodhouse Junction where absolute block regulations continued to apply, and from Woodburn Junction to/from

Above: The unassuming portable building that replaced the original box at Woodburn Junction. *Chris Booth*

Left: The original GCR 1906-built Woodburn Junction signal box is seen here in 1975. The box was double-manned at the time and a Mark 2 Ford Cortina waits for one of the signalmen to come off duty. The track disappearing behind the car is the remains of the former Nunnery colliery branch. *Alan Hazell*

Left: Two camera-shy signalmen inside the box, which originally housed a GCR 81-lever tappet-locked frame, but which was replaced with a 104-lever Railway Signal Co. frame in 1949. *Alan Hazell*

Below: The new panel housed in Woodburn Junction box in 2017, replacing that from 1992. *Chris Booth*

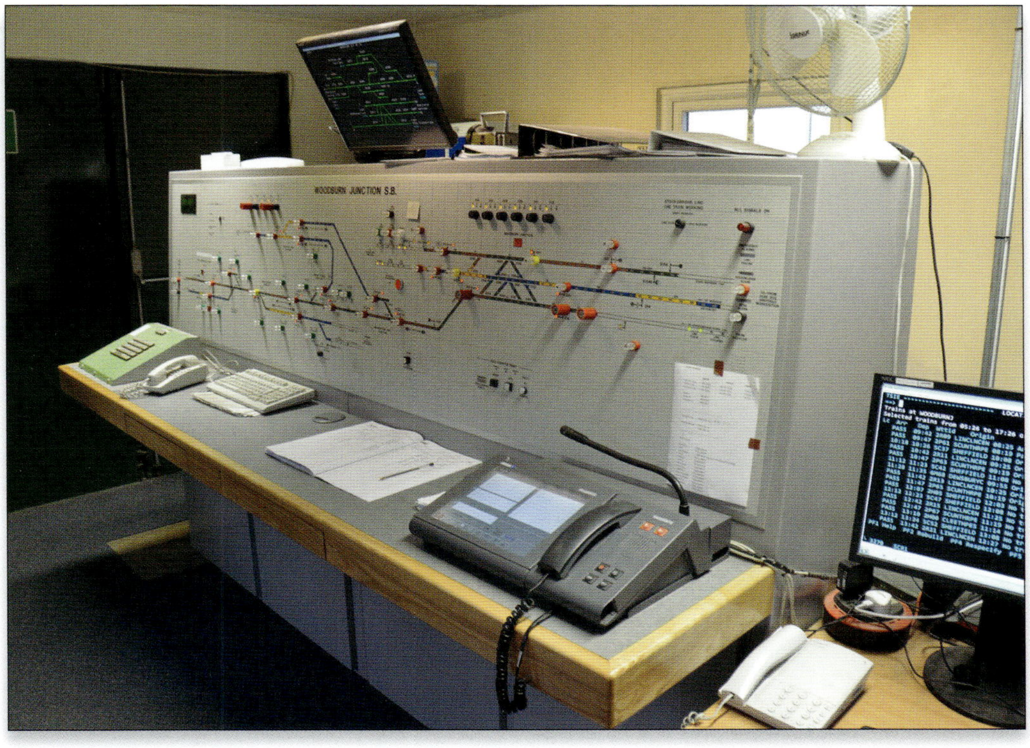

Deepcar where one-train-working regulations continued to apply.

Intended to be a temporary measure pending completion of the resignalling scheme, the "down" and "up" Tinsley Brightside arrival and departure lines between Shepcote Lane and Brightside Junction remained out of use and, although the new panel at Woodburn Junction could control the route, it would never be used again. They were officially decommissioned on Saturday 13th November 1992 with the tracks being lifted in 2007.

Into 1993

January 1993 saw more track realignments and the once large yard reduced to just 20 useable sidings, The overhead electrified lines for the Supertram system between Tinsley South Junction and Woodburn Junction and Meadowhall station were energised at 750 volts DC from 00.01 on Saturday 3rd April 1993, allowing the lines to be used for testing purposes.

The end for the route out of the east end of the yard came on Monday 17th May 1993, when the lines beyond the Sheffield Parkway bridge from Catcliffe Junction to Treeton North and Treeton South Junctions were taken out of use. This also saw the closure of the reception sidings, although an engine line would be retained to allow access to the east end of the depot via the zig-zag route up the bank. The closure of the through route was reputed to be due to the cost of infrastructure renewals and the desire by Trainload Freight to save money. The final train to use the Catcliffe to Treeton North curve was a Didcot to Doncaster working on 17th May 1993, hauled by 37298. From then on, the only access to Tinsley would be from the Shepcote Lane end of the yard. Treeton North Junction was eventually plain lined but the interlocking remains for future reinstatement of the junction should it ever be required.

Table 6 is the final Working Timetable for trains running through the yard from both the Catcliffe and Shepcote Lane ends; after this, access would only be from the Shepcote Lane end of the yard.

The Changing Shape of Shepcote Lane Junction

Below: The junction as built, with the yard to the right of the diagram. *Chris Booth collection*

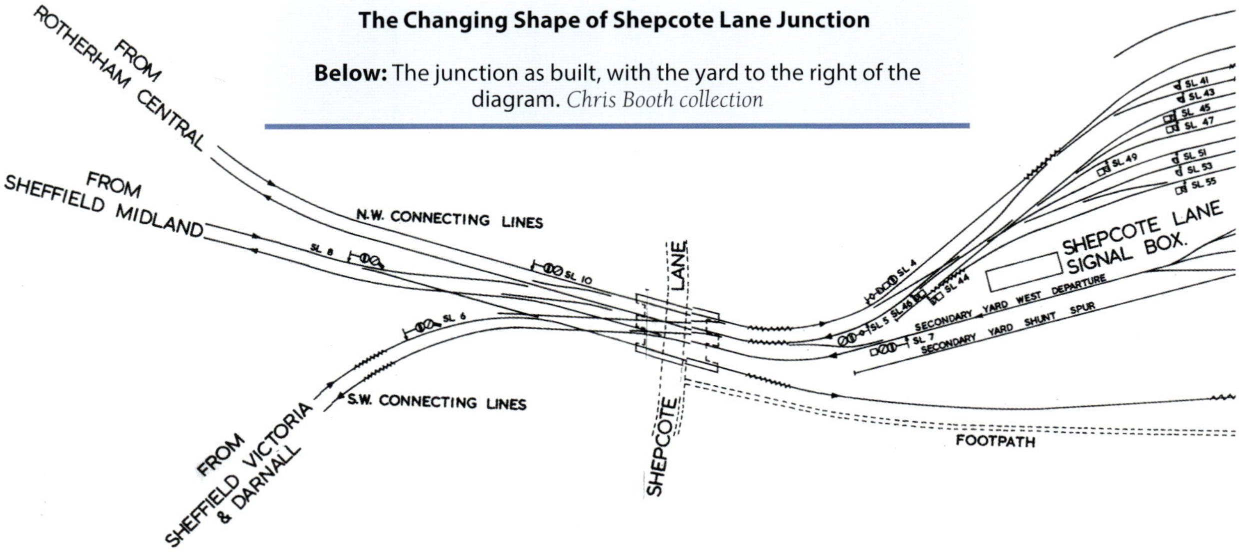

Left: The view from the cab of a loco arriving at Tinsley Yard in 1977. The loco has traversed the route from Brightside (marked on the plan "from Sheffield Midland") and is about to cross bridge 11 over Shepcote Lane and take the line to the diesel depot on the right of the image. This line is shown on the plan just above the word "Footpath". Shepcote Lane signal box can be seen just beyond the pipe bridge. *Roger Wainwright*

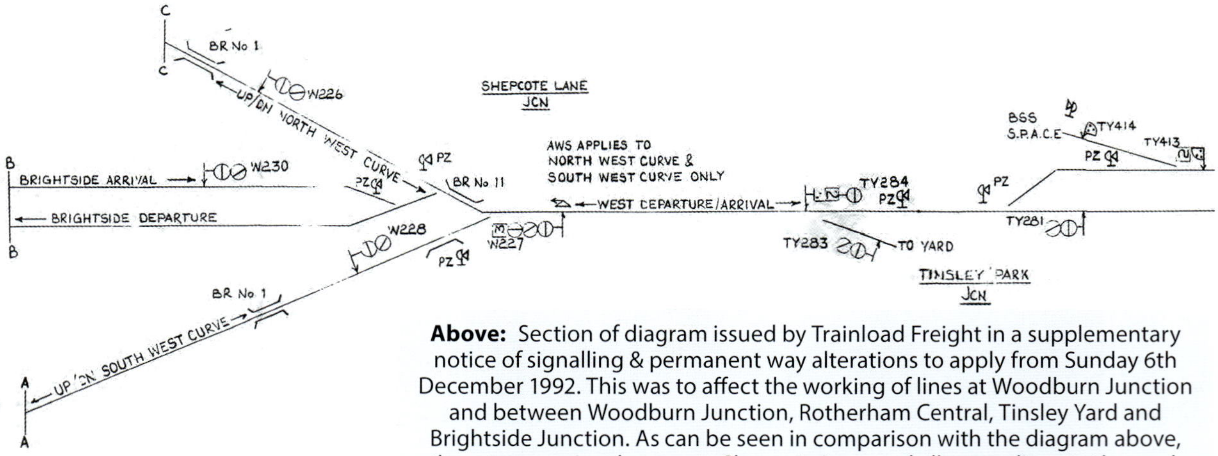

Above: Section of diagram issued by Trainload Freight in a supplementary notice of signalling & permanent way alterations to apply from Sunday 6th December 1992. This was to affect the working of lines at Woodburn Junction and between Woodburn Junction, Rotherham Central, Tinsley Yard and Brightside Junction. As can be seen in comparison with the diagram above, there were major changes to Shepcote Lane and all access lines to the yard. Although re-signalled, the Brightside arrival and departure lines were never used again. *Chris Booth collection*

Left: After resignalling and track rationalisation, this is Shepcote Lane Junction in 2007, as controlled by Woodburn Junction. The signal is W230, which protects the junction from the Brightside direction although, as can be seen, the track from that direction has been lifted. The north west arrival/departure is to the left, with the south west arrival/departure hidden to the right. Both of these are single lines. *Kim Collinson*

Right: A view in the opposite direction to that above. The south west arrival/departure on the left is out of use at this time. The 15mph sign refers to the route towards Brightside, seen diverging from the single north west arrival/departure but which has never been used. The signal left of centre is W228, whilst W228 can be discerned to the right of the 15mph sign. *Dave Stones*

Above: Although displaying Trainload Coal decals, 60090 "Quinag" takes a train of steel from Tinsley Yard along the north west arrival/departure line at Shepcote Lane on 29th May 2001. The pipe bridge at the junction, seen behind the train, allows a direct comparison with previous images of this location. *Stewart Donohoe*

Left: Approaching Broughton Lane, 20003 and an unidentified sister bring a rake of 21-ton hoppers along the "up" goods line in June 1976. The Sheffield & Tinsley Canal is to the left, whilst behind the train is the Tinsley Wire works. *David Ford*

Right: This is the same location during construction of the Sheffield Supertram system. The BR track is now single from Woodburn Junction to Rotherham as the trackbed for the trams is being prepared. The background has changed considerably; all the steelworks buildings have been replaced by the Don Valley Stadium and leisure areas. Notice the steelworks crucible, preserved adjacent to the pylon. *Nial Jordan*

Left: The BR line has now closed and double-track tramway is being laid along the route adjacent to Tinsley viaduct. *Nial Jordan*

Below: This is the Blackburn Valley line in 1976, looking northwards towards Tinsley West Junction signal box and level crossing. The line coming in from Tinsley East Junction can just be seen. Tinsley Viaduct, carrying the M1 motorway, dominates the scene. *Roger Wainwright*

Table 6 - Tinsley Yard 28th September 1992 to 16th May 1993

Down					
6D82	FO	00:30	Tinsley SS	Immingham SS	
6E33	MWFO	19:35	Baglan Bay BP	Hull Saltend	Arr Tinsley 02:56 Dep 04:23 'T'
6E71	SO	01:17	Bescot Yard	Aldwarke BSC	Arr Tinsley 03:46 Dep 03:48 'C'
0V14	MO	04:38	Tinsley TMD	Hull Saltend	Dep Tinsley 04:38 Via Treeton
6E22	TThO	04:30	Spondon	Hull Saltend	Arr Tinsley 06:01 Dep 06:40 'C' 'T'
6E71	MSX	04:03	Bescot Yard	Aldwarke BSC	Arr Tinsley 06:33 Dep 06:44 'C' 'T'
6E49	MSX	03:00	Bescot Yard	Selby Store & Freight Co.	Arr Tinsley 06:42 Dep 08:18 'T' R/R
6E49	MO	03:00	Bescot Yard	Selby Store & Freight Co.	Arr Tinsley 06:50 Dep 08:18 'T' R/R
6E68	TThO	09:00	Peak Forest	Selby Melmerby Estates	Arr Tinsley 10:13 Dep 10:41 'C' 'T'
6D45	TThO	10:32	Aldwarke BSC	Shipley Crossley Evans	Arr Tinsley 10:45 Dep 10:51 'T'
6D53	WO	10:58	Tinsley	Immingham	Dep Tinsley 10:58
6N14	MWFO	12:11	Aldwarke	Stockton Thompson	Arr Tinsley 12:26 Dep 12:59 'C' 'T'
6L43	TThO	13:22	Kilnhurst West	Thames Haven	Pass Tinsley 13:42
0E76	SX	15:13	Tinsley TMD	Holbeck LIP	Dep Tinsley TMD 15:13
6E47	SX	12:07	Cardiff Tidal TC	Scunthorpe	Arr Tinsley 17:34 Dep 19:53
6E15	TThO	16:36	Kingsbury Sidings	Aldwarke BSC	Pass Tinsley 18:33
6M81	FO	20:38	Aldwarke	Workington SS	Arr Tinsley 20:53 Dep 20:55 'C'
6E73	WO	21:50	Earles Sidings	Tees NY	Arr Tinsley 22:51 Dep 22:56 'C' 'T'

Up					
6D50	MX	00:28	Masborough FD	Scunthorpe	Pass Tinsley 00:42
6J50	MO	05:35	Tees NY	Aldwarke	Arr Tinsley 08:17 Dep 08:23 'C' 'T'
0E76	MO	06:27	Leeds FLT	Tinsley TMD	Arr Tinsley TMD 08:29
6D70	SX	08:35	Masborough FD	Scunthorpe	Pass Tinsley 08:44
6J40	WO	07:30	Immingham	Tinsley SS	Arr Tinsley 09:15
0E76	SO	08:02	Holbeck LIP	Tinsley TMD	Arr Tinsley TMD 09:26
0E76	MSX	07:36	Holbeck LIP	Tinsley TMD	Arr Tinsley TMD 09:37
0E33	SO	06:47	Hull Saltend	Tinsley TMD	Arr Tinsley TMD 10:20
6V14	MWFO	08:08	Hull Saltend	Baglan Bay	Arr Tinsley 10:49 Dep 11:19 R/R
6M44	TThO	12:11	Aldwarke BSC	Kingsbury Sidings	Arr Tinsley 12:26 Dep 12:28 'C'
6J45	TThO	13:47	Shipley Crossley	Aldwarke BSC	Arr Tinsley 15:51 Dep 15:58 'C' 'T'
8X06	FO	15:12	Doncaster West Yd	Didcot TC	Pass Tinsley 16:26
6D49	SX	17:00	Masborough FD	Scunthorpe	Pass Tinsley 17:09
6M16	WO	16:13	Tees NY	Earles Sidings	Arr Tinsley 19:14 Dep 19:16 'C' 'T'
6J38	MWFO	16:44	Stockton Thompson	Aldwarke BSC	Arr Tinsley 20:17 Dep 2019 'C'
6M62	SX	21:05	Aldwarke BSC	Bescot Yard	Arr Tinsley 21:20 Dep 21:39 'A'
6J28	ThO	21:00	Immingham	Tinsley SS	Arr Tinsley SS 22:55

'C' Crew change, 'A' Attach only, 'T' Traffic or shunts for other trains to pass, R/R Run round

Dedicated Staff

Staff at Tinsley often showed a remarkable level of dedication to duty. Former Tinsley fitter, Phil Hodgkiss, recalls an unscheduled trip in the middle of the night to ensure a loco was returned to traffic as soon as possible.

"Whilst working a night shift on 25th May 1992 I was working on 37381 after it had received a B exam. During the run up of the air tanks I could hear air escaping from No.1 bogie. After investigation it was found that two brake cylinder seals were blowing by. Unbelievably we didn't have the right seal in stock, but a quick call to Immingham depot revealed they had as many as we could need, so at 03:00 we set off along a very quiet M18 and made it back to Tinsley for 06.00 complete with brake cylinder seal".

Competition Time

In early 1993, local schools held a competition to design a railway poster on the theme of why it is unsafe to play on the railway. The winners were from Catcliffe Junior School and on 20th April 1993 two of the pupils were invited to unveil the name Catcliffe Demon on the side of Tinsley-allocated 47186. The locomotive had been fitted with a diesel engine monitoring system (DEMON), hence the name. The system sent information on the locomotive's technical equipment via a radio link to either Tinsley depot or BR Research at Derby.

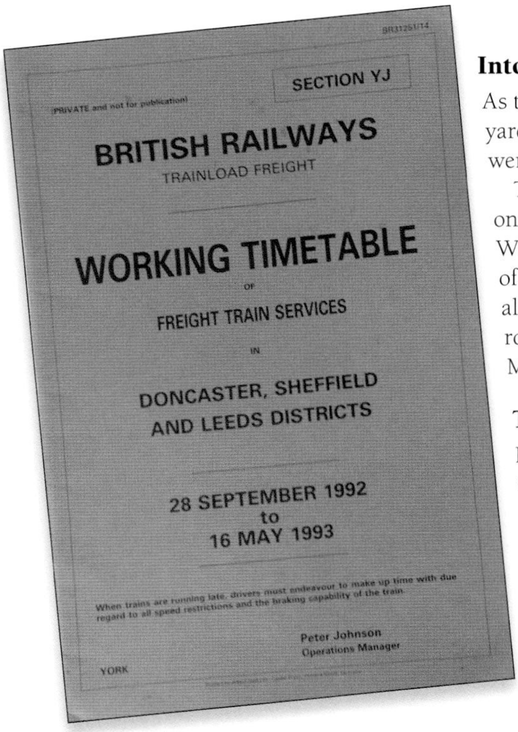
BR31251/14

SECTION YJ

(PRIVATE and not for publication)

BRITISH RAILWAYS

TRAINLOAD FREIGHT

WORKING TIMETABLE

OF

FREIGHT TRAIN SERVICES

IN

DONCASTER, SHEFFIELD AND LEEDS DISTRICTS

28 SEPTEMBER 1992 to 16 MAY 1993

When trains are running late, drivers must endeavour to make up time with due regard to all speed restrictions and the braking capability of the train.

Peter Johnson
Operations Manager

YORK

Into 1994

As the decade progressed, demolition and clearance of redundant buildings in the yard commenced. By February 1994 Shepcote Lane and Tinsley Park signal boxes were no more.

The Sheffield Supertram line from Fitzalan Square to Meadowhall opened on 21st March 1994, sharing the alignment with the former GCR line between Woodburn and the site of Tinsley South Junction. There it followed the route of the former Blackburn Valley Line, adjacent to the M1 Tinsley viaduct, almost to the point where the former GCR route dived beneath the Midland route; it then made a 90 degree turn to reach the tram terminus at the side of Meadowhall station.

The Demise of RfD...

In the years leading up to the privatisation of Britain's railways, the then Conservative Government decided on a franchised model for passenger operations, whilst the freight side was to be broken up and sold off in pieces. The freight companies for sale were Rail Express Systems (RES) and Freightliner, which included RfD and Trainload Freight. In 1994 an effort was made to stimulate greater competition and Trainload Freight was re-organised by geographical areas, rather than by commodity as had previously been the case. Three new businesses were created – TFL South-East, TFL North-East and TFL West, although they were quickly rebranded as Mainline Freight, Loadhaul and Transrail respectively. At this time much of Tinsley's former Trainload Metals traffic was being worked by Loadhaul.

On 13th March 1995, the 100 drivers and train crew (who, by this point, had come under the Loadhaul umbrella) were moved from Tinsley and relocated to a train crew depot at Rotherham steel terminal. This was a simple prefabricated building provided merely to afford drivers somewhere to sign on and read notices etc.

According to staff, the week prior to moving was hectic, as everything was gradually moved to Rotherham although the depot at Tinsley was still running. Equipment not required was left at Tinsley for staff (with permission) to help themselves to, such as the first aid stretchers, (which were stamped LMS), finding their way to the Midland Railway-Butterley. After closure, Tinsley's train crew buildings were soon vandalised. Metal thieves helped themselves to the pipework, with the result that water was running everywhere from toilets and missing radiators. The buildings were soon demolished.

With Tinsley's drivers now based at Rotherham, RfD Saltley-based men had a 6-days-a-week turn to move locomotives between the two depots, with up to five at time moving as 0J03 to Tinsley and 0Z96 back to Saltley. RfD had a contract with Loadhaul for Rotherham men, many of whom were of course ex-Tinsley men, to shunt locomotives around Tinsley depot as required.

...And the Rise of EWS

As the Government developed its plans to sell the freight businesses, management buyout teams formed to acquire some of the companies. At the same time a new business emerged – North and South Railways Limited. This was a consortium headed by US freight operator Wisconsin Central, with additional financing from Berkshire Partners, Goldman Sachs and Fay Richwhite. On 9th December 1995, North and South Railways purchased RES for £24.2 million and, on 24th February 1996, in a move which rather undermined the Government's aspirations for competition, Wisconsin Central purchased all three Trainload Freight businesses for £225 million. The sale of RES included 164 locomotives and 677 postal vans which, when added to Trainload Freight's total compliment of 914 locomotives and 19,310 wagons, gave the new owner an unassailably dominant position in the rail freight market. Wisconsin asserted they had only a 6% share of the total freight market, when road haulage and other forms of freight transportation were taken into account. The four companies continued to trade under their existing names, but in October 1996 Mainline and Loadhaul were merged into Transrail and the combined company renamed English Welsh & Scottish Railway Limited (EWS). Thus, the Rotherham train crew depot came under EWS's control, along with most of the workings in and out of Tinsley Yard, except for any RfD workings and light engines visiting Tinsley TMD.

Until 1995 Tinsley still maintained Class 47s used on the national Freightliner network. As Freightliner and RfD were split for sale, the former handled intermodal trains whilst the latter had mainly less-than-trainload services. In October 1995, 30 of Tinsley's Class 47s were transferred to Crewe Diesel depot for Freightliner's exclusive use, which meant by March 1996, Tinsley was down to an allocation of just 95 operational Class 47s.

The first Class 47 to be repainted for the Freightliner group was 47270, completed at Tinsley and released into traffic in early August 1995 carrying the standard triple grey livery but without decals. The next was 47376 following a light overhaul at ABB Crewe, during which it lost the unofficial Tinsley "Skylark" name it had carried since 21st November 1989. It emerged in the new Freightliner livery of triple grey and red triangle, with the word "Freightliner" near the solebar.

On 25th May 1996 Freightliner was purchased for £5.39 million in a management buyout backed by 3i and Electra Private Equity. As part of the sale the RfD business was required to relinquish some of its Class 47s to Freightliner, as the latter company's locomotives were considered to be some of the poorest still in traffic. RfD was still Government-

Above: A diverted Class 101 DMU has just passed Shepcote Lane No. 9 signal as it drops down towards Broughton Lane Junction in 1987. Broughton Lane Sidings can be seen to the left of the image, with the "up" and "down" main next, followed by the "down" goods, whilst the wagons adjacent to the unit are in Carbrook "down" sidings. *Ted Evans*

Above: Class 37s 37612 + 37605 drop down to Broughton Lane Junction with 1Q13, 08.21 Derby to Derby via Sheffield, Rotherham, Aldwarke, Tinsley, Woodburn, Deepcar, Tinsley and Sheffield on 26th February 2014. The former double-track is now single, as is the Woodburn to Rotherham line. Stagecoach Supertram No. 107 makes its way from Meadowhall to the City; the steelworks in the photo on the left is now history. *Chris Booth*

owned at this time but with Freightliner and RfD now two separate companies, Freightliner issued a tender for future maintenance of its locomotive fleet in August 1995. In the interim Tinsley was still used.

On 24th December 1996, it was announced that EWS was the preferred bidder for the loss-making RfD business. In the run up to the sale of RfD, EWS and another company, "Management Consortium Bid Limited", were both vying for the right to take over RfD and perhaps, had the latter been successful, the history of Tinsley could have been quite different. Freightliner also made a bid for the company shortly before the sale of RfD to EWS was agreed on 12th March 1997.

Throughout mid-1997 there were wranglings between the UK Government and the EU, over subsidies and grants to cover the £168 million track access charges payable by RfD for the Channel Tunnel

Below: 47619 passes Tinsley South Junction with a diverted Newcastle to Poole service on 4th April 1988. The train is on the then double-track route towards Woodburn Junction. The route towards Shepcote Lane and Tinsley Yard has not been used for some time, judging by the rust on the railhead. The building with the red door is the former GCR Tinsley station goods shed; it once had three sidings, one of which ran through the door. Tinsley station was beyond the overbridge and closed on 29th October 1951. *Ted Evans*

Left: Tinsley South again, this time on 7th June 2003. Both the former north west arrival/departure lines, and the line towards Woodburn Junction, have been singled. Stagecoach Supertram No. 109 heads along the double-track tramway towards Meadowhall as 60001 "The Railway Observer" makes its way towards Tinsley with a train of coil from Immingham. The goods shed still exists, albeit hidden by the train. Originally named "Steadfast", the loco was renamed on 23rd February 1997 at Toton. *Stewart Donohoe*

which, along with other payments, came to £242 million over eight years. With RfD still notionally an independent company, Tinsley carried on as before. RfD services at the time included international containerised freight, movement of automotive components and cars by rail, and less-than-trainload freight services for the Ministry of Defence.

The last locomotive to receive RfD livery was 47293 in early November 1997; following its use on a customer's charter from Euston to Daventry on 13th November 1997, the locomotive was named "Transfesa" at Tilbury on 24th November.

The sale of RfD went through on 22nd November 1997. RfD added 1,182 staff and 157 locomotives to an EWS business that then stood at roughly 900 locomotives, 19,000 freight wagons and 7,000 employees. Initially RfD remained a stand-alone business with its own management team. Almost immediately, RfD and EWS locomotives began to be used on each other's services; one of the first instances was on 27th November 1997, when Tinsley's former RfD 47051 hauled the 02.25 Bescot to Eastleigh, formed of 28 VAA vans and PFA flats along with 37516 dead in tow behind the loco. The RfD brand continued until 1st December 1998 when it was renamed English Welsh & Scottish Railway International.

Tinsley Diesel Depot's Final Years

With privatisation looming Tinsley began to take on private work, such as giving an 'E' exam to Pete Waterman's Class 46, D172 "Ixion", thus enabling its return to main line use. In early 1996 Class 50 No. 50044 "Exeter" visited the depot and, although it was the first Class 50 to do so, Tinsley's staff were keen to offer their skills to the locomotive's owner and operators.

Tinsley's allocation in mid-March 1996 comprised 93 Class 47 locomotives in three pools. The first and largest pool was DAET with 74 Class 47s fitted with twin fuel tanks and multiple working equipment; these locos formed the bulk of the RfD fleet.

The second pool was DAST, 12 Class 47 locomotives allocated to less demanding jobs due to their single fuel tank, although a couple, 47348 and 47146, were actually equipped with twin tanks. The pool also contained RfD's last operational Railfreight red stripe-liveried Class 47 in the shape of 47367 "Kenny Cockbird". The final pool was DAYX which held seven stored locomotives including large logo ETH-fitted 47473. Table 7 shows the allocation in March 1996.

Table 7- Tinsley TMD Locomotive Allocation as at 14th March 1996

DAET Pool: Class 47s Fitted with Twin Tanks and Multiple-Working Equipment					
47033 E	47156 T	47217 E*	47280 T	47306 E*	47335 T
47049 E	47186 E	47218 T	47281 T	47307 E*	47338 E
47051 E	47188 T	47219 T*	47284 T	47309 T*	47344 E
47053 E	47194 T	47222 T*	47285 E	47310 T*	47351 E
47085 E	47200 E	47226 T	47287 E	47312 E*	47355 T
47095 E	47201 E	47228 E	47290 E	47313 T*	47360 E*
47114 T	47204 T	47229 T	47292 T	47314 T*	47361 T
47125 E	47205 T	47236 T	47293 T	47316 E*	47362 T*
47144 T	47209 T	47237 E	47297 E	47323 E*	47363 T*
47145 S	47210 T	47241 E	47299 E	47326 E	47365 E
47150 E	47211 T	47245 E	47303 E*	47328 T	47375 E
47152 T	47213 T*	47258 E	47304 T*	47330 T	47378 T*
DAST Pool: Single Tank Class 47s					
47079 E	47276 T	47350 O	47356 O*	47370 E	47540 D
47146 E	47348 E	47525 E	47367 R	47379 T	47555 E
DAYX Pool: Stored Class 47s					
47019 O	47234 E	47291 MU	47302 R	47340 D	47473 L
47207 T	* Class 47s Part of Tinsley's Initial Allocation in 1965				

E – European Livery, T – Triple Grey, R – Railfreight Red Stripe, O – Original Railfreight Grey, D – Dutch, L – Large logo Blue, S – Special.

At this time Tinsley was classified as a level 4 depot, able to undertake locomotive examinations up to category 'E' and component exchange maintenance (CEM) work equivalent to an 'F' exam, which avoided the need to send locomotives to a main works. The depot also perfected its work on wheelset refurbishment; two sets of higher-powered jacks were installed for lifting locomotives to facilitate wheelset, bogie and traction motor changes. Power unit changes and generator overhauls also commenced, something for which the depot had never been designed; they were undertaken with the aid of a road crane. After removal the power unit would be sent to Crewe works for overhaul; wheels were sent to Crewe Electric depot for tyre turning. After refitting, the replacement power unit would undergo extensive testing using the depot's own load bank, which was an ex-steam loco tender.

Modifications to the Class 47 fleet also included the fitting of multiple-working equipment, which was to RfD's own modification. This saw a new jumper housing fitted between the marker lights on the former headcode panel at each end of the Class 47. Other work included the fitting of additional fuel tanks and the removal of vacuum exhausters, as RfD no longer operated vacuum braked trains. Tinsley also undertook a range of smaller work including refurbishing batteries and repairing internal equipment such as fuel injectors, turbochargers and brake valves. Except for superstructure and underframe problems, the depot was able to repair most bodywork problems. The most severely damaged locomotive dealt with was 47307 after one of its cabs was burned out following a collision with a car.

Tinsley's Celebrity Locomotives

As the 1990s progressed, locomotive liveries became ever more diverse and the sea of rail blue began to dwindle. At the same time, Tinsley's embellishments and special repaints increased. Of course, Tinsley was well-known for its celebrity repaints, which it undertook alongside general repaints on its Class 47 fleet. Pet project loco 47145 was booked to go for CEM at Doncaster, but this was cancelled and the loco sent to Crewe instead for a generator repair. Surprisingly it retained its Tinsley-applied special BR blue livery with black cab window surrounds, and on its return to Tinsley its "Merddin Emrys" nameplates were refitted. By the time the loco was noted back in service in January 1992, a further embellishment in the form of a red double arrow symbol had been added below the driver's window.

47375 "Tinsley Traction Depot" was another of the depot's firm favourites. This locomotive was first released into traffic on 3rd December 1965, initially to Immingham depot, but was allocated to Tinsley by the end of that month. After being loaned to Leeds in 1966, it was reallocated there and to various other Yorkshire depots, before returning to Tinsley in May 1980 where it stayed until closure of the depot.

Class 37, 37057, arrived at Tinsley on 16th July 1989 and gained the unofficial name "Viking" on 12th November. After various allocations away from Tinsley it returned in 1992 and by early 1993 was one of just five Class 37s left in large logo blue livery. Tinsley kept 37057 going, along with its unofficial name, until it was placed into the RfD reserve fleet in July 1993.

Painting was carried out at the east end of the shed, the west end being used exclusively for mechanical repairs. RfD's Class 47s had their own "European" livery; the upper band of dark grey had been extended further down the body side, a dark blue roof added, plus bodyside lettering and repositioned numbers. Three light grey Channel Tunnel roundels were added below the driver's window, all of which helped to give a modern but reserved twist on the Trainload Freight triple grey livery. Tinsley chose 47217 to be the first Brush Type 4 to receive the new RfD "European" livery. Originally carrying white TOPS numbers, they were soon changed to black, which was more visible on the grey background. As well as repainting its own Class 47s, in 1994 Tinsley also painted a few of RfD's Crewe-based Class 90s.

In 1996 Bill Foster was the Tinsley TMD depot engineer and, along with his staff, was responsible for ensuring the Tinsley Class 47 fleet was performing satisfactorily. Of the 76 Class 47s in the DAET pool, 56 were required at any one time which was effectively three out of four locomotives. At the time fleet availability was running at 73%. Whilst Tinsley was capable of carrying out exams from 'A' to 'E', in practise its allocation would receive many of the smaller exams, especially 'A' and 'B' exams, at other depots. Locomotives used on Wembley to Dollands Moor services were fuelled at RfD's Dover Town yard fuelling point, whilst Saltley could also undertake 'A' exams and fuelling.

The far-reaching nature of RfD's work meant Tinsley's locomotives visited non-RfD depots for fuel and exams on a contract basis, including Transrail's Cardiff Canton, RES's Crewe Diesel and Mainline's Stratford and Didcot. The last example was used as a convenient base for Tinsley's Class 47s working MoD traffic in the South of England, which was concentrated upon Didcot Yard.

Non-RfD locomotives on Tinsley at this time were rare, although Class 60s had come to Tinsley when new for familiarisation purposes and thus weren't entirely unknown. Transrail also had a contract for the depot to fuel one of its Class 56s which worked into the Sheffield area on steel traffic. Finally, from June 1997, shortly before the sale of RfD to EWS, Tinsley undertook some contract work for Virgin Trains when they began sending their Class 47/8 locomotives there.

Almost immediately after EWS's takeover on 22nd November 1997, the end of Tinsley was in sight; rumours about its possible closure grew ever louder. EWS controlled 90% of the rail freight market and a new order of management, driven by greater efficiency and profit was prevalent. EWS had already reduced the workforce from the 17,000 of 1992 to 6,500 by the end of 1997; they were now striving for nearer to 3,000. When EWS bought the Trainload Freight businesses and secured a renegotiated track access agreement with Railtrack, it introduced a profit-sharing agreement, reduced its staff by 1,800, and was able to cut its shipping rates by over 30%. The next area for rationalisation was its locomotives and maintenance facilities.

Tinsley was a long away from the West Midlands-centred automotive traffic, with Bescot depot better suited for local servicing. Saltley could also offer light maintenance and 'A' exams, whilst Crewe was available for power unit changes and other heavy lifting as required. With Toton and Doncaster being so close in railway terms, EWS's management concluded that they could do without Tinsley.

The End is Nigh for Tinsley TMD

Even at this eleventh hour, Tinsley was still capable of undertaking all manner of work. EWS made use of Tinsley's expertise in Class 47 maintenance by sending some of the former RES fleet there, along with some Class 47s still at work on former Trainload Freight services. Work on the latter included bogie swaps, wheelset changes and repairs to turbochargers and intercoolers. Right at the end came one of its biggest contracts for EWS's competitor, Freightliner, when 47302 was allocated CEM at the depot. Part of the

Below: The final day at Tinsley sees 47145 and 08879 positioned for photographs whilst awaiting the arrival of 47375. *Les Nixon*

overhaul required the fitting of twin fuel tanks and, although the depot had never previously carried out such work, the whole task was carried out within two weeks. Much to the disappointment of staff, the power unit failed to arrive in time to complete the CEM and the locomotive was towed to Crewe for completion. Other jobs, such as a wheelset change on 47315 and a holed fuel tank on Virgin Trains 47839, were each completed inside eight hours.

As EWS began to run the depot down for closure, staff were offered a choice of redundancy or transfer to other depots. According to the Sheffield Star newspaper, a senior manager at Tinsley depot said "We are all devastated to be told about the closure but there was not a lot we could do about it". He went on to say "this is the last repair depot of its kind in the South Yorkshire area and it's definitely the end of an era". He finished with "They said we were high quality and could not be bettered but we were in the wrong place because they said no traffic was going by rail to this area".

In January 1998 Tinsley's staff began to leave for other EWS depots at Doncaster, Toton, Immingham and Rotherham. The remaining staff were concentrated on the 06.00–14.00 shift, with just four people on each of the other two shifts of 14.00–22.00 and 22.00–06.00. The disheartening process of removing Tinsley's equipment had also started, Toton and Doncaster receiving Tinsley's equipment for overhauling turbochargers and brake valves etc. Tinsley's work still continued, and on 9th February 1998 Virgin's 47843 arrived for a 'C' exam, followed by 47814 for fuel dilution.

At the end of February 1998, the remaining 13 scrap locomotives were cleared out, the last two – 47543 and 47520 – being hauled away by 37242. The final two locomotives to enter the depot for attention were 47218 and 47146 which both arrived for 'C' exams on 5th March. Maintenance of the former RfD Class 47s officially continued up until 13th March when the remaining locomotives were transferred en-bloc to Saltley depot.

A TOPS locomotive enquiry for 18th March 1998 saw just 47085, 47125, 47145, 47188, 47299, 47328 and 47351 on shed, all out of service and awaiting transfer or materials, with only 08879 in service on shed. By 26th March only 47145 "Merddin Emrys" was on shed, but even that was out of service. 08879 was still the only locomotive in service; Tinsley was responsible solely for this one loco, which acted as depot pilot. 08413 and 08751 were also stored on depot, but later moved to RFS (E) Ltd at Doncaster for a possible return to work.

The following day Tinsley closed for maintenance and a photocall was held in front of freshly-painted 47145 "Merddin Emrys". Later that day, and most fittingly, 47375 "Tinsley Traction Depot" was sent from Saltley to Tinsley to collect 47145, working as 0J03 11.30 Saltley to Tinsley, arriving 35 minutes early at 12.47. A final photo opportunity was taken with the two Class 47s posed with the final Class 08 on depot, 08879 "Sheffield Children's Hospital". Booked to depart the depot as 0Z96 15.00 Tinsley to Saltley, amid a cacophony of horn blasts and exploding detonators, the duo departed at 13.50 to Saltley, thus bringing an end to the career of the depot. Responsibility for the Saltley out-based shunting locomotives passed to Allerton, whilst 08879 became a Doncaster loco. By the time Tinsley closed its doors, at least 288 differently numbered Class 47s had been allocated there at various times in its life.

On 4th April 1998, a social event to mark the closure was held in the Sheffield railway club. Due to vandalism and metal theft, it was not long before the depot buildings were in a parlous state, which led to their demolition. Rumours at the time suggested the hasty demolition was to thwart any takeover by Freightliner, but perhaps that was just speculation, smoke and mirrors? In any case, by June 2006 the site had been completely cleared for new development and all that remained were the abutments of the rail overbridge used for access. Today the Mercure Sheffield Parkway hotel and a car dealership stand where diesel locomotives once stood.

Should Tinsley Diesel Depot Have Closed? – A Personal Opinion by Alex Fisher

Railway closures, whether it be a line, a station or even a depot, invariably invoke a shrill of discontent from the non-pragmatic. When it comes to the rights and wrongs of Tinsley's closure, there are two key questions: was EWS right to close the depot, and was the Government wrong for overseeing a privatisation process that brought about EWS's dominant position?

If the latter point is taken first, then yes. Despite at the time the Government most probably asserting that it was the best deal for the UK tax payer, they were culpable of allowing EWS to gain too dominant a position in the rail freight market and it wasn't the best deal for competition. Had the management buyout team secured the deal for RfD then Tinsley would certainly have continued for several more years and probably have seen in the new millennium. If Freightliner had secured RfD, then maybe they wouldn't have needed to build their Leeds Midland Road depot and instead maybe Tinsley would have become their main traction maintenance depot. Unfortunately for Tinsley, EWS did secure the purchase of RfD and thus to the first question, was EWS right to close Tinsley depot? And there can only be one, reluctant, answer – yes.

In conclusion, the Government should have made sure that a more competitive split of the freight market was achieved and then maybe Tinsley would have stayed open, but once in EWS' hands, the pragmatic conclusion is that they had no need for it and they rightly had no desire to see it in use by a competitor, and so an ill-fated Tinsley depot was closed.

Yard Clear-up and Creation of Avesta Level Crossing

By early 1998 work had begun to clear up the yard by removing derelict point work and plain-lining the route to and from the SMACC sidings. Additional work removed the disused signalling.

To allow vehicular access to the remains of the yard during its reconstruction period, a temporary level crossing was installed over the single west departure/arrival line, close to the location of Shepcote Lane Junctions on 16th March 1998. Instructions were issued that the barriers were to be kept locked down and the crossing only used during an engineers' possession of the line. Work to commission this crossing was protracted and by November, new instructions were issued that the crossing was for use during daylight hours only, by contractors moving earth. They also stated that before a train was signalled onto this line, the shunter was to proceed to the crossing, lower the barriers and assure the signalman at Tinsley Yard that the crossing was clear. He was then to remain at the crossing until the train had passed clear. It was not until Sunday 26th September 1999 that this crossing was converted into a train-man-operated barrier crossing (TMOB) and made permanent, becoming known as Tinsley Avesta level crossing.

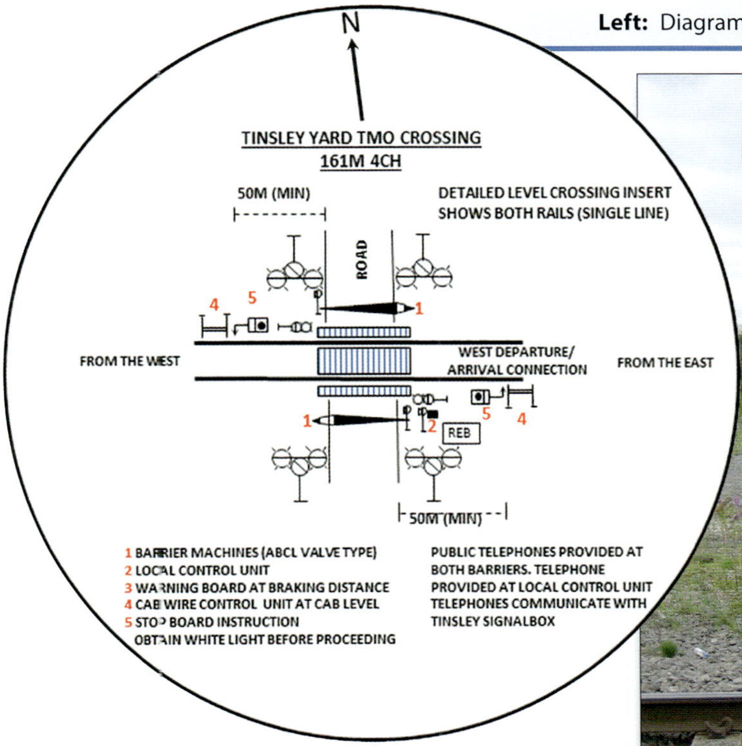

Left: Diagram of the Avesta crossing. *Chris Booth*

Right: Pull cord for operation of the Avesta crossing. *Andrew K. Overton*

Above: 60077 departs from the yard with 6D03, Tinsley to Immingham on 3rd June 2000. The loco is just approaching the train-man-operated Avesta crossing; the driver will lean out of his window and pull the operating cord to lower the barriers. *Stewart Donohoe*

Operation is by means of control wires, which are at cab height on the left hand side of the line on the approach to the crossing from either direction. To initiate the lowering of the barriers, the driver has to pull the control wire, releasing the wire; pulling it a second time stops the lowering sequence and pulling it a third time restarts the lowering sequence.

After the train has passed completely over the crossing, the barriers raise automatically. An instruction states that "If it is necessary to make a propelled movement towards the crossing at the Tinsley end, the shunter controlling the movement must ensure that the movement does not pass the STOP board. Shunting over the crossing is prohibited".

Above: The last loco has departed, the roller shutters are down, and the washing plant is already being removed in this scene on 27th March 1998 – the day of closure. The two tank wagons on the fuel discharge apron were still in situ after the depot had gone; one of them was preserved.

Above: Demolition is well under way on 20th January 1999. The cladding has been removed and, at the west end, the sand towers have been toppled and are awaiting cutting up.

Above left & right: By 21st April the depot had gone, with only water-filled pits and part of the electrical room remaining to show a depot was ever there.

Right: The two tank wagons left after demolition of the depot. Left is ZRW 042174 (ADB 999026), a 22-ton tanker that was scrapped at Booth's, Rotherham in 2001. On the right is ZRO 041204 (A444), the original number of which was 38855. Built by the Midland Railway in 1902 as a New Jersey pure oil tanker, it is now preserved on the Ribble Steam Railway. *All Chris Booth*

Chapter 21

The New Millennium

At the start of the new millennium, the only traffic using the remains of the yard was steel traffic to the Stainless Melting and Continuous Casting (SMACC) plants.

A Brief History of SMACC

In 1992 British Steel Stainless was merged with the Swedish company, Avesta, and renamed Avesta Sheffield. Ten years later another merger, this time with Finnish company Outokumpu, saw the formation of Avesta Polarit. In 2003, British Steel sold its stake in the company, which was again renamed, Outokumpu Stainless.

Historically there were no facilities in Sheffield to roll the stainless steel slab produced by SMACC; slabs were initially sent to rolling mills at Port Talbot or Lackenby for further processing, before being returned to Sheffield.

By the 1980s, the slab was being transported via Grimsby to Bremerhaven (Germany) or Sollac (France) for processing. The rolled steel coil would return, again via Grimsby to Sheffield.

This situation continued until the formation of Avesta in 1992, when traffic was then routed via Grimsby or Goole to Sweden. 4,000 tons per week would be shipped via Goole, and 2,000 tons per week via Grimsby.

With the opening of the Channel Tunnel, steel was briefly transported by rail all the way to Sweden, with a short sea journey on the Kattegut train ferry. This soon ceased due to various problems, and traffic was rerouted via Immingham for onward movement by sea.

Between 1994 and 1997 traffic increased from 180,000 tons to more than 500,000 tons; construction of the Nordic Terminal at Immingham in 1997 served to improve the reliability of the service. By the end of 1999 the timetable for steel trains between Immingham/Grimsby and Tinsley was as in Table 1.

Table 1: Steel Trains at Tinsley in 1999

Headcode	Train	Runs
6J01	00.10 Immingham to Tinsley	SO
6J01	00.22 Immingham to Tinsley	SX
6D33	01.33 Tinsley to Immingham	SX
6D34	05.15 Tinsley to Grimsby Brick Pit Sidings	SO
6D31	05.15 Tinsley to Immingham	SX
6J32	12.10 Immingham to Tinsley	SX
6J34	14.00 Grimsby Brick Pit Sidings to Tinsley	SO
6J35	15.50 Immingham to Tinsley Avesta (SPACE)	SO
6J33	18.42 Immingham to Tinsley Avesta (SPACE)	SX

Above: Whilst waiting to depart for Immingham on 4th May 2000, 60052 "Glofa Twr – The Last Deep Mine in Wales – Tower Colliery" keeps company with an abandoned VVV 4-wheel van amongst the detritus of the decaying yard. *Stewart Donohoe*

Steel Bridge

The highly successful movement of stainless slabs between Sheffield and Sweden became known as "The Steel Bridge".

Each train consisted of 22 wagons. When a loaded coil train arrived at Tinsley Yard, the first half would be moved into the coil build-up area at SPACE (Stainless Plate and Coil Expansion) by the train locomotive, which would then be uncoupled and stable in the yard. Once the wagons were unloaded, the Class 08 based in the yard would shunt the two halves and the second half would be unloaded. The 08 would then take all the wagons to SMACC.

Once the 22 BVAs had been loaded with slab, the train loco would couple up and depart in the evening for Immingham, returning to Tinsley the following day. Additionally, there was a weekly train of stainless slab to Seaforth Dock for export to the USA, at that time moving almost 50,000 tons per year. It departed Tinsley on a Thursday night and returned the following day.

Unfortunately, in 2005 Avesta Polarit closed down its Tinsley rolling plant with the loss of over 600 jobs and the traffic became one way only, loaded trains departing the plant and empties returning.

Joyriding

Tinsley Yard made the national headlines for all the wrong reasons in May 2006, when 19-year-old Daniel Matthews from Rawmarsh was sentenced to 200 hours community service after joyriding with the Class 08 locomotive stabled in the yard.

Climbing into the loco, he started the engine using a spanner and drove it up and down the yard for 40 minutes, changing the points and riding up and down different sidings.

After he had finished, he switched the loco off and cycled away. However, his exploits had been captured on CCTV and he was soon identified after police released footage of him and another man on BMX bikes in the area. Needless to say the British Transport Police condemned his actions as an act of stupidity.

Sheffield District Line Cut Off

The closure of the "down" and "up" Tinsley/Brightside arrival and departure lines between Shepcote Lane and Brightside Junction was intended to be a temporary measure, pending completion of the resignalling scheme in 1992. In the event, the route would never be used again, and was officially decommissioned on Saturday 13th November 1992.

In June 2006, Sheffield PSB's No. 4097A points at Brightside Station Junction (in the "up" & "down" goods line at 161 miles 06 chains) were replaced by plain line. The redundant track at the north end of the "up" & "down" goods line, and No. 1 reception line, leading to the former Tinsley lines, was also removed, thus severing the Sheffield District line from the former Midland line. The tracks towards Tinsley were lifted in 2007.

Broughton Lane

The south west curve from Shepcote Lane to Broughton Lane gradually slipped out of use after 2005, and from 2008 was shown as "temporarily out-of-use" in the sectional appendix.

By now the only access to Tinsley Yard was from the Rotherham direction, via the north east curve and TY319 points, which were secured in the reverse position. Before accepting trains from Woodburn Junction, the Tinsley Yard signaller had first to obtain permission from the shunter to release a "slot", which would allow Woodburn Junction to clear its W226 signal. Trains were then authorised to run via Shepcote Lane Junction and the west departure/arrival line. The Tinsley Yard signaller would clear the position light aspect below TY284 signal to allow trains into the yard. TY283 would be cleared in the other direction for trains to depart back to Shepcote Lane Junction. Because the signal box was now isolated at the far end of the yard, the signalman never saw the trains he was signalling!

Left: With the cooling towers of Blackburn Meadows Power Station and the M1 Motorway Tinsley viaduct in the background, Trainload Coal-liveried 60057 "Adam Smith" takes the single-line curve from Tinsley South to Shepcote Lane on a snowy 4th March 2006. The train is 6J03, Saturdays only Immingham to Tinsley, with return empty BVA cassette wagons for carrying stainless steel. The building just beyond the end of the train is the former GCR Tinsley station goods shed. *Stewart Donohoe*

Right: A few yards further around the curve, and BR blue-liveried 56004 approaches Shepcote Lane on 28th July 1997. The train consists of BBA-F wagons, which were used for steel traffic prior to the development of BVA wagons. 56004 was involved in a collision with 31192 + 31314 at Immingham on 30th July 1982 and was sent to Doncaster Works for repairs. It received replacement Doncaster-style cabs, making it stand out from the other Romanian-built examples. Withdrawn from service on 30th June 2005, it was scrapped at C.F. Booth's in Rotherham on 13th July 2006. *Stewart Donohoe*

Left: Steel to SPACE: After the journey from Sweden, a train of rolled coils trundles along the north west arrival line into Tinsley Yard behind Loadhaul-liveried 60038 on 5th May 2001. The train has been split, with one half placed in the siding in the distance. The rest of the train will be drawn clear of the points at the bottom of the image, before setting back into the SPACE sidings. Two Class 08 pilots, the right one of which is 08879 "Sheffield Children's Hospital", await work outside the shunters' cabin in the remains of the express freight sidings. The water tower from Tinsley Park Colliery days still stands proud. *Stewart Donohoe*

Right: A view over the derelict yard on 15th June 2002. Mainline Freight-liveried 60044 "Ailsa Craig" draws a rake of BVA wagons forward before setting back into the SPACE sidings. Apart from a derelict VVV van, the only wagons in sight are steel carriers. The three adjacent to the coil wagons are BJA bogie steel wagons, built by RFS Doncaster in 1990 with hoods; these were not a success and were stored in the yard for a few years. The main yard is becoming overgrown with silver birch saplings and the air of dereliction is prevalent.

The barrel-vaulted roof of Sheffield City Airport can be seen on the right skyline. Opened in 1997, the airport closed on 30th April 2008. *Stewart Donohoe*

Above: After bringing in half of its train, Loadhaul-liveried 60007 waits as its train is unloaded in the SPACE site on 30th June 2001. After unloading, the resident Class 08 will bring the second half into the yard, while the train engine waits in the main yard. The 32-ton overhead crane makes light work of removing the coils, which will be stored on the adjacent concrete apron.

60007 was one of five class members to receive full Loadhaul livery, the others being 60008/25/38/59. *Stewart Donohoe*

EWS invested £2 million in a fleet of 79 specialised BVA bogie wagons, which were able to transport steel slab or rolled coil on removable cassettes, which made for easier handling at the ports.

The wagons were built by Kockums in Sweden and introduced during 2000–01, with EWS naming Class 60 locomotive 60038 "Avesta Polarit" on 2nd March 2002 to celebrate the event.

Signalling Changes

On Monday 15th June 2009, lightning struck the relay room next to Woodburn Junction signal box, causing major damage to the wiring, batteries and circuit boards etc. Although spares were very thin on the ground temporary repairs were effected, although over the next few months the box suffered regular power failures, sometimes on a daily basis.

On Thursday 22nd October 2009 Woodburn Junction box again suffered a major failure, resulting in drastic measures to get the signalling system operational again. S&T engineers stripped Tinsley Yard signal box of all useful parts and disconnected its panel control and indicator system (its sole function by then was to give a "slot" to Woodburn Junction and to clear one or two signals). All train movements from Tinsley East Junction to Tinsley Yard then required the use of a pilotman, as some track circuits were disconnected and W226 signal was out of use. This duty fell to Sheffield mobile operations managers (MOMs) and often proved problematic as the MOMs were regularly called to emergencies elsewhere. Tinsley Yard signal box was virtually redundant but remained manned for another eight months by a single 12-hour shift, which must have been soul destroying for the signallers!

Woodburn Junction box was still suffering from power failures; with just one working set of computer chips left in the panel there was potential for disaster if they had blown. Things eventually improved following replacement of most of the equipment and the engagement of a company to make back-ups of all chips on site.

Tinsley Yard signal box finally closed on Saturday 12th June 2010 at 11.20, although the official closure was listed as 13.30 on Sunday 19th June 2010. On 1st August 2010 vandals set fire to the box, severely damaging the operating floor and ensuring it could never reopen again.

Above left & right: The sad remains of Tinsley Yard signal box after the fire on 1st August 2010. *Both Mick Charlesworth*

New Methods of Operation

The inefficient use of a pilotman continued for another four months until Monday 4th October 2010, when one-train-working regulations were introduced between Tinsley South Junction and Tinsley Yard, under the supervision of the Woodburn Junction signaller. The first train to use the new method ran on Tuesday 5th October, effectively releasing the Sheffield MOMs from pilotman duties.

Above: The south west curve from Shepcote Lane to Broughton Lane was disused for a length of time and became quite rusty, as can be seen here. This is the view of the bridge over the Sheffield & Tinsley Canal, looking towards Shepcote Lane on 4th February 2009. *Andrew K. Overton collection*

Above: The temporary wooden train staff, sitting on top of the box in which it is normally kept locked, at W205 signal. *John Webster*

A homemade wooden "train staff" was produced; drivers picked up the staff from a locked cabinet situated at W205 signal at Tinsley South Junction, before obtaining permission from the signaller to enter Tinsley Yard. On leaving the yard, the driver would replace the train staff at W205 signal.

Track circuit block regulations were withdrawn from that section of line, but the track circuits were maintained and continued to indicate in Woodburn Junction box, allowing the signaller to see the location of the train. Woodburn's W316 points at Shepcote Lane Junction remained secured in the normal position, and Tinsley Yard TY319 points remained secured in reverse. The now unused TY283 and TY284 signals were switched off and bagged over with large "Xs" to indicate to drivers that they were out of use. In January 2013 the remains of Tinsley Yard signal box were cleared of all asbestos content by specialist contractors and the shell of the building demolished. By 23rd February 2013 it had gone and just a concrete pad remains to show where it once stood.

New Stone Terminal

In connection with the rebuilding of Sheffield's road network, a road-stone and concrete manufacturing plant was built at the Shepcote Lane end of the yard, more-or-less on the site of the express freight yard.

To allow trains to access the yard from the Woodburn Junction direction, the south west curve from Broughton Lane Junction to Shepcote Lane Junction was reopened. New signal box instructions were issued to Woodburn Junction on Sunday 2nd September 2012, and on Monday 3rd September a locomotive running as 0Z99 was used to traverse the curve five times for testing of signals and track circuits.

With two routes into the yard once again, working by train staff was withdrawn and another new method of working introduced. The Woodburn Junction signaller would stop trains at either W205 signal on the north west curve or W212 signal on the south west curve, and obtain permission from the Tinsley shunter for the train to proceed. The driver would then be instructed to proceed towards the Avesta level crossing and await the shunter's instructions.

Once in Tinsley Yard, all movements were controlled by DB Schenker (EWS's successor) ground staff and trains between the yard and SMACC were equipped with an audible tail lamp. This lamp combined the function of a conventional battery electric tail lamp with a warning tone to alert anyone around that the train was being propelled.

The reopened route was only lightly used until timetabled trains commenced. On 1st October 2012, 20301 + 20302 were on route-learning when the leading Class 20 developed an engine problem. The south west and north west curves were used to reverse the engines. DB Schenker also occasionally took advantage of the restored route to run steel trains out of Tinsley Yard via Woodburn Junction and Sheffield instead of via Rotherham.

The stone terminal was initially operated by Freightliner and was served by one stone train per weekday from Bardon Hill Quarry. The first train ran as 6Z48 at 07.00 on Wednesday 6th June 2013 and back as 6Z49 around 13.00.

During 2014 trains were running as 6Z48, 03.19 Bardon Hill to Tinsley, returning as 6M01, 11.19 Tinsley Yard to Bardon Hill.

From 5th January 2015 the contract was won by GBRf, and the trains began to run in one of two paths from either Croft Quarry or Bardon Hill, as shown in Table 2.

Table 2: Stone Trains at Tinsley in January 2015

Code	Origin	Destination	Arrive
6E48	01.28 Bardon Hill	Tinsley	06:28
6M01	10.51 Tinsley	Bardon Hill	
		OR	
6E98	02.41 Croft Quarry	Tinsley	06:23
6V09	10.51 Tinsley	Coton Hill	

By March 2019 only the Bardon Hill traffic was running, 6E48 departing Bardon Hill at 03.04 and arriving Tinsley at 06.42. The return working was operating as 6M83, 10.51 Tinsley to Bardon Hill.

Today's Steel Traffic

Very little steel traffic in the Sheffield area goes out by rail nowadays, although Tinsley still sees a small amount of traffic. In 2018 the remaining sidings at Tinsley are used by the trains shown in Table 3.

Table 3: Tinsley Steel 2018

Headcode	Train	Runs	Conveys
6M12	22.50 Tinsley to Seaforth Container Terminal	ThO	Loaded slab
6E14	14.05 Seaforth Container Terminal to Tinsley. Arrive 18.59	FO	Empties
6J03	02.54 Immingham Nordic Terminal to Tinsley. Arrive 05.57	TWThO	Loaded slab/ bloom/billet
6D03	19.49 Tinsley to Immingham Nordic Terminal	MTWO	Loaded coil
6D03	13.45 Tinsley to Immingham Nordic Terminal	SO	Loaded coil

Sheffield International Rail Freight Terminal

The remains of Tinsley Yard saw major changes in 2006 when plans were put forward for construction of 92,900 sq m (1 million sq ft) of rail-linked distribution warehousing on part of the site. The £70 million project was a joint venture between EWS and Helios (a private development company), to be known as Sheffield International Rail Freight Terminal (SIRFT).

Whilst the redevelopment was taking place, the only sidings available for operational use were Nos. 14 to 18, with No. 18 siding only used as a light loco through road. The remainder of the yard was classed as out-of-use, with No. 19 siding clamped at both ends and siding Nos. 20 & 21 removed. The remaining sidings were lifted and a new, much smaller, five-road yard was laid to the north side of the site, between the massive warehouse buildings and the former yard by-pass line. Two sidings, one of which was on a concrete pad, were situated at the rear of the buildings to allow trains to be unloaded. Unfortunately, these buildings saw no immediate use and were in danger of becoming another "white elephant" until late 2013 when a lease was signed by retailer Marks & Spencer for the two adjoining warehouses, amounting to 647,000 square feet of space. At the time of writing, no goods have yet been brought in by rail – a real shame when all the time and effort to provide a rail connection is considered.

Lights Out

Between September and December 2006, all the redundant lighting towers in the area to be redeveloped were switched out and removed, apart from those situated between siding Nos. 15 & 16, which remained active. The lighting towers were taken down using a 40 tonne Komatsu PC340 long-reach tracked excavator, fitted with a rotating hydraulic shear. This machine cut into the lighting tower at the mid-point and, with the hydraulic shear still attached, lowered the upper section of the tower to the ground. The bottom section was then cut free from its base and lifted clear, the whole tower then being cut into transportable lengths and removed for scrap.

New Route to SMACC

Construction of the warehouse truncated the former access to SMACC, with its complicated switchbacks via

Left: Stone for Sheffield: After unloading its train of stone, 66605 draws its wagons alongside the Sheffield International Rail Freight Terminal (SIRFT) to run round the train and form 6M01, 11.19 Tinsley Yard to Bardon Hill on 27th November 2013. In the background, 60099 acts as "super shunter" to shunt the BVA wagons in the foreground, which will form a train to Immingham later in the day. *Chris Booth*

Right: 66605 runs round its train alongside the huge SIRFT warehouses. The reduction in size of the yard is evident. Two sidings are adjacent to the warehouses, one of which is in the concrete apron. The other five sidings are all that remain of the yard. The north west arrival lines snake around the outside of the yard remains, but by now they are out of use. *Chris Booth*

Left: 66605 stops adjacent to the remains of the main yard control tower and power pack building. The size of the SIRFT can be appreciated, taking almost the full width of the former yard area. *Chris Booth*

the hump underpass. A new line from the remains of the yard ran up a steep incline from the Shepcote Lane end to SMACC; this really tested the mettle of the pilot locos using the route!

Steam in the Yard

In 1913, metallurgist Harry Brearley successfully developed the first stainless steel in Sheffield. To mark the centenary in June 2013, Outokumpu staged a three-day celebratory event at their stainless steel works at Tinsley. DB Schenker arranged for Merchant Navy Class locomotive 35028 "Clan Line" to be on site. The loco was conveyed by 66174 as 5Z35 from Stewarts Lane via Willesden, the North

Right: Stone for Sheffield: As Stagecoach Supertram No. 111 runs alongside on a Meadowhall to Middlewood service, 66605 drops down the single line south west curve from Shepcote Lane Junction to Broughton Lane Junction with 6M01, 11.19 Tinsley Yard to Bardon Hill on 27th November 2013. The single line from Woodburn to Rotherham is adjacent to the loco. The steelworks that formerly lined the route have all now gone. *Chris Booth*

Right: A busy moment at Broughton Lane on 27th November 2013. After being held at the signal in the above image, 66605 now heads towards Woodburn Junction and passes Stagecoach Supertrams 104 on a Middlewood to Meadowhall service, and 116 returning from Meadowhall to Middlewood. The area to the left is now part of the Centertainment complex, in an area that was once a hive of industry. *Chris Booth*

Left: Due to a landslip in early 2014 at Unstone, between Sheffield and Chesterfield, all "up" trains were diverted to run from Sheffield via the "old road" to Chesterfield. The stone train was also subsequently rerouted to run via the north west curve to Rotherham, where it ran round before taking the "old road" to Chesterfield. 66598 rounds the curve and approaches the signal at Tinsley South Junction with 6M01, 12.01 Tinsley Yard to Bardon Hill. Fortuitously, at the same time 66506 passes on the single line with 6E45, 10.20 Hope to Drax power station on 25th February 2014. *Chris Booth*

and West London Lines, and the East Coast Main Line to Doncaster, then south via Rotherham to Tinsley Yard. Over the course of the event, the steam locomotive hauled stainless steel ingots weighing up to 600 tons, and performed demonstration runs over a 1 in 90 gradient hauling 450 tons. The locomotive performed admirably throughout, in front of invited guests and a few intrepid enthusiasts outside the fence boundaries!

Right: Steam in the yard! Merchant Navy 4-6-2 35028 "Clan Line" performed a series of run-pasts with BVA wagons on Saturday 15th June 2013. *Stewart Donohoe*

Below: GBRf took over the contract for stone traffic. 66770 drops down the grade from Shepcote Lane to Broughton Lane Junction with 6M01, 10.51 Tinsley to Bardon Hill on 30th March 2015. *Chris Booth*

Above: The same train is seen earlier as 66770 pulls through the stone terminal to take 6M01, 10.51 to Bardon Hill on 30th March 2015. The disused north west arrival line is in the foreground. The stone terminal is built on the site of the express freight and departure sidings. *Chris Booth*

Chapter 22

Tinsley Open Days

In common with many other BR depots, Tinsley held "open days" for enthusiasts, which were also a good way to raise funds for charity. Tinsley's very first open day was held on 15th June 1980, which took place in the servicing shed only. According to contemporary reports, the event attracted more than 25,000 visitors from all over Britain and, as part of the open day, a special booklet priced £1 was produced. DMUs were used to shuttle passengers from Sheffield station to the open day, and a 10-coach railtour, "The White Rose", ran as 1Z10, London King's Cross to Sheffield via York, with "Deltic" 55018 "Ballymoss" as motive power.

F & W Railtours ran a second railtour called "The Tinsley Terror", using locomotives 20167, 20174, 37158, 37210, 47267 & 47318 on a route from Newton Abbot via Birmingham to Darnall, a station closer to Tinsley than Sheffield itself.

Above: LMS 4F 0-6-0 4027 and two coaches await the crowds on 15th June 1980. Behind is HST power car 43107, with 76033 behind that. A Class 37 and a Class 45 are inside the servicing shed. *Roger Wainwright*

Below: With master & slave unit 13001 in the foreground, the crowds surround the servicing shed. *Chris Booth collection*

Left: 47174 and 55013 "The Black Watch" await the crowds in the early morning sunshine. Unfortunately, the open day was marred by the unforeseen theft of several works plates and engine parts from locomotives on display.
Roger Wainwright

Below: In conjunction with the open day, a shuttle service operated between Sheffield Midland station and Tinsley Yard. Here the shuttle is seen passing the servicing shed.
Ian Cracknell

Table 1: Locomotives on Display 15th June 1980

Diesel	03129, 08523, 08543, 08866, 13001, 13002,
	20017, 20034, 20066, 20078, 20112, 25101, 25253,
	31227, 31292, 31303, 37107, 37226, 40085, 40183,
	45009, 45012, 45019, 45021, 45027, 45028, 45029,
	45030, 45039, 45043, 46020, 46041, 47168, 47319,
	47372, 55013, 56007, 56008, 56078.
HST	43107, coaches 42197 & 41158.
DMU	E50009, E50010, E50043, E56003, E56019, E56030.

Steam	MR "Spinner" 4-2-2 No. 673, 4F 0-6-0 No. 4027
Electric	76033
Locos in Yard	
Diesel	08022, 08492, 08510, 08749.
	20048, 20057, 20131, 20133, 20139, 20186, 20212,
	31168, 31181, 40187, 40194, 45006, 45030, 45041,
	47174, 47312, 47331, 47412.
	2 Industrial locomotives

20th September 1987

An open day for families of staff was held at the TMD on 20th September 1987, although not open to the general public. The locomotives on display are listed in Table 2. Of particular note was the Class 26, which was en route to Vic Berry's Scrapyard in Leicester for examination of asbestos content. Apart from this and the Class 58, the other locos were regulars at Tinsley depot.

Table 2: Locomotives on Display 20th September 1987

08141, 08210, 08244, 08308, 08335, 08500, 08509, 08678, 08691, 08749, 08857, 08870, 08871.
20030, 20064.
26025.
31158, 31165, 31415.
37009, 37068, 37185, 37241, 37682, 37685.
45049, 45107, 45134, 45144.
47122, 47145, 47148, 47279, 47287, 47290, 47372.
56135, 58034.

29th September 1990

Tinsley's third open day was held on 29th September 1990 to celebrate the depot's silver jubilee. A colourful commemorative plate was issued, which sold well at the event.

The BRC&W Type Three Preservation Group's railtour "The Sulzer Rose" used 33042 & 33207 to haul 4TC sets 8001 & 8014 from Eastleigh to Sheffield, where they waited whilst the tour participants visited the depot. The train arrived at the north end of Sheffield having travelled from Eastleigh via Winchester, Staines, St Albans, Toton, Tapton Junction and the "Old Road", so the locomotives were at the south end of the consist for the return journey via Derby.

DMU shuttle services operated from Sheffield comprised of Metro-Cammell two-car Class 101 DMU Nos. 54056 + 51420, and Class 141 Nos. 141106 + 141107. Table 3 lists the locomotives on display.

Table 3: Locomotives on Display 29th September 1990

Diesel	D2199, 06003, 08509, 08749, 08857, 08878, 08880, 08919, 09008, 09013.
	20029, 20043, D7672 (25912), 31547, 37013, 37019, 37029, 37059, 37063, 37070, 37107, 37178, 37185, 37198, 37203, 37414.
	45133, 47107, 47120, 47121, 47144, 47145, 47200, 47209, 47229, 47237, 47270, 47293, 47305, 47308, 47332, 47351, 47370, 47375, 47376, 47378, 47412, 47439, 47447, 47485, 47600, 55015, D9000 (55022), 56002, 58050, 59001, 60013.
Steam	LMS Black 5 4-6-0 No. 44932

The commemorative plate specially commissioned for the open day. *Chris Booth collection*

Right & below right: Freshly repainted into "Dutch" livery, 31547 was on show alongside Foster Yeoman 59001 "Yeoman Endeavour". *Both Chris Booth*

Below: Advance notice of the open day was placed in the popular railway press of the time, this example being from Rail Enthusiast magazine. *Chris Booth collection*

TINSLEY
1965 1990
25th Anniversary Open Day
Saturday 29 September 1990
GATES OPEN TO THE PUBLIC at 10 a.m.
CLOSE at 5 p.m.
TRAVEL VIA SPECIAL SHUTTLE FROM – SHEFFIELD TO TINSLEY (VIA FREIGHT ONLY LINES.)
FEATURING: BRITISH RAIL LOCOMOTIVES & ROLLING STOCK
LOCOMOTIVE NAMING CEREMONY
OTHER ATTRACTIONS • CAB VISITS • MODEL RAILWAY • MINIATURE STEAM RAILWAY • SALES STANDS
SHUTTLE SERVICE AVAILABLE TO SOUTH YORKSHIRE RAILWAY
ADMISSION
£1.50 ADULTS £1 CHILDREN & SENIOR CITIZENS
ADVANCE TICKET APPLICATIONS PLEASE SEND S.A.E. + CHEQUE OR P/O. TO: TINSLEY OPEN DAY, BRITISH RAIL, WOOD LANE, BRINSWORTH, ROTHERHAM S60 5RP
ALL PROCEEDS TO CHARITY

Saturday 27th April 1996

Tinsley's final open day saw around 11,500 people visit the depot which, by then, was under Railfreight Distribution control. The crowd exceeded the open day committee's expectations and the 10,000 printed tickets were all sold by 13.00.

The event raised over £40,000 and it was hoped that, after deduction of costs, around £30,000 would be donated to local charities. The main beneficiaries were Sheffield Children's Hospital, Rotherham Hospital's accident and emergency unit, and the cancer unit at Weston Park Hospital.

More than 50 locomotives were on display. Star attractions included six electric locomotives, 47145 in deep blue with Railfreight Distribution logos and 47817 in its first public appearance in the distinctive Porterbrook livery of white and purple. Table 4 lists the locomotives on display.

Table 4: Locomotives on Display 27th April 1996

Diesel	08879, 20042, 20188, 37605, 45060, 47019, 47049, 47053, 47145, 47194, 47205, 47209, 47210, 47213, 47217, 47219, 47226, 47234, 47276, 47285, 47287, 47290, 47291, 47307, 47316, 47323, 47326, 47351, 47360, 47361, 47375, 47379, 47525, 47786, 47817, 50007, 50031, 50044, 55015, 56004, 56100, D172, D1842.
Electric	86604, 87101, 89001, 90133, 92041, E27000.
For Scrap	47096, 47102, 47190, 47214, 47249, 47288, 47321, 47325.
In Yard	08512, 08587, 08662.

During the day Class 08 No. 08879 was named "Sheffield Children's Hospital" in a ceremony conducted by Richard Jarman of Sheffield Children's Hospital, the loco also gaining a cabside teddy bear plaque.

Credit for the successful open day was given to depot engineer Bill Foster, and the open day committee of Phil Hodgkiss, Pete Wylie, Neil Poppleton and Kevin Elliot.

Two railtours ran to the event. Hertfordshire Rail Tours' "The Tinsley Humper" ran as 1Z37 from King's Cross to Sheffield, where participants had the choice of either visiting the open day or staying on the train for a second railtour; "The Pennine Perambulator" went from Sheffield, around Manchester and back to Sheffield, hauled at various times by 31462 + 31468, 37010, 37372 & 47704.

The second train was Pathfinder Tours' "Tinsley Open Day Specials" which ran from Westbury to Sheffield. Two additional mini tours were similarly available, to Wakefield Kirkgate behind 59205 and to Doncaster behind 59203.

Having participated in the open day, the Fifty Fund, owners of 50044, made use of lifting facilities at the depot to change its No. 6 traction motor on 29th April. The work was completed in a day and the loco returned to the Severn Valley Railway with 50031 and 55015.

Above: 45060 "Sherwood Forester".

Left: Three Class 50s lined up for the camera – 50007, 50031 and 50044.

Left: Four electric locomotives in the form of 27000 "Electra", 86604, 90133 and 87101 stand outside the west end of the depot. *All Chris Booth*

Left: Tinsley's affinity with Sheffield Children's Hospital had led to the naming of 09008 after the hospital at the 1990 open day. The locomotive carried the name for three years before being transferred away from Tinsley. The nameplates were placed in store, but then applied to 08879 at the 1996 open day. The loco also underwent a special Harold Winder & Phil Hodgkiss-inspired repaint and gained a small cabside plaque of a teddy bear, in a ceremony conducted by Richard Jarman of Sheffield Children's Hospital.

Left: The 1996 open day also saw 47817 on show in the distinctive Porterbrook livery of white and purple, the colours being reversed on the other side of the locomotive. 47807 was also adorned in these colours. 47817 was rebuilt as 57311 "Parker" and worked for Virgin Trains, before later being acquired by Direct Rail Services.

Below: As Tinsley was a Railfreight Distribution depot, naturally there were several Class 47s on display in that livery. 47326 "Saltley Depot" is looking resplendent at the east end of the depot. Note that the loco sports a "Yorkshire Rose" cabside plaque.

Withdrawn in January 2004, the loco met its end at the Stockton yard of T.J. Thompson in October 2006. *All Chris Booth*

Above: Visitors were allowed access to the cab of 47323 "Rover Group Quality Assured", which was undergoing repairs inside the depot. This loco was one of the recipients of an unofficial name, in this case "The Jostinot". After privatisation 47323 became part of the Freightliner fleet and was withdrawn after fire damage. It would be cut up in 2003.

Right: An engine-lifting demonstration took place involving 47234. This loco would eventually be rebuilt and become Virgin Trains' 57315 "The Mole". It later received a repaint into Arriva Trains Wales blue, before transferring to West Coast Railway Company and receiving maroon livery.

Above: Pathfinder Tours ran two mini tours in connection with the 1996 open day. The first went to Wakefield Kirkgate behind National Power 59205. On its return with the stock, the second went to Doncaster as 1Z59 "The Riviera Limited", behind 59203 "Vale of Pickering", seen here alongside the South Yorkshire Supertram lines near Broughton Lane, en route to Sheffield. *All Chris Booth*

Chapter 23

On Shed

Tinsley (41A) opened for business in September 1963 and soon became a permanent fixture on the itinerary of many enthusiast societies' trip calendars. Travelling alone, the BR shed directory was the enthusiast's friend when visiting depots, as it contained step-by-step directions to the location of each depot. To visit Tinsley from Sheffield Midland station, the instructions were to catch the No. 4 bus from outside Castle Market, alight at the Three Magpies public house in Brinsworth, then walk along Brinsworth Road to Wood Lane bridge.

Once there, the only access to the shed would usually be to view from Wood Lane bridge and the embankments at either end of the shed. Those lucky enough to be on official visits would be able to see locos inside the depot buildings; for everyone else it was a matter of a good pair of binoculars!

The following is a snapshot of the locomotives seen on official visits by the Pennine Railway Society, records from the Shed Surprise and WNXX websites, and personal visits, including by one of your authors, Chris Booth. Some of these trips were undoubtedly just viewings from the embankments but, whatever they were, here is a look at what could be found on a day in each year of the life of the depot, through the notebooks of several enthusiasts from 1965 to 1998. Maybe you were there?

Sunday 3rd January 1965 (Alex Scott)

D29, D40, D67
D1521, D1537, D1538, D1543, D1550, D1772
D3703, D4028, D4029, D4044
D5685, D5690, D5691, D5692, D5845, D5851
D6749, D6798, D6799
D8023, D8051, D8054, D8057, D8064, D8067
D8606, D8609

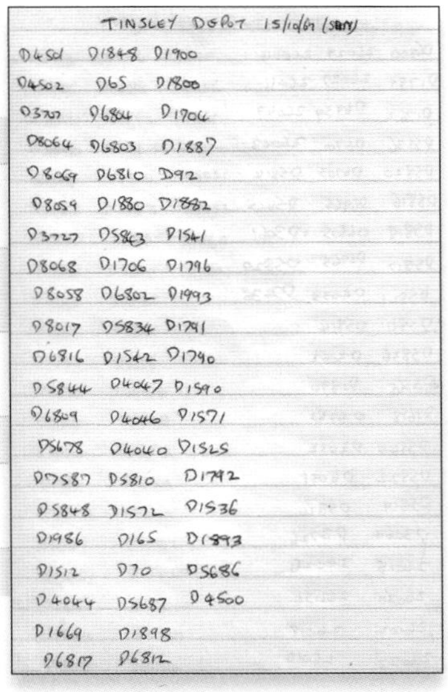
TINSLEY DEPOT 15/10/67 (SUN)
D4501 D1848 D1900
D4502 D65 D1800
D3707 D6804 D1704
D8064 D6803 D1887
D8069 D6810 D92
D8059 D1880 D1882
D3727 D5843 D1541
D8068 D1706 D1796
D8058 D6802 D1993
D8017 D5834 D1791
D6816 D1542 D1790
D5844 D4047 D1590
D6809 D4046 D1571
D5678 D4040 D1525
D7587 D5810 D1742
D5848 D1572 D1536
D1986 D165 D1893
D1512 D70 D5686
D4044 D5687 D4500
D1669 D1898
D6817 D6812

Sunday 12th June 1966 (Clive Greedus)

D83, D119
D1553, D1572, D1706, D1773, D1798, D1867, D1868, D1869, D1873, D1879, D1882, D1891, D1893, D1974
D3253, D3452, D3702, D4035, D4046, D4562, D4500, D4501
D5176, D5204, D7598, D7606, D7609, D7610, D7607, D7647
D5528, D5541, D5549, D5536, D5588, D5854
D6750, D6752, D6754, D6797, D6800, D6804, D6805, D6809, D6811, D6812, D6814, D6815, D6963
D8015, D8016, D8018, D8019, D8058, D8059, D8063, D8064, D8065, D8066, D8069, D8132

Sunday 6th November 1966 (Clive Greedus)

D56
D1542, D1571, D1574, D1575, D1798, D1865, D1876, D1968, D1868, D1878
D3685
D5534, D5630, D5681, D5824, D5845
D6710, D6796, D6807, D6810, D6817, D6905, D6803
D7604, D7631, D7633, D7648
D8013, D8014, D8015, D8017, D8018, D8019, D8020, D8058, D8059, D8064, D8065, D8066, D8068, D8069

Sunday 15th October 1967 (Clive Greedus)

D65, D70, D92, D165
D1512, D1525, D1536, D1541, D1542, D1571, D1572, D1590, D1669, D1704, D1706, D1742, D1790, D1791, D1796, D1800, D1848, D1880, D1882, D1887, D1893, D1898, D1900, D1986, D1993
D3707, D3727, D4040, D4044, D4046, D4047, D4500, D4501, D4502
D5678, D5686, D5687, D5810, D5834, D5843, D5844, D5848
D6802, D6803, D6804, D6809, D6810, D6812, D6817, D6816
D7587
D8017, D8058, D8059, D8064, D8068, D8069

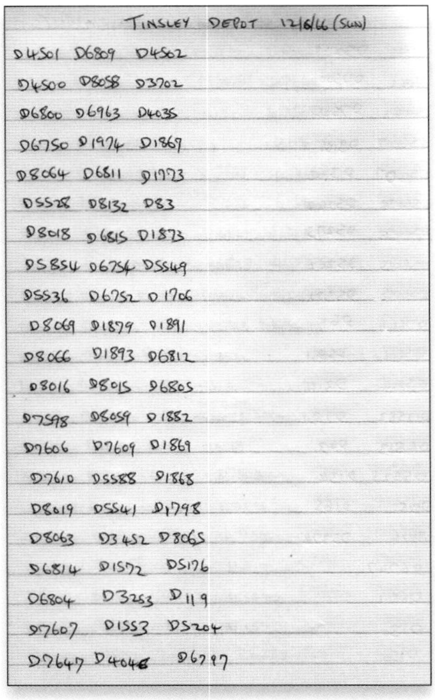
TINSLEY DEPOT 12/6/66 (SUN)
D4501 D6809 D4502
D4500 D8058 D3702
D6800 D6963 D4035
D6750 D1974 D1867
D8064 D6811 D1773
D5528 D8132 D83
D8018 D6815 D1873
D5854 D6754 D5549
D5536 D6752 D1706
D8069 D1879 D1891
D8066 D1893 D6812
D8016 D8015 D6805
D7598 D8059 D1882
D7606 D7609 D1869
D7610 D5588 D1868
D8019 D5541 D1798
D8063 D3452 D8065
D6814 D1572 D5176
D6804 D3253 D119
D7607 D1553 D5204
D7647 D4046 D6797

Sunday 18th August 1968 (Fred Kerr)

D44, D123, D149, D170, D192
D1514, D1515, D1519, D1520, D1521, D1570, D1578, D1703, D1753, D1768, D1790, D1793, D1878, D1879, D1881, D1886, D1986
D3288, D3703, D3707, D3920, D4034, D4035, D4043, D4500, D4501, D4502
D5192, D7533
D5541, D5577, D5582, D5584, D5622, D5684, D5805, D5807, D5808, D5812, D5820, D5824, D5826, D5832, D5849, D5850
D6702, D6703, D6705, D6707, D6716, D6718, D6806, D6807, D6810, D6811, D6814

Monday 23rd June 1969 (Leonard Ball)

D46, D109, D178
D1531, D1533, D1541, D1567, D1794, D1803, D1875, D1899, D1981, D1982
D3251, D3288, D4029, D4034, D4037, D4044, D4500, D4501, D4502
D5561, D5678, D5683, D5807, D5812, D5813, D5814, D5818, D5849
D6704, D6717, D6802, D6805, D6807

Sunday 2nd August 1970 (Chris Booth)

110
1504, 1513, 1520, 1552, 1566, 1568, 1607, 1622, 1739, 1794, 1830, 1983, 1987
3253, 3662, 3707, 4028, 4035, 4500, 4501, 4502
5182, 5231, 5268, 5295, 7522
5534, 5538, 5541, 5552, 5555, 5559, 5567, 5585, 5637, 5676, 5678, 5687, 5802, 5818, 5823, 5827, 5832, 5837, 5844, 5848
6781, 6800, 6801, 6802, 6806, 6809, 6813, 6829, 6860, 6863, 6868
8008, 8035, 8312

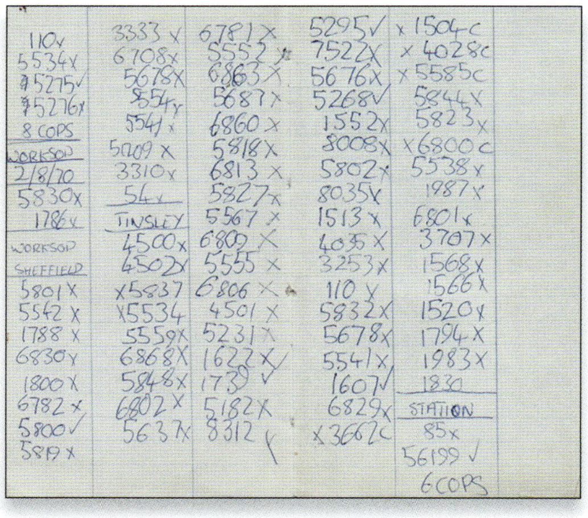

Sunday 23rd May 1971 (Chris Booth)

141
286
1631, 1634, 1785, 1793, 1971, 1994
3283, 3289, 3702, 3707, 4047, 4028, 4500, 4501, 4502
5540, 5552, 5558, 5565, 5567, 5584, 5585, 5686, 5822, 5824, 5830, 5835, 5837, 5844
6788, 6789, 6790, 6805, 6809, 6810 6812, 6813, 6718, 6823, 6861, 6867, 6869
8019, 8025, 8034, 8046, 8048, 8073, 8091
DB975018+ DB975019

Saturday 20th May 1972 (RMweb)

1532, 1570, 1575, 1688, 1772, 1784, 1788, 1793, 1794, 1977, 1982, 1984, 1985
3032, 3251, 3253, 3254, 3289, 3308, 3405, 3707, 3897, 3969, 4030, 4034, 4057, 4500, 4501
5233, 7517, 7611, 7622
5523, 5538, 5558, 5560, 5562, 5567, 5577, 5594, 5607, 5660, 5804, 5808, 5819, 5820, 5839, 5841
6720, 6739, 6749, 6782, 6784, 6785, 6788, 6824, 6826, 6869, 6981
8033, 8128, 8305, 8306, 8315

Sunday 4th March 1973 (WNXX)

68, 69, 83, 131
1101, 1109, 1773, 1779, 1863
3015, 3032, 3035, 3037, 3288, 3289, 3293, 3310, 3405, 3574, 3659, 3897, 3981, 4034, 4039, 4047, 4500, 4501
5169, 5171, 5174, 5228, 5229, 5233, 5252, 7562
5548, 5819, 5852
6724, 6736, 6800, 6805, 6808, 6824, 6827
8020, 8023, 8059, 8060

Saturday 16th March 1974 (RMWeb/Vince Minto)

3330, 3405, 4045, 08024, 08033, 08075, 08218, 08460, 08538, 08801, 08862, 08866, 08878, 08880, 13001, 13002, 13003
8003, 20049, 20058, 20129, 20132, 20208, 20209, 20215
5287, 25012, 25016, 25017, 25022, 25078, 25084, 25092, 25150
5830, 5834, 31140, 31242, 31300, 31307, 31413
37086, 37088, 37101, 37103, 37305
365
91, 46021
1723, 1753, 1885, 47042, 47044, 47053, 47109, 47174, 47297, 47305

Note that TOPS numbers had begun to appear, the renumbering commencing in January 1974.

During August 1974, an intrepid gang of enthusiasts began a week-long odyssey visiting installations and locations, with the help of an East Midlands Rail Rover. This is the part of the tour that took in Tinsley on the 19th and, in their words...

"The tour began with a classic overnight train, namely 1E03, the 00.05 St Pancras–Sheffield from Luton, with 45108 at the head of a good few Mark I compos and parcels stock. Limited doss followed, 45111 took us forward from Nottingham via Derby as booked."

"The highlight of Monday was a couple of hours on the road bridge spanning Tinsley Yard. Despite it being during the miner's holidays, the amount of freight and variety in power in a real industrial setting was awe inspiring to us Southerners unused to steel plants and foundries spewing out noise, smoke and above all atmosphere! One of Tinsley's unique Class 13 paired gronks busied themselves shunting over the hump, the whole time, 08244 and 08861 were substitutes for a missing 13001. We weren't expecting to see Class 76

electrics here but two came in on empty steel traffic to be replaced by noisy 37s or 40s for the run further East. The adjacent shed saw a few loco moves too, but the locals advised that you 'wouldn't get round during the week, easy at weekends though'."

19th August 1974 Tinsley Yard pm (approx 2hrs)

08075, 08218, 08244+08861 (class 13 substitute), 08266, 08504, 08749, 13002
20003+20130, 20060+20204
25015, 25101
31127, 31167, 31302, 31304
37003, 37107
40006
47044
76021, 76048

On day seven of the tour, 25th August 1974, the enthusiasts made the bus journey from Sheffield to Tinsley again, where this time they managed to gain access to the depot. This is what they saw:

36, 46, 103, 131, 135,
08033, 08075, 08076, 08115, 08208, 08218, 08223, 08266, 08335, 08386, 08504, 08507, 08729, 08877, 08880, 08887, 13002, 13003
20003, 20004, 20022, 20023, 20058, 20061, 20062, 20069, 20127, 20130, 20132, 20170, 20202, 20208, 20209,
25004, 25013, 25014, 25018, 25021, 25025, 25033, 25085, 25091, 25108,
31260, 31302, 31305, 31308, 31318,
37017, 37101, 37103, 37115, 37119, 37122, 37126, 37169
40155
46014, 46031, 46051, 46053
1874, 47138, 47166, 47304, 47489

Saturday 11th January 1975 (Andrew Booth)

08033, 08075, 08076, 08183, 08218, 08219, 08223, 08240, 08266, 08801, 08813, 08860, 08861, 08862, 08866, 08880, 13001, 13002, 13003
20008, 20025, 20049, 20056, 20058, 20059, 20066, 20069, 20074, 20127, 20128, 20133, 20202, 20210, 20211, 20228
25001, 25012, 25019, 25020, 25021, 25026, 25078, 25085, 25108, 25150,
31134, 31180, 31234, 31254, 31275, 31278,
37077, 37112, 37118, 37136, 37139, 37140, 37209
30, 64, 121, 45131
47049, 47145, 47167, 47169, 47173, 47180, 47275, 47283, 47284, 47307, 47316

Note that three Class 45s still retained their original pre-TOPS numbers. 30 would become 45029 in April 1975, 64 became 45045 in February 1975, whilst 121 became 45069 in January 1975, which must have been after the date it was seen here. Note also the remarkably similar composition of locomotive classes between 25th August 1974 and 11th January 1975, both with 16 x Class 08s, 10 x Class 25s, 5 and 6 Class 31s respectively, 8 and 7 Class 37s respectively and finally each with over a dozen Type 4s.

Saturday 21st August 1976 (Fred Kerr)

On this date, permission was refused for a visit due to the arrival of the first Class 56s. However, Fred sneaked in to take photos of the Class 56s and, amongst the many other locos on shed, noted the following:

08076, 13002, 13003
20128
37108
40071
46034
47313, 47423
56001, 56002, 56003, 56004

Sunday 1st May 1977 (Fred Kerr)

08022, 08335, 08507, 08813, 08862, 08880, 13001, 13002, 13003
20022, 20119, 20034, 20056, 20058, 20059, 20078, 20214, 20129, 20133, 20130, 20212, 20144, 20145, 20151, 20185,
25121, 25134
31221
37014, 37107, 37108
40146
45130, 46011, 46037
47130, 47531, 47601
56003, 56005, 56008, 56009, 56013, 56014, 56016, 56017, 56018, 56019, 56020, 56021, 56022

Saturday 2nd December 1978 (Pennine Railway Society)

08015, 08208, 08223, 08233, 08510, 08523, 08543, 08729, 08866, 08878, 13002, 13003
20004, 20009, 20021, 20023, 20032, 20054, 20056, 20065, 20078, 20129, 20130, 20133
56003, 56013, 56023, 56032

Thursday 18th January 1979 (WNXX)

08015, 08022, 08024, 08223, 08260, 08266, 08287, 08335, 08485, 08510, 08523, 08861, 08866, 13001, 13002, 13003
20001, 20008, 20010, 20019, 20034, 20056, 20058, 20061, 20098, 20128, 20129
25129
31142, 31149, 31158, 31175, 31227, 31243, 31263, 31268, 31275, 31302, 31315, 31407
37074, 37121, 37226
40007
45008, 45010, 45016, 45020, 45021, 45022, 45028, 45031, 45035, 45038, 45061, 46035
47044, 47110, 47174, 47177, 47305
56002, 56003, 56012, 56014, 56018, 56020, 56022, 56023, 56032

On Sunday 4th March 1979 the depot was visited by the Pennine Railway Society and a lone enthusiast. These two lists show how many locos were seen by the Society, as opposed to those by the individual. You will note that some locomotives tally, but others don't, so which one is correct? Probably both, as they could have been at different times of the day, so locos would have departed, whilst others could have arrived. But where were the Class 25s the Pennine Railway Society saw but the individual didn't? Some locos would undoubtedly have been inside the depot, and maybe the 25s were between other lines of locos that hid them from the view from the embankments? It goes to show how being a member of a society with access to shed permits paid dividends for those wanting to 'cop' more locos!

Sunday 4th March 1979 (Shed Surprise)

08024, 08539, 08810, 08861
20005, 20025, 20056
31147, 31268
37135
45021, 45030, 45061, 45124
47047, 47175, 47501
56007, 56019, 56022

Sunday 4th March 1979 (Pennine Railway Society)

08024, 08087, 08053, 08223, 08810, 08861, 08880, 13001, 13002, 13003
20005, 20008, 20021, 20023, 20035, 20056, 20103, 20127, 20128
25118, 25119, 25147, 25171, 25268, 25276, 25302, 25307
37019, 37022, 37025, 37030, 37135, 37169
40100, 40126
45021, 45023, 45030, 45033, 45035, 45039, 45048, 45061, 45124
47047, 47175, 47455, 47501
56003, 56004, 56007

Tuesday 1st January 1980 (WNXX)

08015, 08244, 08266, 08335, 08485, 08523, 08729, 08749, 08813, 08866, 08877, 08879, 13001, 13002, 13003
20001, 20011, 20017, 20023, 20032, 20054, 20055, 20056, 20057, 20058, 20064, 20076, 20144, 20145, 20154, 20210, 20214
31203, 31227, 31276
37134, 37136
40063, 40124
45007, 45012, 45038, 45036, 45047, 45048, 45105
47165, 47168, 47174, 47175, 47294, 47542
55009
56006, 56009, 56013, 56017, 56018, 56025, 56031, 56032, 56034

Note, "Deltic" 55009 "Alycidon", which was on depot for attention to fire damage on its load regulator. It had arrived on 28th December 1979 and departed on 9th January 1980. Doncaster Works would normally have undertaken this type of work, and it is unclear why it was sent to Tinsley instead.

Saturday 29th March 1980 saw Railway Pictorial Publications Railtours run the "Spotter Special" railtour from Paddington, taking in many different locations. Part of the route, hauled by 45022 "Lytham St Annes" from Derby, took the line from Treeton Junction, through Tinsley and Shepcote Lane Junction, to Woodburn Junction and Sheffield. Whilst passing through the yard the following locos were observed:

Saturday 29th March 1980

08015, 08033, 08507, 08877
20004, 20046, 20058, 20065, 20068, 20076, 20105, 20132, 20144, 20145, 20209, 20212
31270
47168

Friday 24th October 1980 (WNXX)

08033, 08209, 08389, 08507, 08510, 08729, 08878, 13001
20004, 20022, 20052, 20054, 20055, 20058, 20059, 20112, 20133, 20148, 20189, 20208, 20210, 20211, 20213
31147, 31173, 31175, 31221, 31279, 31285, 31289
37064, 37090
40038
45010, 45011, 45015, 45017, 45021(w), 45032, 45039, 45040, 45045, 45048, 45063
47173, 47175, 47374
56016, 56083

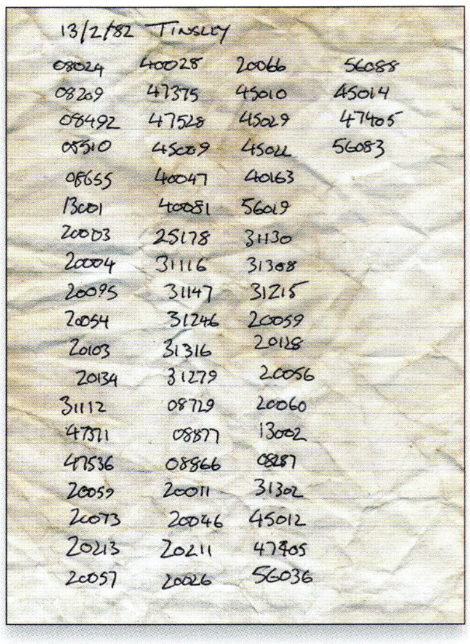

Friday 18th April 1981 (Shed Surprise)

08033, 08208, 08209, 08219, 08287, 08389, 08485, 08492, 08543, 08813, 13001, 13002, 13003
20003, 20004, 20005, 20017, 20021, 20022, 20023, 20026, 20029, 20031, 20046, 20054, 20058, 20060, 20095, 20096, 20103, 20106, 20112, 20129, 20183, 20189, 20208, 20211
25177
31004, 31135, 31315, 31326
37023, 37193, 37237
46049
47279, 47307, 47316, 47319,
56001, 56002, 56031, 56033, 56075, 56079, 56083

Saturday 13th February 1982 (Shed Surprise)

08024, 08209, 08287, 08492, 08510, 08655, 08729, 08782, 08866, 08877, 13001, 13003
20003, 20004, 20011, 20026, 20046, 20054, 20056, 20057, 20059, 20060, 20066, 20073, 20095, 20103, 20128, 20134, 20211, 20213

25178

31112, 31116, 31130, 31134, 31147, 31215, 31220, 31246, 31279, 31302, 31308, 31316

37002, 37121, 37172, 37209, 37215

40028, 40047, 40081, 40163

45009, 45010, 45012, 45014, 45022, 45029

47371, 47375, 47405, 47528, 47536

56019, 56036, 56083, 56088

Sunday 16th October 1983 Stored in Tinsley Yard (Shed Surprise)

DMU: all condemned

M50340, M50342, M50346, M50348, M50349, M50353, M50354, E50372, M50385, M50395, M50526, E50543, E50585, M50764, M50788, M50793, M50815, Sc51013, Sc51015, Sc51016, Sc51025, Sc51026, Sc51033, Sc51036, Sc51038

Sc51042, Sc51048, M51110, M51115, M51117, M51119, M51124, E51266, E51287, E51295, E51972, M53589, E54415, M55989, M56103, M56104, M56108, M56109, M56116, M56118, M56127, E56137, M56151, M56157, M56159,

M56163, M56165, M56295, E56395, M56432, M56452, M56465, E56466, M59007, M59009, M59015, M59018, M59019, M59021, M59032, M59132, M59133, M59135, M59136, M59138, M59140, M59141, M59143, M59145,

M59147, M59158, M59160, M59162, M59164, M59169, M59174, M59180, M59181, M59182, M59198, M59203, M59288, M59351, Sc59397, Sc59398, Sc59402, Sc59407, Sc59411

Saturday 31st December 1983 (Shed Surprise)

08033, 08051, 08208, 08244, 08335, 08434, 08485, 08492, 08507, 08510, 08523, 08678, 08729, 08782, 08870, 08877, 08878, 08880, 13001, 13003

20021, 20023, 20025, 20029, 20042, 20048, 20054, 20056, 20057, 20059, 20064, 20076, 20088, 20092, 20093, 20095, 20096, 20107, 20112, 20145, 20154, 20168, 20210

31125, 31312

37174, 37263

45012, 45019, 45026, 45038, 46018

47279, 47305

56002, 56005, 56006, 56013, 56021, 56094, 56101, 56102, 56107, 56111, 56124, 56128

Stored in Tinsley Yard

20046, 20098

31323

37013, 37024, 37045, 37048, 37126, 37169, 37226

DMU: all condemned

M50340, M50342, M50346, M50348, M50349, M50353, M50354, E50370, E50371, M50384, M50386, M50395, E50544, E50563

Sunday 4th March 1984 (Shed Surprise)

08033, 08051, 08208, 08244, 08335, 08434, 08492, 08507, 08510, 08729, 08782, 08878, 08879, 08880, 13001, 13003

20010, 20023, 20029, 20031, 20032, 20054, 20056, 20057, 20064, 20095, 20096, 20107, 20112, 20128, 20154

25189

31308

37013, 37024, 37126, 37219

45006, 45007, 45010, 45017

47219, 47408, 47426

56005, 56015, 56016, 56018, 56079, 56091, 56092, 56096, 56105, 56107, 56109, 56129

Saturday 4th May 1985 (RMWeb/Vince Minto)

08244, 08266, 08492, 08507, 08783, 08879, 08880

20009, 20025, 20047, 20054, 20055, 20068, 20078, 20086, 20092, 20098, 20105, 20107, 20112, 20116, 20118, 20150, 20162, 20166, 20176

31231, 31234, 31237

37024, 37030, 37036, 37040, 37130, 37246

45066

47029, 47147, 47275, 47276, 47379

56005, 56013, 56014, 56043, 56093, 56109, 56113

Saturday 8th November 1986 (Shed Surprise)

08244, 08389, 08434, 08492, 08500, 08507, 08510, 08783, 08870, 08878

20064, 20088, 20096, 20107, 20111, 20112, 20115, 20139, 20150, 20152, 20164, 20165, 20175

31437

37128, 37194, 37241

45034, 45041, 45046

47018, 47372, 47373

56094, 56095, 56097, 56098, 56101, 56105, 56106, 56107, 56113

Thursday 31st December 1987 (Chris Booth)

Note the appearance of painted names on the locos.

08141 Manvers, 08210 Beighton, 08244, 08308 Langwith, 08434 Midland, 08436, 08509 Wath ETD, 08510 Canklow, 08678 Staveley, 08749 Great Central, 08857 Darnall, 08870 Millhouses, 08878 Grimesthorpe, 08880 Mexborough

20013, 20088, 20095, 20130, 20135, 20157, 20175, 20195, 20197

31125, 31126, 31142, 31184

37062, 37079, 37117, 37194, 37251, 37285

45012, 97409(45022), 97410 (45029), 97411(45034), 45049, 45052, 45110, 45124, 45141

47012, 47093, 47098, 47143, 47145, 47211, 47279, 47293, 47299, 47309, 47311, 47314, 47360

Tinsley Yard

45051, 45070, 45120, 45127

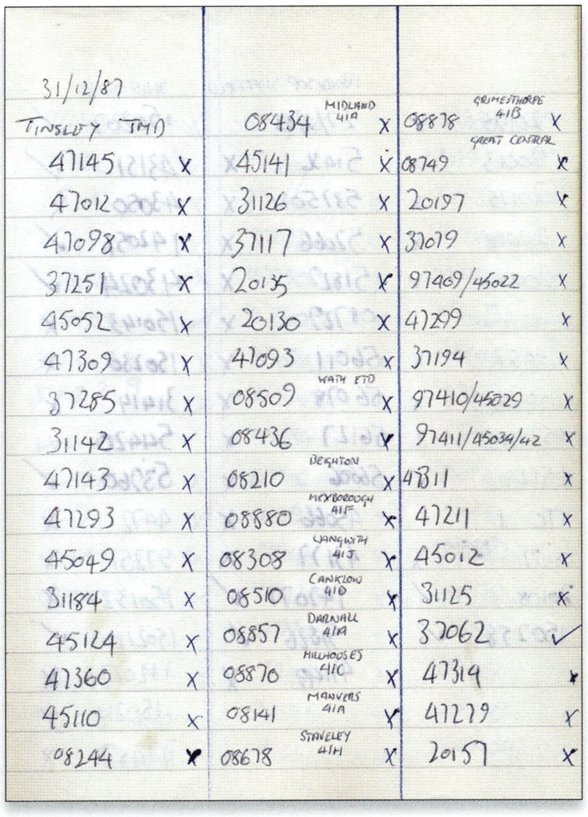

Sunday 4th September 1988 (Shed Surprise)

08210 Tuxford, 08436 Beighton, 08509 Wath ETD, 08678 Staveley, 08857 Darnall, 08870 Millhouses, 08871 Barrow Hill, 08880 Mexborough, 08919 Cadeby
20060, 20092, 20147
37194, 37678
45103 Griffon, 45106, 45115 Apollo, 45128 Centaur, 45141 Zephyr
47104, 47203, 47217, 47290, 47317
56054
Tinsley Yard (all condemned)
08141 Manvers
20060, 20092, 20147
45007 Taliesin, 45012 Wyvern II, 45033 Sirius, 45037 Eclipse, 45046 Royal Fusilier, 45049, 45052 Nimrod, 45107 Phoenix, 45110 Medusa, 45110 Athene, 45120, 45124 Unicorn, 45134 Neptune, 45140 Mercury, 45145 Scylla, 46009 Old Dalby (actually 45104!)
47093, 47130
97409 Lytham St.Annes, 97410, 97411, 97412, 97413

The pseudo-46009 in the yard caused much confusion amongst enthusiasts, as 46009 had been destroyed in a test crash with a nuclear flask on the Old Dalby test track on 17th July 1984!

Sunday 10th December 1989 (Shed Surprise)

08509 Wath ETD, 08749 Great Central, 08871 Barrow Hill, 08878 Grimesthorpe, 08919 Cadeby, 09008 Aldwarke, 09013 Shepcote
20025, 20030, 20046, 20118, 20156
31146, 31286
37054, 37679
47009 Guillemot, 47146, 47298 Kestrel
Snow Plough: ADB965309

Monday 7th May 1990 (Shed Surprise)

08509 Wath ETD, 08880 Mexborough, 08919 Cadeby, 09008 Aldwarke, 09013 Shepcote
20178-cond
31132, 31260-cond, 31280-cond, 31282
37009 Typhoon, 37109, 37238 Spitfire MKII
47012 Magpie-cond, 47014-cond, 47097, 47098 Oriole, 47105 Goldcrest, 47118, 47236, 47280 Pedigree, 47283, 47331, 47377
Tinsley Yard (all condemned)
08141 Manvers
20126, 20226
45007 Taliesin, 45012 Wyvern II, 45033 Sirius, 45037 Eclipse, 45046 Royal Fusilier, 45052 Nimrod, 45106, 45120, 45124 Unicorn, 45128 Centaur, 45130, 45134 Neptune, 45140 Mercury, 45141 Zephyr, 45145 Scylla, 45150 Vampire, 46009 Old Dalby (45104)
47017, 47412
97409 Lytham St.Annes, 97410, 97412, 97413 Amethyst
ADB968024

Sunday 20th January 1991 (WNXX)

08509, 08749, 08880, 08919, 09008, 09013
20176, 20198
31285, 31405
37079, 37110, 37298, 37425
45135 (Peak Rail Preserved)
47018, 47120, 47145, 47152, 47214, 47220, 47289, 47323, 47332, 47447, 47605

Wednesday 25th December 1991

Christmas Day 1991 fell on a Wednesday. According to one visitor to the depot who was "just passing - and it had to be done", there were a lot with the same opinion! These are the notes of one WNXX poster

08141, 08509, 08876, 08879, 09008, 09013
31459
37029, 37053, 37058, 37114, 37261
47006, 47096, 47102, 47125, 47195, 47220, 47302, 47313

Whilst this is what Alan Waterhouse noted on the same day:

08141, 08509, 08749, 08876, 08877, 08919, 09008, 09013
31407, 31459
37029, 37053, 37058, 37114, 37198, 37203
47006, 47096, 47102, 47125, 47195, 47220, 47302, 47313

There are slight discrepancies between the two; maybe some missed locos that the others saw!

Sunday 12th January 1992 (Shed Surprise)

08141-cond, 08509, 08749, 08876-cond, 08919, 09008
37015, 37114, 37190
47006-cond, 47102-cond, 47188, 47195-cond, 47204, 47207, 47323, 47375, 47378
56098
60046
Tinsley Yard (all condemned)
20126, 20226

45130

47017, 47412

Saturday 20th March 1993 (Chris Booth)

08879, 08880, 08919

37032, 37209, 37248, 37373

47063, 47095, 47096, 47102, 47146, 47190, 47195, 47293, 47301, 47325, 47335, 47349, 47361

Sunday 24th April 1994 (WNXX)

08389, 08661, 08691, 08879, 08880, 08919

37238

47096, 47102, 47142, 47221, 47249, 47288, 47310, 47317, 47321, 47325, 47345, 47347

Saturday 7th October 1995 (Andrew Booth)

08379 Earles, 08389

47033 Royal Logistic, 47050 (minus cab), 47052, 47060, 47079, 47096, 47102, 47190, 47214, 47249 Sea Hawk, 47288, 47286 Port of Liverpool, 47321, 47325 Red Rum, 47339, 47358 Ivanhoe, 47375 Tinsley Traction Depot

Friday 16th August 1996 (WNXX)

08879, 08905

47095, 47096, 47102, 47125, 47146, 47186, 47190, 47214, 47222, 47249, 47291, 47303, 47309, 47316, 47321, 47325, 47330, 47344, 47356

Sunday 30th March 1997 (Pennine Railway Society)

08512, 08879

47033, 47102, 47222, 47144, 47190, 47321, 47096, 47214, 47310, 47325

Sunday 8th March 1998 (WNXX)

47520, 47543

On 27th March 1998, 47375 "Tinsley Traction Depot" collected 47145, posed for photographs with 08879, then left as 0Z96, 13.50 Tinsley to Saltley, the final departure from Tinsley depot.

Below left: The view from the path. Trodden by many feet over the years, this was the path that the vast majority of enthusiasts visiting the depot would have taken. *Brian Hancock*

Below: 47328 at Tinsley depot on 21st June 1993. Built at Brush and delivered as D1809 on 27th January 1965, she was renumbered 47328 in April 1974, 47396 on 21st March 1994, and back to 47328 on 27th July 1995. She was withdrawn in December 2003 and scrapped at C.F. Booth, Rotherham in September 2005. *Keith Long*

Chapter 24

Personal Recollections of Tinsley

Anyone with an association with Tinsley Yard, whether as ground staff, office staff or train crew, all have their memories of their time at Tinsley. Some have put pen to paper to record some of their recollections and have been kind enough to share them with us in their own words.

James Greaves – Shunter

"At the age of 18 I had outgrown my position as train recorder, with a grade of junior railman at Woodhouse Junction, and was sent to be a shunter at the mighty Tinsley Marshalling Yard."

"Although when I arrived Tinsley was on the decline and had seen better days, there was still plenty of activity so you could appreciate just how busy it had been in its past. I was given the title of leading railman in the main yard, which was deep in the hub of the yard itself. I recollect being given a basic induction and a shunting pole and told to shadow a senior railman, who was to become my regular mate."

"The place was intimidating with its constant chatter of locomotives ticking over and wagons steadily rolling down the hump system. The mess room was full of characters from carriage & wagon staff, train preparers, train crew waiting for their trains to be marshalled and other shunting staff. Behind the mess room was the brake bank (or 'brake kip') which was basically a hill with guards vans stabled on it. Once a train was ready a brake van would be rolled by gravity onto the rear of the train, and this was where the guard carried out his duties. A train would depart via TP13 signal outside Tinsley Park signal box and the guard would roll the brake onto the back of the train and attach before giving the driver a hand signal to depart."

"Locomotives would arrive all the time to be attached to their prepared trains on the 49 roads (sidings) that swamped the landscape. One of my jobs was to set a route using hand points and fetch the locomotives in and attach them to prepared trains. I recall just how dangerous it was nipping in between the moving wagons, slinging the screw or instanter couplings on the hooks."

"The rules required you to walk back towards the main hump tower, pin down brakes (which could be 80–90 wagons away) then walk back and attach the couplings and associated brake pipes. This was just not practical and it became the norm to nip between the moving wagons; although dangerous, it was the only practical way you could carry out your job. I learned the skills associated with this role very quickly using the pole to move wagons by barring the wheels with the shunting pole."

Above: As viewed from the main yard control tower, what could possibly be a Class 9F 2-10-0, departs from the yard with a mixed train via the main yard east departure on 12th August 1965. Note the brake vans on the brake kip to the left of the train. The furthest one will probably be rolled down the incline to be attached to the rear of the train when it comes to a stand, before heading off towards Catcliffe and Treeton. *National Railway Museum/Science & Society Picture Library DM354*

Above: An instanter coupling between wagons, set to the "long" position. Turning the centre link to the "short" position would close-couple wagons and help to prevent "snatch" as the wagons moved. *Model Rail Forum/Creative Commons Licence*

Below right: As viewed from a diverted passing passenger train, English Electric Type 3 No. 37067 is attached to a rake of oil tankers in the express freight yard at Shepcote Lane on 19th May 1978. This loco is now 37703 and can be found at the Bo'ness & Kinneil Railway. *Arnie Furniss*

"The method of short coupling instanter couplings by using the stick soon became second nature."

"Of all the jobs I've done on the railway I do have to say I loved this job; I had never been so fit working with first class, down to earth, Sheffield & Rotherham salt of the earth people, meeting drivers from all over the network. It was a very dangerous job but in a strange way very satisfying."

Dramas

"One of the most dangerous situations I faced was when the Dowty retarder system failed. I recall a massive ferry van running away at around 10mph with nothing going to stop it from moving, the only logical thing to do was shout on the local tannoy 'ferry van right away Shepcote Lane'. This thing did run all the way to Shepcote Lane signal box, running through several sets of points but it was the norm, the points suffered no damage; they were compressed air points that just needed the signalman at Tinsley Park to alter them back to the corresponding position and they would be fine. The pilot driver and I would take the 350 shunter down to Shepcote Lane and bring it back – no real drama."

"Derailments were a common theme in busy yards; it didn't happen every day but it certainly occurred regularly. We just grabbed the re-railing ramps and lined them under the pair of wheels off the rails, then pulled them back on with the pilot."

"I remember being asked by the yard supervisor to fetch a pair of locomotives in the yard. The hand points were always bouncing off but you just gave them a kick and they soon fitted up. This particular day I did exactly as I normally did, pulled the points over and kicked them up and waved the driver in; the next thing was a bang, clatter & bang. I saw the driver bouncing up and down in his seat – the leading bogie was well and truly derailed. I walked back to get to the locomotives and got the biggest telling off of my life from the driver of 31164, lambasting me for not making sure the hand points were fully up. These Frodingham men took no prisoners when they vented their anger at you, but this was Tinsley and these things happened."

"Tinsley was a notorious place for being on sodden ground; it was built on a marsh ground, so I'm led to believe, which probably explained why there was so much movement in the track. [In actual fact it was due to sodden ground caused by a combination of bad drainage and oil leakage from the Dowty equipment – Author]. *The wooden chocks fell out all the time holding the gauge together; you never saw any P-way staff maintaining it."*

You Couldn't Make it Up!

"I vividly remember the overhead stanchions were still in situ when I started there in 1981, but the wires had been cut down due to the closure of the Woodhead Route. I recall a guy fitting earth straps to the rail from the stanchion. I remember asking him what he was doing. He replied 'I've been fitting these straps for years', I remember saying the wires had gone, but he replied 'I've been doing this job for years, the contract to put them on is still ongoing until we finish them'. That was good old British Rail which we all remember through rose tinted spectacles."

Class 76s

"As a child I was always fascinated by the Class 76 electric locomotives and both my brothers were traction trainees learning to drive them at Rotherwood. Anyone knowing anything about railways will tell you these things were way ahead of their time, but that's another story!"

"I recall at Tinsley this particular day we were busy marshalling trains together, when this rake of withdrawn Class 76 electrics

Right: Driver Eddie Kisby grins from the cab of 40033, which had arrived into Tinsley Yard with a train from the Hope Valley Line and is now ready to return to Guide Bridge. *Roger Wainwright*

Below right: 20076 + 20064 wait to depart Tinsley with 76030, 76025, 76034, 76051 and 76027 as 9X35, 07.23 Tinsley Yard to C.F. Booth's scrapyard on Tuesday 15th March 1983. The driver was Les Barnes, with relief driver Roger Wainwright. These are the Class 76 locos from James' story. Note that the overhead stanchions are still in situ, perhaps being fitted with earthing straps by the guy in James' story! *Roger Wainwright*

arrived from Reddish outside the yard, hauled by a Class 40 locomotive. It was stood outside in the reception sidings quite a while, whistling away, but the Guide Bridge driver was getting impatient and decided to bring them in while we were shunting. I didn't see what happened but there was an almighty bang and the pilot loco collided with the Class 40, and when I walked down the pilot loco was listing around 45 degrees and derailed. The Class 40 was still on the track but the traction motor was exposed and damaged. Eventually the Class 40 was pulled back and taken to the TMD shed; shortly after the packing vans arrived and started re-railing the stricken 08 shunter."

"My day was up but I recall the next day that these beautiful Class 76 electrics were stood there awaiting their fate at the locomotive graveyard in Rotherham (C.F. Booth's). I decided to have a good look around them, truly amazing locomotives, everything was brass & copper, which in today's scrap metal market would make them worth thousands of pounds. I remember switching on the main breakers in the engine room (didn't have a clue what I was doing) – the whole locomotive lit up like a Christmas tree, the compressor started working and all the cab instrument panel lit up. These beautiful machines had been stood for best part of two years, but the build quality was amazing; even the handbrake near the secondman's seat was made out of brass. It was a sad day when they went on their way to C.F. Booth's scrap yard for the inevitable."

"My memories of Tinsley will always be fond ones but it was clear it had seen better days when I left to fulfil my ambition to become a signalman. The place buzzed 24/7 with diesel smoke and engines ticking over, something us old railwaymen will never forget."

Walter Ford – Relief Senior Railman

"I worked at Tinsley from 1970–71, as a relief senior railman, when I had a position in the express freight sidings. Here we prepared trains in Nos. 1 to 20 roads at the west end, and Nos. 1 to 4 in the express sidings. This involved pulling trains out into the express sidings to go south to Toton, Washwood Heath, Bescot and South Wales, via Treeton. It was a very busy job at the time."

"I moved from the express sidings to the reception sidings to work as a cutter, which involved working with a partner cutting trains up into sections, ready for them to be hump shunted into their respective sidings in the main yard and secondary yard."

"This was heavy work; one of us uncoupled wagons whilst the other collated the wagon information to ensure the wagons would go into their correct sidings. We used to take the collated information into the small 'Addo' cabins in the reception sidings, to send the information via the 'Addo' machines to the main control tower."

"Another job I held was in the main control tower as an operator and supervisor. On snowy days I also acted as a snow shoveller to clear the points and keep the yards moving."

"I also acted as a chargeman on the reception sidings until the hump and the reception closed in 1984. After closure of the

Left: Class 76 cab controls, as seen in one of the locos brought into the yard in James' story.

Below: After travelling from Tinsley Yard via Catcliffe, Treeton North Junction and Masborough South Junction to Holmes Junction, 20076 + 20064 propel 9X35 back towards C.F. Booth's at Holmes Junction on Tuesday 15th March 1983.
Both Roger Wainwright

Above: A cutter at work in the reception sidings in September 1965. Here he is taking the number of the wagon, along with the details of its destination and loading from the wagon label.
Chris Booth collection

Right: An example of a wagon label. *Ted Evans*

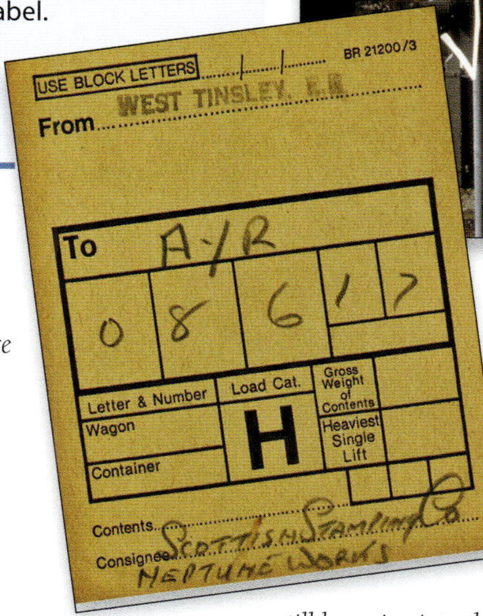

USE BLOCK LETTERS BR 21200/3
From WEST TINSLEY E.R.
To A/R
0 8 6 1 7
Letter & Number
Wagon
Container
Load Cat.
H
Gross Weight of Contents
Heaviest Single Lift
Contents
Consignee SCOTTISH STAMPING Co NEPTUNE WORKS

reception sidings, we were relocated to the former platelayers hut which was adjacent to the closed main yard tower. Our job then was to let trains into and out of sidings 16–21 which were connected via a new section of track. Shortly after this we were moved to the west end of the yard."

Decline

"The yard was then in decline and we were only there to see trains for the steel works. A train would arrive in the morning with steel coil, which was placed in the SPACE sidings. When the wagons were unloaded, we would take them, six at a time, into the SMACC sidings. Here they would be loaded with slab and this procedure would carry on all day until the rake of wagons were all unloaded and reloaded, after which the train would be prepared for departure to Immingham, to be shipped to Sweden. Twice a week there was also a train to the Seaforth container terminal at Liverpool."

"Even when Tinsley was in full working order it was never used how it was supposed to have been. It was designed to allow trains to depart eastwards from sidings 1–15, whilst still humping into the other sidings; however, this was never done whilst I was there."

"I recall one incident after privatisation when we were about to take strike action. As such, management sent staff out from the Doncaster offices to learn our job, one of them being the female secretary of the main man at Doncaster! Where it took our staff 50 minutes to complete a train turn around, it took them 3 hours to do the same job!"

Left: As a Class 25 passes Tinsley Yard signal box and enters the reception sidings with a rake of fitted 16-ton mineral wagons, so 13001 shunts another train over the hump in July 1977.
Allan Trotter, Eastbank Model Railway Club

David Higdon – Yard Design

"My involvement with BR(ER) began in the fifth and final year of my Liverpool B.Arch. degree in 1960–61. I was undertaking a research project on the potential for glass reinforced plastic (GRP) factory-produced lightweight buildings and spotted a news item about a new GRP relay room at Pitsea on the LT&S electrification. The CCE's office very kindly sent me details of the building, after which I gained a good assessment on my submission."

"The CCE's architect (H.H. Powell) kept in touch and, after asking whether I'd be interested in working in the CCE's research & development (R&D) group, summoned me for interview at King's Cross and attached a white First Class cardboard ticket!"

"The CCE's long grimy stock brick office block was actually the east side of the King's Cross station structure and served to stabilise the cast iron train shed arches. And recently built within the cab road arcade was a mezzanine structure housing the architect's office, accessed through glass doors from the head end of platform 1 (This has all since been replaced by platform 0)."

"The interview went well and I was introduced to Reiner Volhard, the R&D group leader who also brought in Arthur Quarmby, the young architect who'd designed that GRP relay room."

"I returned to Liverpool, having been offered a job with a salary of £800. I didn't realise it at the time but this was £40 more than BR's pay scale divide between travelling First and Second Class. It was to prove very embarrassing at times working with long-serving experienced railwaymen."

The Chief Civil Engineer's Office, King's Cross

"Fascinatingly Victorian in atmosphere, nevertheless the building seemed charged with cutting edge ambition. Most offices had coal fires in cast iron grates fed by a trolley circulating continuously to drop off re-filled brass coal scuttles. Many drawings dated back to the 1840s and 1850s, with mind-blowingly beautiful ink draughtsmanship on linen, all fully water colour rendered. Assistants were expected to routinely amend these with updates of trackwork and alignments, and revisions to lineside buildings. Many specialist items of drafting apparatus dated back to Victorian times: mahogany boxes of exquisitely carved wooden railway curves, extraordinarily complex planimeters and pantographs – all to be signed for in ancient ledgers with the mid-Victorian railway company names on their bindings, guarded by equally long-serving clerks."

"On the other hand, in our R&D group (of about 12 assistants) we might get summoned at short notice to a presentation by the Bristol Aeroplane Company, of their proposal for a plastic railway coach from the fusing together of two identical injection mouldings produced in a 70-foot long pair of male/female concrete moulds (I remember that one in particular). Chief officers would seek our opinions afterwards about what had been proposed."

"We had some interesting alternative ideas, e.g. Dowty made some headway with a project for network-wide wagonload distribution propelled entirely by computer-controlled lineside plungers (as used at Tinsley) that either speeded up or slowed wagon wheels! However, this was deemed a step too far!"

"Being in London, many of the older group members had come to King's Cross from other railway offices, in particular London Transport and the Southern Region. Both offices used 'in-house' design teams and had a reputation as innovators, whereas the Western and North Eastern used outside consultants. We each were assigned a range of projects we were expected to research then discuss with the others, orchestrated by Reiner the R&D group leader on Friday afternoons. I particularly enjoyed a couple of collaborations with BR Research at Derby: one about ameliorating the effect of noise on lineside settlements – a current issue once again with HS2."

Jewel in the Crown

"At the start of the 1960s the Sheffield District with its specialised steels etc. was at that time very much the 'jewel in the crown' from the point of view of BR revenue generation – it seemed natural to invest heavily there, not only on Tinsley but also on station improvements at Sheffield Victoria, Elsecar & Hoyland etc."

"Tinsley was my main job. I've already written about the long model constructed along the mezzanine office to determine the position of the control tower [See Chapter 2 – Author]. My longer-term task was to devise a practical way of building the range of yard buildings. The two ex-Southern group members argued persuasively for concrete, after long experience of Exmouth Junction pre-cast component usage."

Polish Permanent Way Engineers

"One distinctive feature of the CCE's office in 1961 was that the permanent way team working on future capital projects such as Tinsley (and the Peterborough realignment) all appeared to be middle-aged Polish émigrés. There were rooms in the upstairs offices that were wholly Polish-speaking. Within them a 'Franz Kafka' atmosphere of intensely complex industry prevailed, occasionally interrupted by someone slipping off his drawing board stool to make the fire up from the brass scuttle. I wonder whether it was them who named the proposed inclined line of guards' brake vans at Tinsley a 'brake kip'."

Tinsley TMD

"The TMD was a separate job within our group, pre-dating the rest of the yard buildings. The job architect was Tony Hall, who'd already completed Finsbury Park to service the Deltics then being commissioned. The pitched roof form to ER diesel depots was decided upon by A.K. Terris the CCE, who personally insisted upon this style of roof section for all the new ER diesel depots, to assist natural dispersal of diesel fumes – (BR(ER) was the first region to be entirely diesel and electric. (I'm not sure the pitched roof was entirely justified by the physics of diesel fumes, though the portal frame section certainly made for a more weatherproof construction). Under A.K. Terris, Roger Waters was very much into research-lead innovation and I was in the CCE's R&D group that co-ordinated with BR Derby central research and also with LT."

John Miller – Rotherwood Guard

"It's 2am and I sign on at Rotherwood and meet the driver. We are given a Class 76 loco and we climb on board & travel light locomotive to Tinsley Yard. On arrival I drop off the loco, go into the shunters' cabin and pick up the TOPS list and the shunter tells me which road the train is on, and that it's all ready to go. So, I flick the handpoints, guide the loco onto the train, couple up the loco, hand the driver the driver's slip, release the handbrakes on the leading three wagons (making sure they matched the train list), and left the driver to draw out while I went up onto the bank (brake kip) and got the brake van ready. The brake kip was where the brake vans were stored to wait their next duty and was inclined to allow them to roll by gravity towards the rear of a waiting train."

"So, after the train drew out I released the brake, free-wheeled the van onto the back of the train, coupled up, gave the driver the green light, and off we went."

A Slow Journey

"Now this driver (who better remain nameless) never messed around – he just went out and back as quickly as he could, which I loved as it usually meant an early finish to the shift. However on this day, as we got to Wadsley Bridge, I was thinking 'much slower mate and we will be on overtime (must be following something, I thought).' We CRAWLED to Oughtibridge and on past Deepcar all the way to Penistone. I was thinking 'what the hell is going on? – there's no leaves on the line (so the loco was not slipping)' but on we went trundle-trundle all the way to Woodhead and into the tunnel. I remember being stood on the brake van veranda thinking 'shall I get out & shove?'"

"We got over the top of the grade inside Woodhead tunnel and by the time we exited the tunnel we were flying, and I mean scary speed, even to the point where I was applying the brake van brake to try and slow us down."

"Through Torside, Crowden and Hadfield, I'd never been so quick and I thought maybe the loco had a fault that had rectified and the driver was making up time. We slowed a bit for Dinting and the viaduct, then Godley and we arrived at Dewsnap sidings where the train terminated."

Surprising!

"My brake van, however, had come to a grinding halt outside the signal box on the main line at the east end of the yard. I was on the brake van veranda looking what was going off when the signalman slung the box window open & shouted `You the Guard?' I nodded, at which he replied: `Bloody idiot, you have brought TWO TRAINS IN ONE' ... `Eh?' I replied ... `Your train is in the yard, the loco's at the far end of the siding and your train's too bloody long; you're double the length, double the weight....double everything!'. `Oh sh**' I replied, `What you want me to do?'. `Nothing' said the signalman, `the shunting staff are uncoupling the train and a Class 40 is coming onto the rear to pull you out and put the rest into another road!'. `OK' I replied."

"Well, this took an hour to complete, after which we went back light loco to Rotherwood. The driver said only one thing on the way back ... `PULL ANYTHING THESE 76s WILL!' – how blooming true!"

"The outcome of it all was that a shunter at Tinsley had NOT uncoupled between the rear wagon of my train and that of the next train, and so we took all the wagons to Dewsnap!"

Brian Hanley & Roger Wainwright – Relief Drivers

Retired Tinsley relief drivers, Brian Hanley and Roger Wainwright, recall a few anecdotes of their time at Tinsley. Brian remembers one train that he used to work when newly transferred to Tinsley in 1964. This conveyed oil from Hull for Colwick and Annesley, routed via the GCR lines from Doncaster and Rotherham to Darnall, then via the GCR connection to the LD&EC line at Beighton.

The train was 4M26, 01.30 Hull to Annesley which ran on Mondays excepted. It was timed to pass Rotherham Central at 02.57 and depart Tinsley East Junction at 03.03. Because of the gradients between Broughton Lane and Darnall, Tinsley drivers would take a loco to Tinsley East Junction where they would buffer up to the rear of the train of tanks, which was usually hauled by a WD 2-8-0. They then banked the train to Darnall. The train was timed to pass Attercliffe at 03.08 and Darnall at 03.13. From there, the train would be allowed to carry on towards Woodhouse East Junction, passing at 03.25 but the banking engine would drop back and follow to Rotherwood, where it was put into the sidings before returning to Tinsley. The train was booked to arrive at Annesley at 04.43.

Right: The steel portal TMD building takes shape at Tinsley in early 1964. *Chris Booth collection*

Top: 76046 awaits further duties at Rotherwood on 1st October 1978.

Above: 76022 + 76023 burst out of Woodhead Tunnel on 18th November 1980 with 8M17, the 11:18 Monckton Coke Works to Northwich, which the locos would work from Wath as far as Mottram.
Both Chris Booth

Spink Hill

Roger recalls an incident whilst at Spink Hill, on the former line from Beighton Junction to Langwith Junction, with a Class 56 waiting to couple to a train of HAAs. The guard had walked around the train and taken the brakes off the wagons; they started to move before the loco was coupled. Needless to say they collided with the loco and sheared the buffers off the 56, which then had to be returned to Tinsley as a failure!

Roger also recalls one driver collecting horse manure from the fields as they were waiting for their train to be made up. He put this in bags and placed it on the loco footplate, much to the annoyance of his compatriot on the footplate!

A colleague of Brian's, Driver Maurice Butcher in charge of another train at Spink Hill, had another episode of a runaway there. Sometimes the train would be collected directly from Westthorpe Colliery, other times from the sidings adjacent to the closed Spink Hill station, where it had been spilt into about 15 wagons per road.

On one particular day the loco buffered up to the wagons in the first road and the guard went to couple the train up, with the intention that the driver would draw forward and set back onto the second rake of wagons.

However, whilst walking round the first line of wagons, in a *"bright idea"* moment he also released the brakes of the second rake as he walked along. Needless to say once all the brakes were off, the wagons began to move and pick up speed, heading out of the yard and along the branch towards Killamarsh! It is said that they achieved a speed of about 60mph past Glovers flour mill at Killamarsh before the signalman at Beighton Junction had an indication in his box, looked up and saw them approaching, just as a train had passed on the main line! Before he could react, the wagons became derailed on the trap points and caused an almighty pile up! That was one for the Tinsley breakdown crews to deal with.

Another anecdote Brian recalls involved the Class 08 pilot loco from Tinsley, heading to Glovers flour mill at Killamarsh to shunt some wagons. The driver collected the staff from the signalman at Beighton Junction (Killamarsh signal box had closed by this time) and set off to the mill, only to find that the permanent way staff had removed the track in the yard!

Alan McKenna – Telex Clerk

"I started my career on the Railway in April 1971, aged 19 coming up to 20. My Dad was a railwayman and I had never intended to follow in his footsteps, but was getting married that year and the job was offering a salary of £630 pa, which was £100 more than I was getting at the time. I remember, after I had been offered the job at the medical in Sheaf House Sheffield, being asked by the medical officer what job I had been offered. When I told him he said that's OK because did you know you are colour blind and would not be able to work on the running line. Never knew that I was until then!"

"Anyway, on to Tinsley. My job title was Telex Clerk Grade CO1. After a five-week training course at Eversholt House in London, I was ready for the job. We had a small office in the west end building which was situated next to the departure line for roads 40 to 46, which were electrified for the Bo-Bos which ran through Woodhead to the Manchester area. Next to our office was the main yard supervisors' office. We had like a serving hatch so we could talk to the supervisor and his assistant without going round. This building also housed the mess room. Also attached to this building was the tappers cabin."

"We worked three shifts – 0600–1400, 1400–2200 and 2200–0600 – but because there were two of us on the night shift it meant working nights every other week. Although we only worked Monday to Friday we did work Saturday mornings on the day shift. I always found the changeover from days to nights and vice versa quite difficult. Even worse when we were working 12 hours. We didn't have a relief clerk so any holidays, sickness or lieu days were covered by 12-hour shifts."

"The job of telex clerk was really a forerunner to TOPS although of course we didn't know this at the time. Our team consisted of four telex clerks backed up by six (I think) number takers. Our job was to send details of the make-up of trains (i.e. wagon type, load, destination, departure time etc.) to other terminals. We only sent these details to other major terminals. These would include Washwood Heath, Warrington, Bescot, Dewsnap, Tees, Tyne, Toton, Ashburys, Grimsby, Doncaster, Hull, Lincoln, Godley Junction, Mottram, Healey Mills, Whitemoor, Temple Mills, Severn

Above: 47319 arrives at Westthorpe Colliery with a set of empties for the colliery. These will be loaded to form a train to West Burton or Cottam power stations. *Roger Wainwright*

Tunnel Junction and probably one or two more that I can't remember."

"The number takers tended to come and go although two that I can remember were Kevin Chatfield and Cyril Livingstone. Their job was to go out after a train was prepared and read out the wagon label details over a radio link which we recorded on to a small reel-to-reel tape recorder. The Telex clerks would then type these details out on to a punched tape which was then sent through the Telex machine to the various places."

"Cyril lived nearby at Catcliffe and when he was working nights it was common knowledge that he had been in The Plough pub since about six o'clock most times. You could always tell when he had drunk one too many as instead of reading out the label details he would sing them!"

"The rechargeable batteries for the radios weren't very good and often went dead very quickly leading to lots of crackle and difficulty in getting the messages through to us. Cyril got wise to this and would often make crackling noises in the radio and blame the battery thus saving him going down the train. Despite this he was very likeable and everyone turned a blind eye to his drinking until one night we lost him and he was found on his back in the six foot singing to himself. From then on he was transferred to the job of storekeeper."

1970s

"During the '70s we had about 125 trains departing every 24 hours. Guards didn't have to prepare their own trains. There was a dedicated team of train preparers (TP) who did this. Each TP had a set number of trains to prepare each shift. Most of them were on permanent 12 hours but that was no hardship. I think the Telex clerks and the supervisors were the only ones who physically worked 12 hours. Two TPs who I was very good friends with were Peter Green and Albert Hunt. They used to work together thus finishing the job much quicker, a 12-hour day shift for them finished when they had prepared their allocation which was usually about 15.00. This included about an hour for a cooked breakfast about 10.00 and a walk down to the Plumpers public house on Shepcote Lane for dinner."

Below: Spink Hill in 1977. Westthorpe Colliery internal shunter reposes in the shunt neck as two rakes of HAA wagons wait in the sidings. This view is from the cab of a loco that is about to cross to the former "up" line. This was originally part of the LD&EC Beighton branch with Spink Hill station in the distance, beyond which is the tunnel of the same name. *Roger Wainwright*

Above: Secondman's view of the "End of One Engine in Steam" board affixed to the signal at the Beighton Junction end of the Westthorpe branch. This was the former LD&EC Beighton branch from Langwith Junction, by now only serving Westthorpe Colliery and shortly after being singled. Over to the left is the "Old Road" with Beighton Junction in the distance. *Roger Wainwright*

"The tappers next door had an easy job. They used to sit in their cabin playing cards most of the day and usually only ventured out when someone asked them to check a flat tyre. They did 12-hour shifts most of the time as well."

"As well as sending out consists of the trunk trains we would also receive telegrams from Sheaf House with details of out-of-gauge loads, bullion trains and similar. These had to be passed on by us to the signal boxes concerned as they had details of the special bell codes to be used. I once got a 'form 1' as I told every box except for Treeton South as I thought it didn't concern them. The signaller held the train up (rightly) until he received the information. I was still in my probation period and was quite worried. However I had a good union man who got the 'form 1' rescinded, proving that I had not been given any training on the handling of these."

"I don't remember the year we moved out of the main yard to a new building next to the admin block on Wood Lane but would guess about 1973. This was when we started to look at pre-TOPS which meant the Telex messages we sent to the other yards had to be in a certain format."

"Again not sure about the date, but probably a year later, our building was extended and after a two-week training session which took place in a railway carriage on the top shed, we started with TOPS proper and I was re-graded to CO2. Our office became Tinsley AFC (area freight centre). The job was new to everybody and many railway staff treated us with suspicion, thinking that we would do away with jobs (lots of jobs did disappear in the end including mine but it was nothing to do with TOPS)."

TOPS

"TOPS was very exciting and everyone I know remembers it as being the best job ever. It was very manic as we were working on real time and every train had to be reported both in and out including local trips and engineers' trains and light engines. Reports were sent to us from all of the Sheffield area, some by telephone and some by fax which was very new in the 1970s. Think the railway was the first industry to use fax by a long way."

"We worked a lot of 12-hour shifts as I remember. I can also remember quite clearly taking home my snap more than once and I wasn't the only one. Every wagon and every loco was represented by a punched card which was filed according to location; Deepcar, SFT, Broughton Lane, Orgreave, Rotherwood and many others. Looking after the yard file was my favourite as, although it was the busiest, every wagon was filed in an individual compartment depending on which road it was in. To me it was like playing trains."

"Things moved on, and around the end of the '70s we were told that we no longer needed to use punched cards as TOPS was going cardless. This was a major panic for some as they wondered what we would do without a physical card for each wagon etc. Actually, it worked great."

"Saturday afternoons and Sunday were always quiet as the only movements were locos, engineers' trains and newspaper trains (this was before we started to report passenger trains in the later stages of the system). Late Saturday morning and Saturday afternoons usually saw a handful of trainspotters watching the locos come on and off the depot. Often they would knock on our door asking if we knew what locos were on the shed. Most of us would print them off a loco list (for a fee!)."

"On the weekend nights we were the only ones around that part of the yard (single manned) and it could be a bit lonely at times. However on the plus side there was plenty of wildlife prowling around, particularly foxes. One of the worst nights was that night once a year when the clocks went back. The clocks were controlled from a central point and at 02.00 would wind back an hour. I seem to have worked more than my fair share of those!"

"We had a newspaper train that left Manchester at something like 01.45, due in Sheffield at 02.30. The TOPS computer wasn't happy when you tried to show it arriving at 01:30 (before it had left!)."

"Within the AFC the personnel was made up of four or five TOPS clerks, three supervisors (yard supervisor, area supervisor and can't remember the title of the third but he was in charge of the other two). We also had a travelling chargeman whose job it was to drive to some of the remote locations (Brookhouse and Westthorpe Colliery spring to mind) to report on the movement and consist of trains from these locations. One of the chargemen, Geoff (forgotten his surname) had been a butcher in a previous life and used to bring cuts of meat that he had prepared and sold to us at a cut price."

"During the winter of 1978/79 we had very bad snowfalls. I remember on one occasion me and a work colleague, Fred Anyon, walked down to Sheffield station (about 2 miles). It was chaos, no trains running at all. Eventually we managed to hitch a lift on a loco going to Tinsley depot, booking on for our shift about 2 hours late. It was an afternoon shift and nothing much moved all day. Can't remember how we got home that day but on another occasion the snow started mid-shift and by about 20.00 nothing was moving at all. Mick Barrett who was an enterprising signalman in Tinsley Yard box persuaded someone at the depot to run a loco for us. It dropped me off near Heeley bridge and a short walk home!"

Above: In a scene replicated most days at Tinsley, 20115 + 20002 have just arrived in the yard with 6T30 trip from Deepcar on 26th September 1985. *Michael Rhodes*

The 1980s

"In 1983/84 the TOPS offices at Barrow Hill, Rotherham Masborough and Wath were all closed down and incorporated in Tinsley. By this time I was a grade CO4. All the personnel from the three depots were relocated into our office and then jobs allocated on a seniority basis. I was redundant and spent about 6 months spare. This meant coming into work to see if there were any jobs to do – there hardly ever was – and then going home again after a few hours."

"The only day of the week I had a regular job was delivering wages to signal boxes on Thursdays. Eventually I was told to do a job in Sheaf House managing an admin office – something which I had no idea what I was doing and hated every minute. It was only a temporary job and within six months I had secured a job in the booking office at Sheffield and my career took a new turn. I eventually reached the dizzy heights of retail training manager but retired early in 2009; however, I always look back at my time at Tinsley with great fondness."

Above: A typical Saturday at Tinsley. 37127 stands in front of 37096, keeping company with 31229 and an unidentified pair of Class 20s. *Roger Wainwright*

Chapter 25

Tinsley Today

In 2015 the 50th anniversary of the opening of Tinsley Yard was not celebrated in any way. We hope this book will, in some way, redress this. To examine what remains of the yard, we begin with some aerial photos taken by John Webster in June 2015 – which are in stark contrast to those seen earlier in the book. Parts of the yard are returning to nature and motor vehicles dominate the landscape – such is progress.

Below: Looking towards Catcliffe from above Wood Lane bridge. The site of the TMD is now devoted to car sales, whilst in the distance is a hotel. The long concrete pad that can be seen centre bottom was the footprint of Tinsley Yard signal box, while the site of the reception sidings is now a small forest!

Bottom: Wood Lane bridge is top left, with the remains of the main yard control tower in the centre, now just a pile of concrete rubble. The area of trees is the site of the hump and servicing shed. *Both John Webster*

Above: Now looking towards Shepcote Lane, we see the huge bulk of the SIRFT and the remaining sidings. The remains of the main yard control tower are to the bottom left. *John Webster*

Special Instructions to the Signaller at Woodburn Junction

Working of Trains into Tinsley Yard

You must contact the Tinsley Yard shunter/PIC to obtain permission for a train to proceed towards Tinsley Avesta TMO level crossing stop board, before setting the route from W205 signal to W226 signal. W226 signal position prevents a freight train from stopping at the signal due to a steep rising gradient on approach; therefore an aspect control will be applied to W205 that will prevent the route up to W226 signal from clearing unless W226 signal is off. Confirmation from the shunter or driver that the movement in is clear at Tinsley Yard complete with tail lamp is not required.

You must contact the Tinsley Yard shunter/PIC to obtain permission for a train to proceed towards Tinsley Avesta TMO Level crossing stop board, before setting the route from W212 signal to W228 signal. Confirmation from the shunter or driver that the movement in is clear at Tinsley Yard complete with tail lamp is not required.

The signaller must record in the Train Register Book the times that:

- Signal W212/W205 is cleared for an inward movement
- Signal W227 is cleared for an outward movement

Signalling

The only signalling in place today governs access into the yard, under the control of Woodburn Junction signal box. This box received a new TEW SM48 panel in May 2017, in conjunction with resignalling around Tinsley East in preparation for tram-train operation between Meadowhall and Rotherham Central. The resignalling saw the removal of the former Brightside arrival and departure lines (former Sheffield District Railway lines) from Shepcote Lane to Brightside.

The remains of the yard are operated by DB Cargo (formerly DB Schenker), with GBRf working stone traffic. Like any yard, a safe system of working is in place. Special instructions apply to trains working between Broughton Lane/Tinsley South and Tinsley Yard; the latest versions, brought into operation from 30th June 2017, remove the need for all drivers to stop at signals to receive instructions about entering the yard, provided the signaller has received prior authority from the DB Cargo Person in Charge (PIC) or the shunter.

September 2017 saw a minor amendment to these instructions insofar as if there is no DB Cargo shunter on site, the first freight operating company operative to arrive should contact the Woodburn Junction signaller and take charge of the yard as PIC. If that operative is the driver of the first train he can then proceed, checking that all routes are set for the direction of travel and stopping short of any obstructions.

When a DB Cargo ground staff team member (GSTM) subsequently books on duty, they should enquire of the Woodburn Junction signaller as to who holds the PIC authority of the yard and relinquish that person of the responsibility accordingly.

DB Cargo also operates a safe system of work (SSOW) for the Tinsley complex, part of which reads:

Person in Charge

The DB ground staff team member (GSTM), when occupying the yard, will be the PIC of all train movements in all areas of the Tinsley complex, excluding the stone terminal. When the Tinsley GSTM books on duty, they are to phone Woodburn signaller to take PIC responsibility of the Tinsley complex. All movement must then be authorised through the DB PIC. When DB have not taken PIC responsibility, the PIC is the first FOC staff to be present on site. They must phone Woodburn signaller to take PIC responsibility of the Tinsley complex. If it is a driver that arrives on site first and assumes responsibility, then they are to proceed at caution, stopping short of obstructions, and stopping to check and set all hand points are set for the direction of movement.

When DB GSTM books on duty, they will contact the current FOC PIC (as defined by the Woodburn signaller) and relinquish them of PIC duties.

The Tinsley complex consists of 5 locations. These are SMACC, SPACE, Tinsley Yard, warehouse, stone terminal (this SSOW does not cover the stone terminal – GBRf managed site).

Line Speeds:

- 5 mph hauling & propelling in Tinsley Yard & from/to the west head shunt.
- 10 mph hauling only on the bank between west & east head shunt. (no propelling movements authorised up or down the Bank).
- 5 mph hauling & propelling from east head shunt to but not including SMACC loop.
- 5 mph hauling & 3 mph propelling in all SMACC complex (including loops).
- 5 mph hauling & propelling from Tinsley Yard to gates at SPACE (and vice versa).
- 3 mph on SPACE pad.
- 5 mph hauling & propelling within Tinsley Yard and warehouse sidings.

Tinsley Yard – TOPS Stannox 25200

Tinsley Yard is accessed from the "up" & "down" Tinsley freight line. It consists of five sidings, which are all open-ended and controlled by hand points. There is a gradient that falls towards the west end of the yard (way in/out). There is a road crossing at either end of the five sidings, and an access road along the side of No. 5 siding. There are no signals in this yard.

No. 1 road to be used for run-round purposes only;
No. 3 road must be managed and remain clear for shunting purposes;
No. 5 road to be used for wagon repairs;
NB: There is a STOP board when departing the yard.

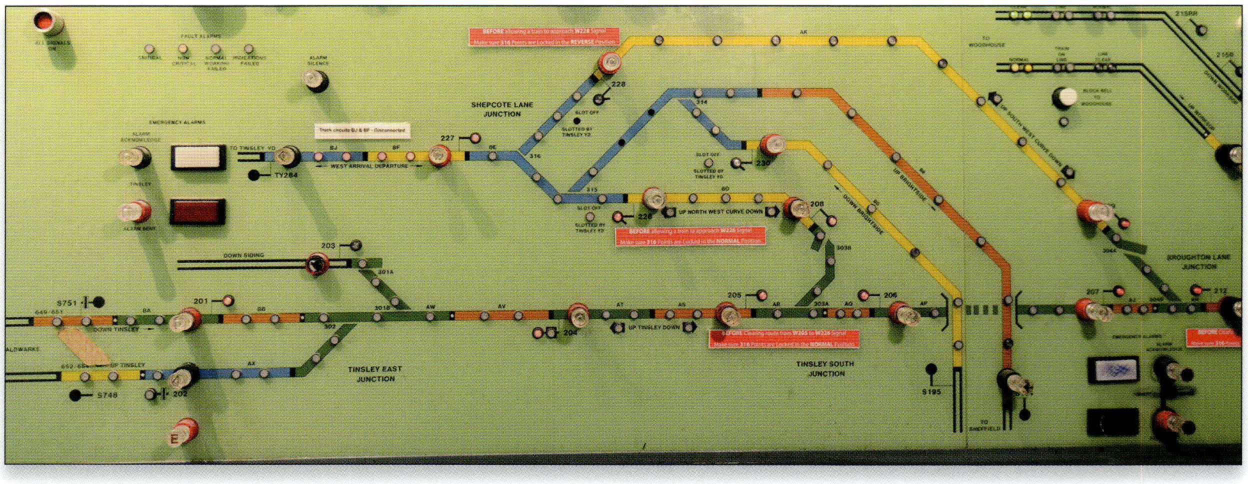

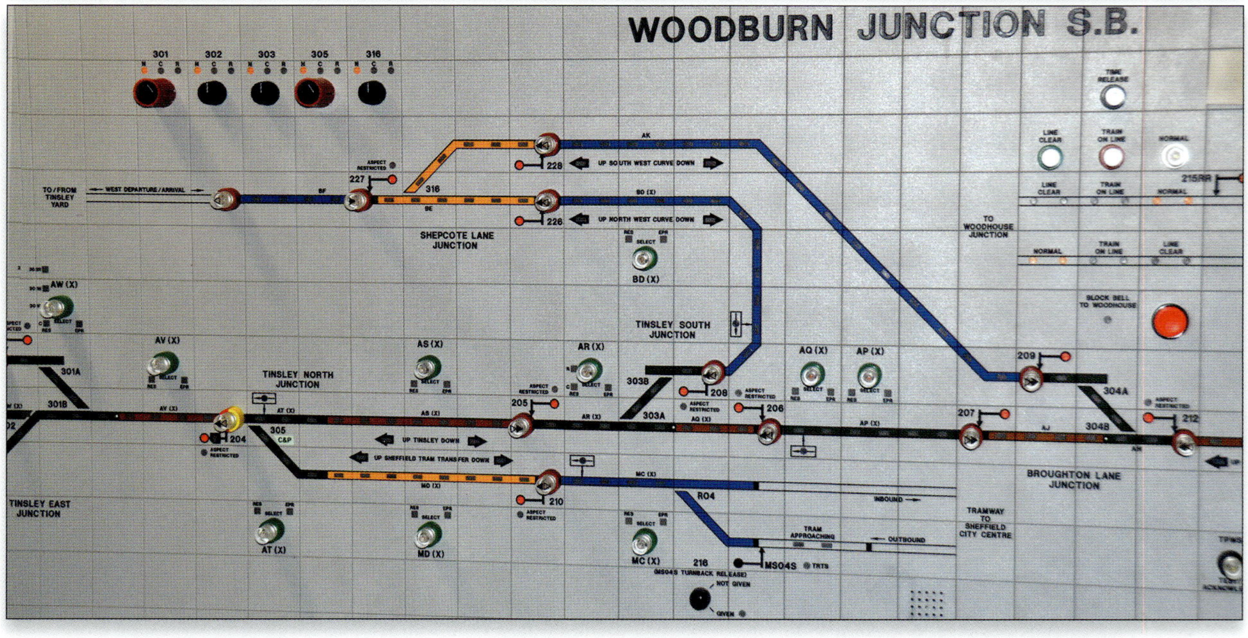

Top: The original Woodburn Junction panel showing the Broughton Lane/Tinsley South Junction to Shepcote Lane Junction area, and the lines to Brightside Junction going off the bottom of the image. This route was never used. *John Webster*

Above: The new panel in Woodburn Junction showing that the Brightside lines have been excluded, thus simplifying Shepcote Lane Junction. The tram-train route is at the bottom of the panel, Meadowhall to the right, Rotherham to the left. *Chris Booth*

SPACE – TOPS Stannox 25208

The site consists of a single line leading from the west arrival line. The siding (unloading/loading pad) is protected by a security gate which is electrified. The DB Cargo PIC must unlock and open the gate before allowing train movements through it. The pad has a single white line painted to the left hand side of the siding, indicating a limit of clearance for the train. Another painted line runs adjacent to the crane rail. This is to warn staff of their proximity to crane movements.

SMACC – TOPS Stannox 25212

Site rail access is via a single line connected to the west end departure line in Tinsley Yard. The route from the yard to SMACC's arrival/departure line is unsignalled and hence controlled by the DB Cargo PIC, and is known as the bank. It has a head shunt at the east and west end and a run-round loop at the top of the bank. There is a road crossing at the bottom of the bank, protected by road signs only. There is a severe gradient on the bank that falls east to west.

Beyond the run-round loop is the SMACC complex. Access is protected by an electrified gate. The single line from the loop passes through the gate and splits into the RMS siding and the pad leading to the bay. In the SMACC complex, there is an open unprotected road crossing used by heavy plant machinery. Extreme caution must be exercised when working in the SMACC complex. These gates form part of the site security and must be kept closed and locked to maintain security.

Warehouse

The warehouse is two sidings adjacent to Tinsley Yard; one is a siding in a concrete pad, the other is a run-round siding. They are both open-ended and connect to Tinsley Yard at both ends. Access/departure is controlled by the DB Cargo PIC. Before any movements leave the warehouse sidings, the Tinsley Yard PIC MUST be contacted to authorise the movement.

Stone Terminal

The stone terminal is accessed from the warehouse sidings and is a GBRf managed site. No movements to be made out of the stone terminal without the PIC authority.

Possibly the first and only time a Freightliner Class 70 visited the site was on 12th June 2014, when 70011 worked the stone train into the terminal. Freightliner lost the stone contract to GBRf in 2015, and GBRf Class 66 predominate, with occasional visits by GBRf's only Class 59 No. 59003.

Above: During the last few weeks of Freightliner's tenure of the stone traffic, 66602 hauls 6M01 out of Tinsley for Bardon Hill at Shepcote Lane on 3rd December 2014. *Mick Charlesworth*

Right: Seen on 24th March 2015, GBRf 66770 is in the Aggregate Industries stone terminal and is ready to commence its journey out of Tinsley Yard with 6M01, 10.51 Tinsley to Bardon Hill. The stone will be processed into asphalt for resurfacing of Sheffield's roads by contractor Amey. The SIRFT is to the far left, and in the foreground are the north east arrival and east arrival lines, with the disused access to SPACE bottom. This view would once have looked across the express freight and departure sidings towards the secondary yard. *Stewart Donohoe*

Left: This is 60099 with 6J03 Immingham to Tinsley which, having recessed in the sidings at 06.30, is now being shunted. *Stewart Donohoe*

Below: The train has come out of the sidings adjacent to SIRFT via the track furthest from the train. Having moved in the Catcliffe direction to beyond the points just in front of the loco, it is now propelling towards the camera on what was formerly the north west arrival, the only usable track as the east arrival is covered in undergrowth in parts. *Stewart Donohoe*

Below: The train now propels past the remains of the main yard control tower towards Shepcote Lane. The demolition of the tower was just starting when these views were taken on 4th April 2015. *Stewart Donohoe*

Above: 66120 acts as 0J03 the yard pilot and collects wagons to take to SMACC on 30th March 2015. The lighting tower on the right was one of the last few to survive in the yard. *Chris Booth*

Three photos showing the remains at Catcliffe and Treeton North.

Above: Signals remain at Catcliffe in February 2009. TY204 (left) and TY202 guard the east departure towards Treeton, whilst the east arrival is on the right of the image. The reversing siding formerly lay between the two lines.

Right: The remains of Treeton North Junction, with Sheffield PSB's S305 signal protecting it on the "Old Road".

Below: Looking across Catcliffe viaduct towards Treeton, the track has been disturbed by thieves in anticipation of being stolen. *All Andrew K. Overton collection*

The Main Yard Control Tower Meets its End: By the turn of the century the main control tower was looking increasingly decrepit, plans being made in 2013 to demolish it at the same time as Tinsley Yard signal box was demolished. However, with the discovery of asbestos in both buildings the costs for its removal meant that the funding ran out. The tower gained a further reprieve until January 2015 when plans were again made to demolish it. 24-hour private security was employed to keep trespassers away; demolition began on 2nd April 2015 with completion by 13th April 2015. *All Chris Booth*

Above: With the disused steel works water tower to its right, 60100 shunts BVA wagons adjacent to the stone terminal on 13th May 2015.
Mick Charlesworth

Right: Lord Beeching unveils the commemorative plaque to announce the opening of Tinsley Marshalling Yard on 29th October 1965.
Chris Booth collection

The Future

A future development that could have affected the remains of Tinsley Yard was High Speed Two (HS2), the planned new high speed rail link connecting London with the West Midlands, Manchester and Leeds.

The route originally proposed in the Sheffield area showed it running through the Rother Valley, following the alignment of the "Old Road". The route was then shown to traverse a viaduct over the River Rother flood plain before entering a deep cutting under the A630, across the end of the former Sheffield airport, and then onto an embankment before crossing diagonally across the former Tinsley Yard site, cutting through one of the SIRFT warehouses. A later map showed the route had been moved slightly southwards to cross between the stone terminal and Shepcote Lane Junction.

Sigh of Relief

In July 2017 the final route was announced, which now shows the line following the M1 motorway as far as the M18, then following that motorway northwards, thereby missing out Tinsley altogether – probably with sighs of relief from residents in the area.

Still Activity

Whilst there is still activity to be seen in the yard, albeit very little nowadays, long gone are the days of hump shunting and diesel locos trundling to and from the diesel depot.

Just over 50 years since Beeching opened the yard, will its remains see its 75th anniversary? We would like to see many more years of operation, but also hope that memories of Tinsley Marshalling Yard will take a long time to fade.

Bibliography

As well as official British Railways files perused at the National Archives at Kew, Sheffield Archives and the National Railway Museum Archives, information has been obtained from the following sources:-

Batty S.R. (1994) *Ian Allan. Railways of Sheffield*
Batty S.R. (1984) *Ian Allan. Rail Centres: Sheffield*
Booth A & Chapman S (2009) *Bellcode. Railway Memories No. 21: Rotherham, Mexborough & Wath*
Chapman S (2013) *Bellcode. Railway Memories No. 27: Sheffield*
Fiennes G.F. (1967) *Ian Allan. I Tried to Run a Railway*
Freeman Allen G. (1981) *Ian Allan. The Eastern Since 1948*
Gough J. (1989) *The Railway and Canal Historical Society. The Midland Railway: A Chronology*
Hurst G. (1992) *Milepost Publications. Register of Closed Railways 1948–1991*
Hawkins, Hooper & Reeve (1989) *Irwell Press. Diesel Depots: The Early Years*
Rhodes M. (1988) *Oxford Publishing Co. The Illustrated History of British Marshalling Yards*
Rhodes M. (2016) *Platform 5 Publishing. From Gridiron to Grassland: The Rise and Fall of Britain's Railway Marshalling Yards*
Shannon P (2014) *Ian Allan Publishing. Speedlink*

Periodicals, Journals and Magazines

Honyman M. (2004, November) *Sheffield's Bridge of Steel, Rail Express 104*
Rhodes M. (2004) *Tinsley Yard Remembered, Rail Express*
Rhodes M. (1997, March) *Traffic Profile: Speedlink, Rail Express*
Shannon P. (1991, February) *Speedlink axed, Modern Railways 98–99*
Sutton P. (1996, June) *Tinsley: Three Decades of Change, Rail Express*
Not credited (1965, January) *The ER's New Diesel Depot at Tinsley, Modern Railways*
Not credited (1998, October) *EWS Shuffles its Shunter Pack, Rail Express*
Not credited (1964, December) *Diesel Developments on British Railways, Modern Railways*
Ford D (1998, May) *The End of the Line for Tinsley TMD, Railway Magazine 53–55*

Below: A view from Tinsley Yard signal box of the reception sidings sees a train either just arriving or about to be hump shunted. Two staff lean on the veranda of the brake van, whilst over to the right a Brush Type 4 makes its way along the engine line. On top of the embankment a line of Class 20s await further duties. *Ivan Stewart*

Appendix I: The Dowty Hydraulic Units

The Dowty Hydraulic Units

Figure 1: Approximately 23,500 small hydraulic units secured to the rails were used to exert a retarding or boosting force on the wagon wheel flanges as they passed over. The diagram shows a representation of this action. As the wheel passes over it pushes the piston down and the resistance to the oil in the cylinder has a retarding effect. Once over, the oil pushes the piston back upwards thus imparting a boosting effect to the wheels if required – this is determined by the speed of the wagon passing over it.

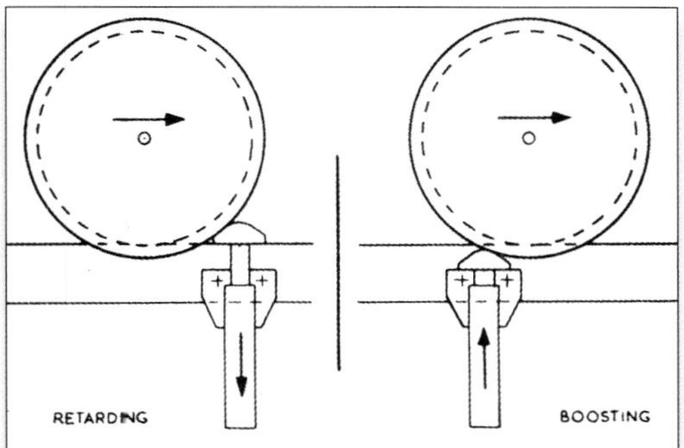

UNIT ACTION
FIG. I

Figure 2: Low-speed retarder units. These self-contained units, operating at 0 ft/sec – 8 ft/sec, were only partially filled with oil. The rapid depression of the piston by a wheel flange caused oil to flow through a speed control valve and consequently caused a pressure difference across it. This lifted the speed control valve against its stop, thereby sealing off the orifices in its periphery. Further increase in pressure due to piston movement opened a relief valve, thus allowing oil to pass out through the holes in the side of the speed control valve. Energy was therefore dissipated from the wheel flange due to the effort required to force the relief valve up against its spring so that the oil may pass through. Should the wagon be travelling at less than the required speed of the unit, as the piston was depressed the oil escaped through holes in the speed control valve and the pressure rise was insufficient to lift the speed control valve against its stop. This allowed the piston to collapse and there was no reduction in the wagon speed.

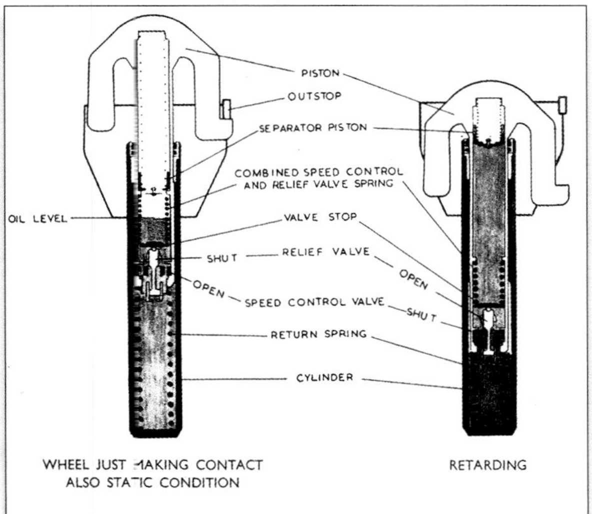

THE RETARDER
FIG. 2

Figure 3: Booster/retarder units were connected to hydraulic pipelines by means of flexible hoses, the power for these being obtained from hydraulic pumping units known as power packs

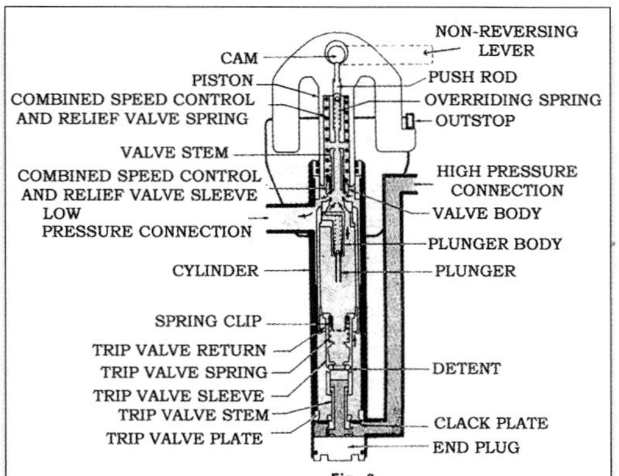

Fig. 3

(A) Retard action – Rapid depression of the piston as the wagon passed over caused the oil initially displaced to flow from the main cylinder to the low pressure line, which closed the speed control valve by lifting it together with the valve body onto its stop. This prevented further flow to the low pressure line and the pressure below the piston rose instantaneously to a value slightly higher than the high pressure line. The clack plate opened and further depression of the piston forced oil into the high pressure line, creating a high resistance to downward travel and an extraction of energy from the moving wagon, thereby slowing it down. If a large number of retarding strokes were made, the quantity of oil pushed back into the high pressure line could cause excessively high pressure and, in this case, the combined speed control and relief valve blows and discharges oil into the low pressure line. Since the speed control valve body lifted and no longer sealed the plunger body, oil was free to escape when the plunger was depressed. Consequently the plunger collapsed into the plunger body, when it contacted the spring cup, and thus did not select the valve sleeve. When the piston started to ascend, the speed control valve would then return to its normal position and oil from the low pressure pipeline would restore the piston to its original position. High pressure oil could not enter the piston due to the clack plate being shut and the other passage via the trip valve sleeve being sealed by virtue of the position of the sleeve.

(B) Booster action – As the wheel flange depressed the piston, the displaced oil escaped from the main cylinder to the low pressure line via the orifices in the speed control valve. The oil flow was insufficient to close the valve against its spring force and the piston was depressed at low pressure, offering little resistance to the wagon. Because the speed control valve did not lift, oil remained trapped inside the plunger body causing the plunger and body to act as a strut when it struck the spring cup. The plunger thus made contact with the spring cup near the bottom of the piston travel and depressed it until it bottomed on the trip valve, which was mechanically moved until the detent was disengaged. The trip valve spring then 'fired' the trip valve sleeve down into the booster selection, where the ports on the trip valve sleeve coincided with those on the trip valve stem. High pressure oil was then admitted into the unit and immediately closed the speed control valve. The full force was then exerted against the wheel, thereby propelling the wagon. Finally, the high pressure was cut off as the trip valve sleeve was lifted by the trip valve during the last fraction of upward piston travel.

Non-reversing lever – this feature was included on low speed booster/retarder units (4ft. / sec.) positioned in sidings. It prevented wagons from being accelerated backwards after recoiling from wagon rakes in sidings. When travelling in the reverse direction, the wagon wheel would depress a non-reversing lever prior to the depression of the piston. This rotated a cam and depressed a push rod which in turn depressed a valve stem via the over-riding spring. This action of the valve stem closed the orifices in the speed control valve which caused an instantaneous rise in pressure when the head of the piston was contacted by the wheel. High pressure within the unit then lifted the speed control valve and the valve body onto its stop against the action of the over-riding spring. The unit then completed the normal retarding action.

Appendix II: More Dowty Installations?

More Dowty Installations?

It is interesting to look at other proposals for the use of Dowty equipment during the 1960s, when it seemed that BR could not decide on the future of marshalling yards. Consequently, there was much correspondence between the British Railways Board and Dowty.

Although Tinsley would be a flagship project for BR, a memorandum from Mr L.W. Ibbotson (chief operating officer of BR) in April 1964 shed doubt on the future of hump yards, and went further to say there was little prospect of further business for Dowty's in the equipping of new hump yards in this country, as it was unlikely that any more hump yards would be built.

However, the memo went on to ask if a start could be made in investigating how far the Dowty equipment could be used economically in existing yards, of which there were two categories – hump yards and flat yards.

Hump Yards

It was stated that it would be necessary for the Dowty equipment to be fitted without major alteration to the hump gradients or layout, since it was unlikely that use of the hump could be sacrificed for the time necessary to make radical alterations. It also said that the Dowty equipment would have to tolerate higher speeds if it were to be installed near the hump in modern yards.

A suggestion was that there may be a case for putting in Dowty retarders below the existing primary or secondary retarders in order to avoid the need for pushing down and correcting wrong shunts, but that the expenditure would have to be justified by the ability of the yard to handle more traffic, thus allowing other yards to close.

Flat Yards

The memo of Mr Ibbotson stated that: *"we are bound to be left for a number of years with numerous flat yards, some double and some single-ended, which have always proved a most uneconomical and intractable proposition. The capacity of these yards is so low that we have to have many more than we would like. In many cases, however, the construction of a major hump yard to eliminate these small yards cannot be justified, especially in the circumstances where the shunting of individual wagons is likely to become a decreasing factor over a period of years."*

The memo carried on, *"I should like to develop a case for Dowty operation with parallel reception lines and an inclined shunting deck. If the wagons could be drawn on to the shunting deck by capstan or mule with the assistance of Dowty boosters, the incline from the shunting deck would give the wagons the necessary start for running into the yard. In this way we might well double or even treble the capacity of a flat yard and do away with the shunting engine altogether."*

A brave idea maybe, but Mr Ibbotson went on to ask if anyone thought there was a future for projects like this. He suggested that the Scottish Region should develop the flat yard idea, leaving the Eastern Region to develop hump yards. He concluded by saying that he had not contacted Dowty over this idea, and he did not think anyone else should unless they thought the idea worth further investigation.

Further memos were sent on the subject, particularly the desire to do away with a shunting loco altogether. It was suggested that a yard should be designed so that the wagons could be moved from reception sidings to departure sidings by gravity alone, controlled at a pre-determined speed by Dowty units. The yard should be planned so that it started with reception sidings, leading to a primary sorting siding fan, followed by a secondary sorting fan and finally departure sidings. The required separation between shunts would be on a time interval basis.

The idea was that, once a train had arrived in the reception sidings, it would be cut into the required shunts then, when ready to shunt, the whole train would move forward by gravity and be controlled by the Dowty units at a pre-determined speed. The first cut would continue forward through the king points into the appropriate siding of the primary fan under Dowty control. The remainder of the train would be bought to a stand on the reception siding at the position of the first cut then, after an appropriate time interval to allow the first cut to clear; the second cut would be released and so forth. This idea would never be taken forward, probably due to its complexity or the costs involved, but there was plenty of enthusiasm for it at the time.

With BR having previously enquired with Dowty about using their equipment in the proposed new yards at Wigan and Feltham, on 5th January 1967 Dowty voiced their concerns with BR about future work as they understood that the Wigan project had been abandoned and the Feltham scheme was in abeyance, leaving them with no work in hand for BR

Mention was made that they had been in conversation with a Mr Woodbridge previously, and were under the impression that they would be involved in the installation of Dowty equipment in certain yards already equipped with conventional retarders. The memo stated that this was exactly the type of project they had in mind as a follow-on to the construction of new yards, as it offered the benefit of the Dowty system at a reasonable cost. They also hoped consideration would be given to equipping hump yards where the retardation of wagons was dependent on the manual operation of brakes by wagon chasers, and even to flat yards.

It was mentioned that the contribution BR had made to the development of the system, including the experiment at Hull, had come to £39,000, but that Dowty had spent much more on the project. They were hoping for business from abroad (although they would later reject requests) but what they really wanted in view of the capital invested by both parties was the exploration of further uses for the system with the aim of winning more home orders.

The response by Mr F.C. Margetts of the BR Board to Dowty on 30th January 1967 was not what Dowty was hoping for. Although it praised the equipment in use at Bescot which had been much improved from experience gained at Tinsley, they went on to say, *"Due to the rapid change of emphasis from conventional to new methods of freight movement, in particular the freight liner and merry-go-round coal trains, along with the continuing increase in full block load working between source and destination, the requirement for marshalling yards is declining. Many have already been closed and more will follow, and in others a reduction in the working load and the facilities required will be possible"*. Was this the first nail in Tinsley's coffin?

Mr Margetts' letter continued: *"I think there is little doubt that we shall not build any new marshalling yards and, against the above background, capital expenditure on existing yards will be difficult to justify. However, the Regions have been reminded that if they have any marshalling yards for which a continuing future can be seen and on which a limited degree of modernisation would bring an adequate return, to bear in mind the possibility of installing Dowty equipment."*

It is clear from further correspondence that BR did not completely abandon Dowty as, in a letter dated July 1967 from Mr G.F. Fiennes to the chief operations officer at the BRB, a reference was made to the possible installation of Dowty retarders at Temple Mills at a cost of £220,000, with an estimated saving of £55,000 on damage to wagons and the use of locomotives for pushing up. This was not taken up by the Board.

Changes

Things started to change in May 1968 when a memorandum stated that Sir George Dowty was refusing to accept firm requests for wagon control equipment from South Africa and New Zealand; the general manager of the railway division had resigned, leaving a junior executive in charge. There were also concerns that the company was ceasing production and development work as there seemed to be no future for the equipment in this country.

This seems at odds with what BR were proposing as, at the time the Eastern Region was experiencing a rise in damage reports to wagons at their Tyne and Tees yards and would be anxious to make a case for the installation of Dowty equipment there, using their own design and development resources, and the experience of Tinsley. It was said that those two yards were amongst the most efficient of the modern mechanised yards, and if they were not considered satisfactory for their continuing role under the new freight plan, it suggested the older less-efficient yards proposed for retention and development would have to be equipped with the Dowty system.

The memorandum went on to say that Dowty had suggested a cheap form of installation for flat and gravity yards was being designed. The simple installation to which Dowty was referring comprised low-speed retarder units in each siding which, because they did not use booster/retarders, did not require power packs or distribution pipe work. A speed setting of 5 feet/second was proposed, but was considered to be too low as it removed too much energy from stiff-running wagons, which would then stop near the entrance to the siding and require *"pushing up"* to clear them. The thought was that 8 feet/second would be preferable for the first 10 wagon lengths of the yard to help eliminate the problem. Another idea of Dowty's was a cheap hydraulic point machine which could be operated from a small console by the head shunter, responsible for the cutting of wagons and uncoupling. This would merely move the points hydraulically instead of by hand lever and could not take the place of an expensive automatic system. Dowty were quoting in the region of £10,000 for a yard with 10 sidings.

There were problems with Dowty's proposals, as staff would still be required to indicate when wagons were clear of the points and, without the use of booster/retarder units, the wagons would still require pushing up if they fouled individual points. The memorandum concluded that, notwithstanding the problems, if there was a way to improve overall efficiency with sufficient economy to service the capital, then it was worth closer inspection. A further letter to Dowty on 9th January 1969 stated that BR was giving further consideration to its marshalling yard policy and that it seemed probable a review would be made of existing partial-automatic and non-automatic yards, with a view to modernisation and improving wagon transit time. However, BR was concerned about suggestions that Dowty was seriously considering whether to remain in the market, and so BR enquired what Dowty's reaction would be if an invitation to tender was issued. A meeting took place at BR headquarters on 31st January 1969, to discuss the potential application of the Dowty system to a number of other marshalling yards. Dowty stated they would be interested in quoting for further orders if requested to do so.

At the time, BR was considering improvements to approximately 20 yards. This was expected to require an average of 7,500 Dowty units per yard, in a five to ten-year installation programme beginning in 1970/71. By now, Dowty had developed a second generation of equipment with an improved plunger head and a means of reducing oil losses, but the company was reluctant to develop these further without an assurance that manufacture on an adequate scale would follow.

David Bick designed a far simpler and cheaper method of control in 1969 based on gas-hydraulic units, which were tested in Tinsley Yard and installed at Scunthorpe New Yard in 1972. A later retarder/booster development utilised compressed air instead of hydraulic oil, but no other proposals for further Dowty installations were ever taken up in this country. Today no equipment exists in use on the national network, although Dowty-style retarders have found use in Europe and in the USA.

Appendix III: Job Descriptions

Job Descriptions

In any workplace there has to be a set routine if the work is to be done safely and efficiently. At Tinsley, much work went on behind the scenes to ensure the yard operated efficiently, and departing trains left on time and with the correct consist of wagons. A document was issued which set out the job descriptions of the operatives and methods of working for the administration staff in the yard; a copy of this document was reissued in 1973 and although quite technical in its terminology, it is as set out below and describes how each job was designated back in the day.

Job Description – Assistant Yard Manager 'D'

1. Responsible for all lines and sidings between Shepcote Lane, Treeton Junction and Treeton South inclusive.
2. Supervision of the yard function including area traffic office, train crews, signal boxes, main yard (MY), and secondary yard (SY) control towers, and all yard staff.
3. Responsible for all aspects of train and traffic working and for liaison with divisional control office.
4. Responsible to assistant to area manager.

Job Description – Traffic Supervisor 'C'

Traffic Planning Yard Operations

Assess current and forward traffic position in Tinsley Yard and its environs by reference to yard state and departure chart, connectional charts, consists, local freight train departures (LFTD) control advices, print-outs, and anticipating clearances.

Where traffic accumulations or shortages suggest that modifications should be made to the planned working, this should be done by any means possible, including the following:

1. Arrange special trains or make cancellations.
2. Arrange alternative loads to those booked.
3. Stable outside yard as necessary after consulting control.
4. Divert block loads on incoming trains to destination station. If not possible divert to MY, SY or EF if feasible, and then arrange clearance of this traffic.
5. Arrange trip loadings with the appropriate control section.
6. Use unbalanced engines and men for clearance of excess traffic.
7. Divert traffic into alternative roads in conjunction with MY west supervisor and SY supervisor.
8. Arrange traffic restrictions for incoming trains and keep book record.

Empty Wagon Orders

Ensure that main yard and secondary yard supervisors are given empty wagon orders each day. Liaise as necessary where wagons can be provided by humping into the appropriate roads. Amend print-outs as necessary, for purpose of meeting wagons orders, providing barrier wagons etc. Request instructions from wagon control, Sheaf House, when there are surplus empties on hand.

Supervision of Receptions

1. (a) The supervisor at all times will have overriding responsibility for deciding upon which reception siding arriving trains should be placed.

   (b) When receptions are congested, supervisor to regulate acceptance of trains, bearing in mind connectional traffic, engines and men's next workings, crews' times on duty and traffic position on main lines outside yard.
2. Consult connectional charts and instruct reception chargemen as to cutting priorities. Arrange for chargemen to be advised of temporarily altering cutting instructions, as when certain types of empties are ordered to changed destinations.
3. Arrange removal of non-humping wagons from receptions, liaising with main tower as necessary. Advise tower supervisor of details of non-humping wagons so that he can arrange to detach before arrival on receptions where possible.
4. Advise tower supervisor when connectional traffic should be taken by train engine to west end.
5. Decide order of humping, bearing in mind connectional requirements as shown on connectional charts, road-holding capacity, and future reception needs.
6. Instruct tower supervisor grade 'A' of alteration to booked workings and orders for additional workings of locos released from train on reception sidings.

ODM Requirements

Receive advice of ODM maintenance requirements and decide when the work can be done. Record standing times created by ODM maintenance/emergency situations. Liaise with ODM charge hands to restore normal working as soon as possible.

General

1. Perishables, specially advised wagons, and vehicles which need marshalling to be advised to the MY or SY supervisor before these have been humped. Liaise as necessary.
2. Maintain current record of traffic restrictions.
3. Supervision of ATO staff.
4. Responsible for daily compilation of time book for ATO staff.
5. Maintenance of traffic log book.
6. Responsible to area yard manager.

Job Description – Charting Supervisor

1. Enter ATI summary from connectional charts to formica strips and attach on ATI section of board.
2. Update strips from connectional charts with amendments following receipt of print-out. Place on board in reception sidings position. Enter details on yard state and departure chart when train is humped (information supplied by tower supervisor).
3. Maintain yard state and departure chart showing amended totals when wagons are removed from, or placed in, the sidings (information supplied by west end, SY and EF sidings).
4. Note ATI for details of wagons arriving in excess of road holding capacity, or beyond the capacity of the booked services to clear. Advise traffic supervisor of diversions which may be necessary. Note probable shortage of traffic for outgoing services. Liaise as necessary with traffic supervisor.

5. Record empty wagon orders and advise all concerned. Amend print-outs when wagons can be humped into appropriate roads. Liaise as necessary with main tower supervisor and west end supervisor. Keep book record of wagons supplied to meet orders and advise wagon control at appointed times.
6. Advise tower supervisor of altered state of sidings at west end of main yard.
7. Advise supervisor at main yard west and secondary yard of engine nos. for departing trains.
8. Operate UHF radio equipment.
9. Give details of wagons in roads to west end supervisors as required.
10. Check outgoing consists against wagon totals reported drawn from yard. Advise Telex clerk of inconsistencies.

Job Description – CO2 Charting

1. Summarise Telex consists, LFTD advices and train wires. Enter details on connectional chart. Note critical connections, monitored traffic, perishables, explosives, vulnerable and specially advised wagons and any other vehicles of importance which must be connected. Advise traffic supervisor as necessary.
2. Summarise print-outs and enter on connectional chart. Note discrepancies between ATI and print-out information and enquire into circumstances. Amend wrong cuts and advise as necessary.
3. Extract details of 'Z' notational traffic for form 'B'.
4. Record incoming trains on arrival sheet and inform traffic supervisor of cancellations and late running trains.
5. Note failure to receive LFTD or consists. Also note errors in LFTD and consists. Advise Telex clerk of omitted and incorrect consists so that the matter can be taken up currently and advise traffic clerk so that fault can be taken up in writing.
6. Check outgoing consists for specially advised traffic. Where failures occur liaise with DMO as necessary re forward movements. Check outgoing consists for monitored traffic.
7. Maintain record of all wagons underload stopped for repairs and process as per instructions.
8. Assist traffic clerk as necessary

Job Description – Traffic Information Clerk

1. Extract details of traffic for local firms from incoming consists and enter on card index.
2. Enter firms' requirements of empty wagons supplied, on card index for that day.
3. Advise local firms of traffic on hand, together with details of exceptional consignments. Obtain information about acceptance problems and analyse firms' difficulties and needs.
4. Liaise with private sidings, Sheffield Fright Terminal (SFT) and outstations re current and anticipated outward traffic.
5. Record current stock at outstations and keep stock information boards up to date.
6. Prepare statement of current stock at 08.00 for use by conference assistant.
7. Deal with enquiries from all points on current transit matters.
8. Monitor consists with special regard for wrong traffic. Take up in correspondence with sending points and divisional office as necessary.
9. Send written confirmation to sending points of incorrect consists and duplicate copy to admin office.
10. Enter engine numbers and crews on count down sheet.
11. Supply pipeline figures to divisional office.
12. Assist as necessary.

Job Description – Telex Clerk

1. Operate Telex machine.
2. Receive and distribute messages as per instructions.
3. Keep train recorders informed of altered workings and late traffic.
4. Pass outgoing consists to charting supervisor so that check can be made against wagon totals drawn from yard.
5. Send corrected consist details to destination points after check by charting supervisor.
6. Make enquiries via Telex machine as required. Send enquiries on non-receipt of consists.

Job Description – Supervisor A, Tower

1. Hump trains in the order determined by TO
2. Record on form 'A' time of arrival and time humped of all incoming trains.
3. Report train arrivals and engine released to Control.
4. Arrange removal of non-humping wagons from receptions. Liaise as necessary with traffic supervisor.
5. Advise traffic supervisor of full roads and of trains which cannot be humped owing to insufficient room.
6. Push roads down as necessary after liaison with traffic supervisor and advise west end supervisor when this has to be done.
7. Advise west end supervisor when traffic is being humped into empty roads.
8. Advise traffic supervisor of catch-ups, and re-hump wrong shunts as directed.
9. Maintain up to date totals of wagons in roads.
10. Carry out instructions for protection of ODM staff.
11. Arrange closing-up movements for cutters when required.
12. Operate yard lighting switches as necessary.
13. Change tapes and spools in "Addo" machines.
14. Cleanliness of tower: maintenance of time book for associated staff.

Yard State (YS) and Departure Charts (DC)

This chart shows all the roads in the main yard and secondary yard sorting sidings in vertical columns, with horizontal coordinates of ten-minute intervals over an eight-hour period.

The charting supervisor enters the total number of wagons in each road. As trains are humped into the various roads, the totals are added to, thus the figures are cumulative against a time scale.

When a train departs from the west end of the yard, the charting supervisor deducts the wagons withdrawn from the total and enters reporting number and new total in the vertical column against the appropriate point in the time scale. The YS and DC thus gives an up to date record of the state of the yard.

The YS and DC is used for the following purposes:

1. Decision making. The booked departure time of a freight train is known, and the number of wagons offering for that train is shown in the appropriate column on the YS and DC. Since the facts are known, it is then a matter for decision as to whether a train should run as booked, or whether the resources (locomotive, train crew) could be better employed dealing with other traffic, covering shortages, or working to an alternative destination.
2. Assessing the yard position. The wagon capacity of each road is known constant, whilst the YS and DC show the number of wagons currently standing in each road. There is thus a clear guide as to the remaining capacity of each road. When it is seen that a road is becoming full, one of the following courses of action has to be taken:-
   a) Arrange a clearance of traffic from the west end.
   b) Divert wagons to an alternative road.
   c) Cease humping wagons for the full road.
   d) Draw load back to reception.

Daily Connectional Charts

The charting clerk maintains a file of connectional charts in a numbered sequence. When a consist or LFTD advice is received, the charting clerk selects the appropriate connectional chart and enters the summarised details upon it.

The connectional chart is then passed to the charting supervisor. The charting supervisor enters the summary of cuts onto a formica strip, and then places the strip on the upper section of the ATI board

Using this information, the charting supervisor advises the traffic supervisor, and the main yard and secondary yard supervisors, of unusual flows of traffic, of traffic beyond the holding capacity of the roads, and, of connections which might be desirable. The connectional chart is then passed to the traffic supervisor.

The traffic supervisor assesses the connectional chart for acceptance, cutting and humping priorities. When the train arrives on the reception line, the traffic supervisor enters the arrival time and reception line number on the card. The connectional chart is then passed to the charting clerk.

When the train has been cut, the charting clerk compares the printout information with the ATI details, and after checking for cutting errors, amends the connectional chart. The charting clerk secures the print out to the connectional chart by paper clip and passes both documents to the charting supervisor.

The charting supervisor examines the connectional chart and makes amendments on the formica strip for road diversions, empty wagon orders, etc. He updates the formica strip to the reception sidings position on the ATI board, amends the print-out as necessary and advises all concerned.

The connectional chart and print-out are then passed to the traffic supervisor, who places them in the appropriate clip, and uses the information to decide the order of humping.

When the train has been humped, the traffic supervisor enters the time humping was completed. He then gives the connectional chart to the charting supervisor, who shows details of missed connections, and at the same time, updates the formica strip details to the YS and DC.

The connectional chart is finally passed to the charting clerk, who completes the *"summary of missed connections"*. The attached print out is stapled to the incoming consist for filing.

The daily connectional chart is an important tool in the work of the ATO. It depicts summarised details of the composition of each terminating train, together with the critical connections. It is thus of great value in determining the acceptance, cutting and humping priorities of each train.

Consists: Inward Trains

A consist is received in duplicate in the Telex office. The Telex operator passes both copies, together with the carbon sheet, through a sliding window to the charting clerk.

The charting clerk ticks off the *"consist and LFTD check list"*. He then summarises the consist in duplicate and enters the details on the appropriate connectional chart. At the same time, he notes critical connections, monitored traffic, explosives, vulnerable, and specially advised wagons which must be connected. He advises the information clerk of any incorrect details on this consist.

The charting clerk passes one copy of the consist to the traffic supervisor, and he retains the duplicate for checking wrong cuts as on the print-outs. When the train has been humped, the duplicate copy is placed in the ATO file, together with the print-out. The traffic supervisor checks his copy of the consist for vehicles which cannot be humped, or wagons that need to be taken to the west end for connectional purposes, or for any other situation which may warrant special action (e.g. block loads). The traffic supervisor then passes the consist to the information clerk.

The information clerk extracts details of traffic for firms within the Tinsley area and enters on the appropriate card files. He also monitors the consist for wrong traffic and incorrect details and takes in correspondence and via Telex as necessary. The information clerk then files the consist for later despatch to the divisional manager at Sheaf House.

A consist contains details of wagon types, wagon numbers, sending and destination stations, consignees, and in some cases the type of traffic. The consist then, is the most valuable source of advance information for the ATO.

Consists: Outward Trains

The Telex operator must endorse the consists with both his own and the train recorder's initials. The Telex operator retains one copy of outgoing for file and eventual dispatch to the divisional manager at Sheaf House. He passes the other copy through the sliding window to the charting clerk who will examine the consist for monitored and specially advised traffic.

When failures occur in the movement of specially advised traffic, the charting clerk should consult the divisional manager at Sheaf House, re the forward movement. The charting clerk endorses the countdown sheet with the train recorder's initials.

The consist is then passed to the charting supervisor who checks the total reported on the consist against the total reported drawn from the yard. In the case of any discrepancy between the two totals, the charting supervisor should advise the Telex operator, who will send a message of correction. The consist is then placed in the TO file.

Consists for outward trains are used for advising destination yards of the composition of train departing from Tinsley.

Local Freight Train Details (LFTD) Advices

The charting clerk receives LFTD advices by telephone and enters the details on the LFTD form. He then makes the entry on the appropriate connectional chart and ticks off the *"consist and LFTD check list"*. The information clerk takes up in correspondence with the sending point in the case of omitted or incorrect LFTD. The LFTD form is then filed together with the print-out.

LFTD is the means by which local yards and sidings give advance information to Tinsley. Communication is usually via the BR telephone service.

Print-outs

A print-out is a summary of the cuts made in the train by the reception sidings staff, and it is produced by an "Addo" machine. The slave "Addo" machine is placed at the side of the charting clerk's desk. When a print-out is received the charting clerk summarises the details and amends the connectional chart, if necessary. He notes discrepancies between consist and print out information and enquires into the circumstances with the reception chargeman. He then amends wrong cuts and advises as necessary.

The charting clerk affixes the print-out to the connectional chart by paper clip and passes the two documents to the charting supervisor. When the traffic supervisor receives the print-out (with the connectional chart) he examines it for light wagons, perishables etc., and then places both documents in the appropriate reception sidings clip. After the train has been humped, the print-out is separated from the connectional chart and stapled to the incoming consist for filing.

Humping Record: Form 'A'

Kept by the tower supervisor, this is a record showing all trains arriving in the reception lines, times of humping and which pilots were used. Delays, Dowty maintenance times and relevant remarks should be entered on these forms.

The arrival times, reception line numbers, and times humped should be reported currently to the ATO.

Details of wagons humped and reception lines empty, must be given to divisional manager, Sheaf House control office at the end of each shift.

Daily Summary of Missed Connections

Maintained by the charting clerk. The reasons for missed connections are summarised under coded headings. This form is used by management to improve the service to customers. Copies of this form are then sent to:-

Regional HQ, York
Divisional manager, Sheaf House
AMO, Tinsley

Weekly Summary of Missed Connections

Book record and graph compiled by the traffic information clerk.

Monitoring of Wagon Load Transits

The charting clerk compiles this form, obtaining his information from inwards and outwards consists, LFTD, and the yard state and departure chart.

This form is designed to sample the service to customers, and it is also a management check on transit times.

For outward traffic the form is compiled in duplicate:
1st copy to DMO FG section.
2nd copy to destination station.
For inward traffic, one copy is sent to the terminal.

Departures "Countdown Sheet"

This is basically a list checking on all the component parts which go together to produce an end product; a train departing on time. The "countdown sheet" is a tool to assist in the making of the most efficient and economical use of resources, compatible with the maintenance of expeditious transit times for wagons.

The "countdown sheet" is:

a) An early warning system of the shortage of power or men for departing trains.
b) Early notice of alterations in type or class of motive power on inwards workings which work out again, thus enabling the preparation of correct loads.
c) A precise method of making judgement on the traffic needed to run a service from Tinsley, or whether the resources available should be utilised to deal with other traffic peaks.

Divisional control advises traffic office of the punctuality performance of trains approaching Tinsley. If a train so advised has an outward working from Tinsley, the sheet is marked in the appropriate column. Thus the traffic supervisor has a further early warning of either possible late start for an outgoing train, or non-provision of resources for the train.

The information clerk enters details of locomotives and crews originating in Tinsley on the "countdown sheet" and the traffic supervisor obtains the information in the case of out and home workings. In the case of shortage of power or trainmen, the traffic supervisor must decide whether to cancel the train, or whether adjustments can be made to other workings, in liaison with the AYM.

The charting supervisor gives details of locomotive numbers to the MY and SY Supervisors, so that the correct load can be prepared for each train.

Consist and LFTD Check List

The consist and LFTD check list is a document designed to monitor received consists, so that omitted and incorrect consists can be identified. Action can then be taken to improve the service. The check list is also used to record the punctuality performance of incoming trains, so that early assessment can be made of the possibility of maintaining connections. Also, when the punctuality performance of each incoming train is known, the ATO will have an early guide, where out and home workings are concerned, of the availability of resources. This check list is also used to record incoming locomotive numbers, so that early advice can be given to train preparers in the case of return workings.

The charting clerk compiles this check list and ticks off received consists and LFTD as the message arrives. He also enters locomotive numbers in the appropriate column.

When any consist or LFTD advice is overdue, omitted or incorrect, the charting clerk initiates action to correct the fault. He advises the Telex clerk so that the matter can be taken up currently, and gives details to the information clerk. The information clerk uses the pro-forma to advise starting points of overdue, omitted or incorrect consists.

Incorrect/Omitted Incoming Consist Pro-Formas

The information clerk indicates faults on a pro-forma and dispatches it to the sending points of omitted and incorrect consists, so that efforts will be made to improve the service. The file copy will be passed to the O&G section to be raised again if no reply is received in a reasonable time.

Card Index for Firms in the Tinsley Area

The information clerk extracts details of traffic for local firms from consists and enters these in the appropriate card index. He also shows the daily empty wagon requirements of each firm and wagons supplied to meet the orders.

The cards showing inwards traffic should be used to keep firms currently advised of traffic en route. Acceptance problems and other relevant matters should be recorded on the cards, and all concerned advised. These cards are intended that current clearances are made, to connect with the best possible forward service.

The card index system is designed to ensure that a close and regular contact is made with all firms in the Tinsley area, thus promoting an improved relationship between BR and the customer.

Morning Conference Report

The 08.00 stock position to be recorded by the traffic information clerk.

On the area stock sheets, traffic which will not be cleared by booked services, and the brake van position, to be shown in red. The following documents to be available for conference:

1. Area and yard stock sheets.
2. "Countdown sheets" (list of trains starting from Tinsley).
3. Daily summary of missed connections.
4. Extracts from the following log books: traffic supervisors, main tower supervisors, secondary tower, train crew supervisor.
5. Train crew staff position.

The morning conference report is designed to enable management to assess the traffic position throughout the entire Tinsley area.

FRS Orders Book

This book is kept by the charting supervisor. He enters the daily wagon orders from the divisional manager, Sheaf House, together with details of wagons supplied to meet the orders. The information clerk extracts details of wagons ordered by, and supplied to, local firms, and enters these in the appropriate card index.

The FRS orders book is intended to enable ATO staff to maintain a constant check on the supply of empty wagons for customers.

Traffic Restrictions Book

Maintained by the traffic supervisor. This forms a current and historical record of all traffic restrictions that affect the Tinsley area.

The purpose of the book is to enable traffic supervisors to take early action to avoid difficulties which may arise from the imposition of traffic restrictions.

Daily Log Book

Maintained by the traffic supervisor. This log is a daily record of traffic arrangements, delays and pertinent matters for management information.

Weekly Time Book

Compiled by the traffic supervisor to record hours worked by all staff in the area traffic office (ATO).

The times staff employed in the ATO report for duty, and leave duty, are recorded in the weekly time book. This is the basic document from which weekly salary payments are calculated. The AYM and the area administrative assistant will make random checks of the book entries.

Stationary Order Book

The stationary order book is a document which enables staff to maintain a supply of necessary items from stores by a uniform and systematic method. The duplicate page enables the stores department to account for all items supplied.

The stationary order book is located with the traffic supervisor in the ATO. This book is solely for stationary items and not locally manufactured forms which are dealt with as a separate situation. In the event of any member of the ATO requiring stationary, he should inform the traffic supervisor of his requirements. The book should be submitted monthly to stores, and the items requested will be supplied according to availability.

Locally manufactured forms peculiar to the office are specially tailored to that office only, and therefore at least

Left: Specially painted for the "Three to the Sea" railtour from Sheffield to Brighton, run by the Class 20 Locomotive Society on 2nd May 1987, 20030 "River Rother" and 20064 "River Sheaf" repose on the 'proto' road at Tinsley, with a Class 37 and three other Class 20s. *Roger Wainwright*

one months' notice must be given to the staff/admin office in writing, stating the type and numbers required.

Area Information Board

Maintained by the information clerk. Using a chinagraph pencil, the information clerk records details of current traffic stocks throughout the Tinsley area, showing brake vans on hand at each point. The information will be displayed for use of the ATO staff, the AYMs and management.

Whereas the detail on the board is altered currently in line with received information, the whole of the detail on the board is checked for accuracy at 08.00 hours each working day, and becomes the source for compilation of the morning conference report.

The Information Board is a constant visual reminder of accumulated traffic.

Wagons Stopped for Repairs or Loading Defects, or Involved in Mishaps

Tick messages in respect of the above wagons will be phoned from source to the charting clerk, who should be given the following information:-

a) Type of wagon, and wagon number
b) Label date
c) Sending station, destination, and consignee
d) Defect, if known

The charting clerk will enter these details in a book and will arrange for a Telex message giving the above information to be sent to:

a) Sending station
b) Destination station
c) Divisional manager, Sheaf House
d) Customer (where appropriate)

Where the use of the Telex is not possible, a telegram should be sent via the BR telecom lines.

When the wagon has been repaired, the charting clerk is advised of the forward service so that Telex messages or telegrams can be sent to the same points as when the wagon was stopped.

The charting clerk should keep the information clerk advised of such stoppages, and he should also ensure that the wagons are not subject to undue delay.

Below: On a typical day in 1978, 13003 has just finished shunting a cut of wagons over the hump and now has the brake van to dispose of. *Ivan Stewart*

Appendix IV: List of Abbreviations

AC	Alternating Current
ASLEF	Associated Society of Locomotive Enginemen and Firemen
ATI	Advance Traffic Information
ATO	Area Traffic Office
BICC	British Insulated Callender's Cables
BPGV	Burry Port & Gwendraeth Valley
BR	British Railways, later British Rail
BRB	British Railways Board
BREL	British Rail Engineering Limited
BSC	British Steel Corporation
BTC	British Transport Commission
C&W	Carriage & Wagon
CCE	Chief Civil Engineer
CEGB	Central Electricity Generating Board
DC	Direct Current
DMD	Diesel Maintenance Depot
DMU	Diesel Multiple Unit
ECML	East Coast Main Line
EE	English Electric
ER	Eastern Region
ESC	English Steel Corporation
EWS	English Welsh & Scottish Railway
FOC	Freight Operating Company
FPL	Facing Point Lock
GBRf	GB Railfreight
GCR	Great Central Railway
GE	Great Eastern
GER	Great Eastern Railway
GF	Ground Frame
GN	Great Northern
GNoI	Great Northern Railway of Ireland
GPL	Ground Position Light
GRP	Glass-Reinforced Plastic
GSTM	Ground Staff Team Member
HST	High Speed Train
IFS	Individual Function Switch
LD&EC	Lancashire, Derbyshire & East Coast Railway
LMR	London Midland Region
LMS	London Midland & Scottish Railway
LNE	London North Eastern
LNER	London & North Eastern Railway
LNWR	London & North Western Railway
LT&S	London, Tilbury & Southend
LTM	Line Traffic Manager
MCD	Miniature Control Desk
MGR	Merry-go-Round
MML	Midland Main Line
MPD	Motive Power Depot
MR	Midland Railway
MS&LR	Manchester, Sheffield & Lincolnshire Railway
MSW	Manchester, Sheffield & Wath
NCB	National Coal Board
NER	North Eastern Region
NMR	North Midland Railway
NRM	National Railway Museum
NSE	Network SouthEast
NUM	National Union of Mineworkers
NUR	National Union of Railwaymen
NX	Entrance-Exit
PA	Public Address
PIC	Person in Charge
PSB	Power Signal Box
R&D	Research & Development
RfD	Railfreight Distribution
RNAD	Royal Naval Armaments Depot
RSCo	Railway Signal Company
S&LJR	Sheffield & Lincolnshire Junction Railway
S&RR	Sheffield & Rotherham Railway
S&T	Signalling & Telecommunications
SDR	Sheffield District Railway
SFT	Sheffield Freight Terminal
SIRFT	Sheffield International Rail Freight Terminal
SMACC	Stainless Melting and Continuous Casting
SPACE	Stainless Plate and Coil Expansion
SSOW	Safe System of Work
TC	Track Circuit
TCB	Track Circuit Block
TLF	Trainload Freight
TM	Traffic Manager
TMD	Traction Maintenance Depot
TOPS	Total Operations Processing System
TRA	TOPS Responsibility Area
UCC	United Coke & Chemical Company
UHF	Ultra-High Frequency
WCRC	West Coast Railway Company

Index